W9-CGM-801

Science Wars

Science Wars

Andrew Ross, Editor

Duke University Press   Durham and London 1996

© 1996 Duke University Press
All rights reserved
Printed in the United States of America on acid-free paper ∞
Typeset in Melior by The Marathon Group, Inc.
Library of Congress Cataloging-in-Publication Data
appear on the last printed page of this book.
The text of this book originally was published without the
chapters by Roger Hart, N. Katherine Hayles, Richard C.
Lewontin, and Michael Lynch; with unmodified versions of
the chapters by Steve Fuller, Ruth Hubbard, and both pieces
by Andrew Ross; and without the index, as *Social Text*
46/47, vol. 14, nos. 1 and 2 (Spring/Summer 1996). The
essay by Richard C. Lewontin originally appeared in
Configurations, vol. 3, no. 2 (Spring 1995): 257–65.

Contents

Andrew Ross Introduction   1

Sandra Harding Science Is "Good to Think With"
16

Steve Fuller Does Science Put an End to History,
or History to Science? Or, Why Being Pro-science
Is Harder than You Think   29

Emily Martin Meeting Polemics with Irenics in
the Science Wars   61

Hilary Rose My Enemy's Enemy Is—Only Perhaps
—My Friend   80

Langdon Winner The Gloves Come Off: Shattered
Alliances in Science and Technology Studies
102

Dorothy Nelkin The Science Wars: Responses to
a Marriage Failed   114

George Levine   What Is Science Studies for and
Who Cares?   123

Sharon Traweek Unity, Dyads, Triads, Quads,
and Complexity: Cultural Choreographies of
Science   139

Sarah Franklin Making Transparencies: Seeing
through the Science Wars   151

Ruth Hubbard Gender and Genitals: Constructs of
Sex and Gender   168

Richard Levins Ten Propositions on Science and
Antiscience   180

Joel Kovel Dispatches from the Science Wars   192

Stanley Aronowitz The Politics of the Science Wars
202

N. Katherine Hayles Consolidating the Canon   226

Michael Lynch Detoxifying the "Poison Pen Effect"
238

Roger Hart The Flight from Reason: *Higher
Superstition* and the Refutation of Science Studies
259

Richard C. Lewontin A la recherche du temps perdu:
A Review Essay   293

Les Levidow Science Skirmishes and Science-
Policy Research   302

Andrew Ross A Few Good Species   311

Contributors   321

Index   325

Introduction

Andrew Ross

AT the end of July 1995, Republican moderates in Congress broke ranks to help defeat a brutal package of anti-environmental legislation. The Contract with America juggernaut had run into its first engine trouble after almost seven months of all-terrain joyriding. The consolidated voice of punditry pronounced that voters had made it known that health and environmental safety was one regulatory issue where the anti-fed evangelicals were "going too far." Political pressure quickly reversed the vote, but the damage was already done: it was stymied in the Senate, and a number of other bills began to run into heavy weather. The significance of this moment is worth dwelling upon. The Gingrich Congress had announced that it was putting almost sixty years of post–New Deal consensus politics to the sword by reviving ideology. There was little reason to doubt its word in the first six months of legislative activity. But here, in the first major coalition vote won by Democrats in the new Republican Congress, it looked as if ideology was ceding to the assumed consensus mood of the electorate. A line had been drawn at the edge of the wetlands. The legislation itself was no small potatoes; it would have eviscerated EPA control over the poisoning of soil, water, and air by industrialists, removing the "excessive constraints" to their business practices and sending a clear signal that the polluter-pays principle was no longer in effect. But why did environmental legislation stem the Republican onslaught where other, equally radical, deregulatory measures had swept through Congress with conservative Democrat support?

Let us suppose that an electoral consensus about health and environmental safety really does exist and that it marks a meaningful limit to the current congressional war on public interest. The growth of that consensus surely owes much of its resilience to a rise in popular technoskepticism. The middle classes, in particular, are increasingly scientifically informed about the hazards of advanced industrial life and are ill-disposed to see the risks multiply. Pollution, unlike poverty and hunger, (ultimately) spares no one, and so the politics it begets has high legitimacy. But middle-class anger only goes so far. The truly decisive moment of environmental politics comes only when the ecological consequences of industrialization can no longer be economically externalized. That moment is not yet upon us (it has been further deferred by the global free-trade agreements), but its social and cultural symptoms have long preceded it, giving birth to a social movement that deploys science against the industrial threats generated by science.

This contradiction—using science against science—speaks faithfully enough to the Janus-faced development, development associated as much with destructive as with productive forces, of technoscience in the modern period. In the popular mind, the allegiances of many of the sciences to elite interests—military, corporate, and state—and to the cause of superindustrialism have underscored the perception that they are far from democratic in practice. As for those with vested interests or managerial obligations, it is increasingly apparent that the risks generated by industrial life may come to threaten the scientific process of production itself, a phenomenon first acknowledged by industrial elites in their own 1974 Club of Rome report. Poisonous threats to life are ultimately threats to property interests and to natural resources—air, water, minerals—that industry has long counted as free. The rise in technoskepticism, then, parallels a crisis in industrialization that is often mistaken for a crisis of great environmental magnitude. Commentators like Ulrich Beck have explained this conundrum by theorizing a new phase of modernity—reflexive modernization—where modernity today everywhere confronts the effects of the earlier, primary modernization that we recognize as industrialism. That moment of self-confrontation involves a radical critique of the very scientific rationality that served as the vehicle of industrial society. For Beck, then, the growth of this self-critical temper is not counter-Enlightenment but rather the triumph of the skepticism by which rational scientific inquiry is formally turned upon its own foundations and methods. This reflexive turn is the basis for modernity's continuing development under the aegis of risk politics, where the management of technoscientific risks becomes a major factor of power as well as a new opportunity for scientific expansion.[1]

Readers may quibble with the apparently simple replacement, in Beck's paradigm, of the axial law of accumulation (and/or the production of scarcity) with the principle of the overproduction of risks (and/or the production of security). Can't they be mutually affirming? How can a slippery, semiconceptual quality like risk come to be the primary criterion that defines social structure? At the very least, Beck's location of technocritique within the mainstream of modernization moves it away from its customary whereabouts: the countercultural margins populated by Luddist stereotypes. What used to be characterized as knee-jerk technophobia is now an everyday response to the risks and threats generated by advanced industrialization. The critique of science is normalized and domesticated, exercised in a hundred little ways. It forms the basis for massive public anxiety about the safety of everything from the processed food we consume to the steps we are making toward a biologically engineered future, often described, in popular shorthand, as "tampering with nature." Though knowledge about the risks surrounding us is unevenly shared, the risks are a part of everyone's daily experience of modern life.

Embedded in this normalized technophobia is a nugget of information that is no longer news: science does not have a monopoly upon rationality! Most of the contributors to this issue would probably acknowledge this, although they would have their own ways of putting it. And yet there is still something scandalous, and of course irreligious, about saying such a thing. The reserves of guilt that are summoned up by our superegos betray the basis in faith of so much of our own collective assent to the authority of Science and to its dogma of progress through unratified experiment. But why should the burden of proof lie with the designated skeptic? In a world racked up to a level of nuclear, biogenic, and chemical overdevelopment that cannot insure against its own annihilation, let alone guarantee its own sustainability, technological ethics more than entitles the skeptic to ask for evidence of rationality. Who can deny that we live in an open environment that has been utilized, on both a micro- and a macromanaged scale, as an experimental laboratory for life-denying substances and processes? And yet we still worry about being mistaken for frothy-mouthed apocalyptics, pilloried by science boosters for cheerleading the flight from reason and truth into the welter of disorder that is a mirror-image of the world's end.

Some may take comfort in the idea that there are competing claims on rationality. The sober relativist assures us that, in reality, no one is unhinged and that we all operate within the framework of our own patched-together rationality. The Western laboratory scientist observes the institutional rules, follows the local procedures, and accedes to the general "value-free" belief-system as logically as the Chinese bare-foot doctor or the rainforest shaman functions within their own cultural environments. Nor, we are reminded, is this laboratory scientist a singular culture type; each of the stable sciences, and each of their specialty subfields, has its own domain of technical consensus, a closed circle of opinion that makes sense only in relation to the particular instruments and genres of data analysis developed for that field.

But let us not mistake the latter as a description of diversity. If anything, it is a recipe for technocratic expertise, all the more credible when supported by an adequate apparatus of persuasion, whether in science or in government. Yet the remoteness of scientific knowledge from the social and physical environments in which it will come to be measured and utilized is as irrational as anything we might imagine, and downright hazardous when it involves materials that can only be properly tested in the open environment. The unjustified conferral of expertise on the scientist's knowledge of, say, chemical materials, and not on the worker's or the farmer's experience with such materials is an abuse of power that will not be opposed or altered simply by demonstrating the socially constructed nature of the scientist's knowledge. That may help to demystify, but it must be joined by insistence on methodological reform—to involve the local experience of users in the research process from the outset and to ensure that the process is shaped less by a manufacturer's interests than by the needs of communities affected by the product. Such methodological reform will lead from cultural relativism to social rationality.

Nor can we be satisfied that the "successes" of the lab scientist, the Chinese doctor, and the rainforest shaman are equally relevant and adequate to the cultures they serve. The power and authority of the Western scientific method has a global reach. Outside the West, the material impact of its applications is often inappropriate to local cultures and environments, and their need for extensive technical-support systems (backup parts, pesticides, interface technologies, training, and so on) creates relations of dependency between provider and recipient. No less relevant is the political impact of technoscience— manifest in the pressure to adopt Western ways. So, while the destructive history of technology transfer (in the Green Revolution) has long been officially acknowledged, the funding of subsistence science in Third World nations, such as India, for example, continues to be dwarfed by the investment in military, nuclear, and space science— the "big science" projects that count in the competitive global game of national prestige. Again, the relativist can show us how Western technoscience becomes acculturated and syncretized in other parts of the world but cannot insist that alternatives to this one-way process should be encouraged. Once it is acknowledged that the West does not have a monopoly on all the good scientific ideas in the world or that reason, divorced from value, is not everywhere and always a productive human principle, then we should expect to see some self-modification of the universalist claims maintained on behalf of empirical rationality. Only then can we begin to talk about different ways of doing science: ways that downgrade methodology, experiment, and managerial needs in favor of local environments, cultural values, and principles of social justice. This is the way that leads from relativism to diversity.

As for democratic rationality, we are hard put to see where it enters into those sectors that are currently viewed as the benchmark criteria of success and progress in our business civilization—capital growth on the one hand and technological innovation on the other. Protests about the concentration of elite decision making in these sectors are derailed by invoking the rule of "international competitiveness" or kindred threats.

Critics of capitalism have long diagnosed its irrational tendencies (contradictions, crises, and inefficiencies) and questioned the morality of its exploitative creation of surplus value. In response to a labor theory of value, market apologists point to the subjective or marginal utility of commodities as the ultimate, if not the primary, source of economic value. Such theories, which vaunt their neutrality in the name of science, appeal to a market principle of the artificial scarcity of goods, not to be confused with the reality of material shortage in a depleted natural world that is the habitual outcome of monetary growth.[2] As recognition of the possibility of ultimate scarcity deepens, many analysts have come to see the resulting contradiction between capital accumulation and the degradation of resources as the only ultimate limit to growth.

Do the sciences recognize or serve any of these economic rationalities involving nature and market conditions? Officially, of course, they do not, and, according to conservative defenders of the faith, they should not. But the reality is quite different. With regard to method, the value-free ideology of *Wertfreiheit* may well have been conceived initially as a hedge against the intrusion of authoritarian dictates, from both the church and the monarchic state. In some fundamental respect, scientific objectivity was thereby associated with the rising economic forces of free trade. Following Boris Hessen's epochal 1931

London lecture on the relationship between Newton's *Principia* and the needs of the emergent English mercantile bourgeoisie, historians of science have expanded on the indebtedness of empirical science to market interests. Feminist critics have also described how the androcentric rationality of the scientific method has served not only to exclude women professionally but also to reinforce their social subjugation through its subordination of the private, subjective realm of experience with which they were socially identified. Earlier in this century, *Wertfreiheit* was invoked for immunity from ubiquitous political incursions, including both Fascist (Aryan) and Stalinist (Lysenkoist) science, but increasingly it has been used to fend off social criticism and to protect professional contracts with the corporate-military state. As Robert Proctor has argued, the principle of value-free science had especial appeal in the "end-of-ideology" period of the Cold War and, for the critics of science, over the next two decades it came to signify political quietism in the social sciences and selling one's skills to the highest bidders in the natural sciences. If scientific knowledge is not beholden to market forces, why is it that geologists know more about oil-bearing strata than other strata, or that virologists know more about viruses that attack tobacco than about other strains?[3] If stable sciences really are objective fields of knowledge and inquiry, why have so many—seismography, oceanography, and microelectronics, to name a few—evolved directly from military research and development as part of the spin-off system that is habitually cited to justify the benefits to society of the vast military budget? As Sandra Harding points out in her contribution to this volume, "If you want to do modern agribusiness, modern technosciences can help; if you want to maintain a fragile environment and biodiversity, those sciences, so far, have been of little assistance."

On the other hand, significant sectors of the sciences are increasingly devoted to dealing with the risks and threats generated by the industrialized phase of science's transformation of nature. Scientific development is a source of environmental problems for which science provides new market solutions. Waste management and environmental security are industrial sectors in their own right. Scientific echelons of the ecology movement are now fully professionalized and industrialized. Even segments of the military's vast surveillance infrastructure have been retrofitted to serve as ecological warning systems. Are these developments continuous with science's service to state and business elites, or do they represent a crisis in the relationship of service, in tandem with the crisis of industrialization itself? With such questions at the forefront of the moral, if not the operational, philosophy of science, it is hardly surprising that the ideology of objectivity and truth is under renewed scrutiny or that its defenders are rallying.

The recent call to arms has taken the form of what some have called the Science Wars, a second front opened up by fresh conservative recruits lured by the successes of their legions in the Culture Wars. Seeking explanations for their loss of standing in the public eye and for the decline in funding from the public purse, conservatives in science have joined the backlash against the (new) usual suspects — pinkos, feminists, and multiculturalists of all stripes. The epochal 1993 congressional decision to pull federal funding for the superconducting supercollider project affected these scientists in much the same way that Waco affected the militia movement. It was interpreted as a sign that the Cold War contract between government and scientists could no longer be taken for granted. Had the cozy postwar agreement, whereby scientists' political neutrality was traded for Department of Defense (DOD) dollars, survived the publicity storms over Indochina and Star Wars only to fall afoul of the new penny-pinching mood in Congress? More likely than the treachery of the Feds, the supercollider decision was in line with shifts in the techniques of industrial management, for which the complex hierarchy of the high-energy physics lab had become a superannuated model. "Big science," in other words, was in trouble, not because of sustained ideological criticism, but because it was out of sync with the new downsizing and decentralizing tendencies of business organizations.

In any event, science boosters and patriots wounded by the supercollider decision did not have to look far for scapegoats. In the search for nation spoilers, the Culture Wars had already produced a cast of culprits. The sequel, bankrolled and coordinated by the same right-wing groups (National Association of Scholars) and foundations (Olin), has probably already come to an op-ed near you, and we have suffered another season of asinine anecdotes about feminist algebra, queer quantum physics, and Afrocentric molecular biology, backed up by warnings that these people are battering rams, no less, behind which the barbarian flood of counter-Enlightenment irrationalism and pseudoscience lies seething in wait to claim the citadel.

The churlish tone of this backlash was set by Paul Gross and Norman Levitt's book *Higher Superstition: The Academic Left and Its Quarrels with Science* published in the Spring 1994. In spite of the authors' claim that they are not "stalking horses for social conservatism." *Higher Superstition* belongs fair and square to the choleric tradition of Alan Bloom, William Bennett, Roger Kimball, Hilton Kramer, and Dinesh D'Souza. Presented as a wake-up call to unsuspecting scientists, it identifies and caricatures "science bashers" in the same promiscuous fashion as those before had fingered the defilers of their Great Books tradition: "The relativism of the social constructionists, the sophomoric skepticism of the postmodernists, the incipient Lysenkoism of the feminist critics, the millenialism of the radical environmentalists, the racial chauvinism of the Afrocentrics" (252). Gross and Levitt's effort generated its share of coverage in the scholarly media and drew cutting responses from the ranks of those who demolished *The Bell Curve*. This early attention was followed by a series of lavishly funded, highly publicized conferences intended to mobilize a broad coalition from the natural sciences, the social sciences, and the humanities. The best-publicized conference, "The Flight from Science and Reason," hosted in June 1995 by the New York Academy of Sciences clearly laid out the agenda of linking together a host of dangerous threats: scientific creationism, New Age alternatives and cults, astrology, UFOism, the radical science movement, postmodernism, and critical science studies, alongside the ready-made historical specters of Aryan-Nazi science and the Soviet error of Lysenkoism. The organizers had made an effort to include liberals and leftists—Wendy Kaminer, Bogdan Denitch, and others—among the pledged opponents of irrationalism. While some of the latter speakers suggested that the rise of the religious right might be a greater threat to rational inquiry than critiques from the left, the general level of argument relied on caricature. Alternating between thinly disguised attacks on the very existence of an academic left (which, in the organizers' minds, has all but taken over the humanities and social sciences) and ill-equipped attempts to explain the existence of the myriad alternatives to the orthodox scientific credo, the conference often degenerated into name-calling. The audience, primarily composed of science professionals and journalists, might well have expected some competent, even self-critical analysis of the reasons for both the flourishing of New Age movements and pro-democracy criticism of establishment science. Instead, all they got was a crude call to arms, leveled against folk devils within the "cultural elite," that ran lockstep in style if not in influence with the pit bull moralizing of the Buchanans, Doles, and Gramms of the moment. While the public appeal of a crusade against this newly conspiratorial "antiscience movement" is obscure, the local intent was quite plain: to get scientists (who haven't been sorely vexed by curricular displacements of T. S. Eliot by Toni Morrison) involved in the academic P.C. wars.

Of course, this is not the first time that a hue and cry has been raised about the decline of science's authority or its imperilment on all sides by the forces of irrationalism. Such complaints are endemic to the official discourse of the science establishment. But more than one commentator has observed that in science's long contest with religion, the titular status of dogma has changed sides from one authority to the other. Indeed, in this volume Steven Fuller argues that the *secularization* of science, resulting from the increasing decline in state funding, will be a crucial test of popular credibility in its tenets. That many famous modern scientists, especially physicists, have been mystics barely raises eyebrows. That a large number of North Americans today sustain a belief in creationism while living within a technologically advanced society where scientific rationality is the civil and economic religion is considered a clear and present danger to civilization. For Harvard physicist Gerald Holton, author of *Science and Anti-Science* and co-organizer of the "Flight from Science and Reason" conference, "antiscience" sentiments are an ominous and "ever present" reminder of "the Beast that slumbers below."[4]

Demonology of this sort does nothing to explain the substance and mass appeal of alternative forms of rationality. Asking how and why science—academic, corporate, and military—responds only to elite needs and interests might provide more persuasive answers for the flourishing of popular faiths in fringe places. For the most part, however, class-soaked pronouncements about the return of the Dark Ages among the ill-educated masses are intended to reinforce the myth of scientists as a beleaguered and isolated minority of truth seekers, armed only with objective reasoning and standing firm against a tide of superstitions. The disparity between this image and the world of actually existing science is staggering. For well over a century, the Golden Age image of science as a sequestered craftlike pursuit has been undermined by the wholesale proletarianization of scientific labor in commercial production. The vast majority of professional scientists today are industrial workers, producing local, technical knowledge, not publishable research (not to mention the service echelons of the scientific workforce—lab technicians and assistants, manual workers, maintenance and cleaning staff, high school teachers, and so on). Research and production science is dominated by the leading weapons, chemicals, biotechnology, energy, and microelectronics industries. Access to basic science is still the driving motor of corporate capitalism despite the recessionary cutbacks in the scientific workforce. And contrary to all the fine talk about the enlightened pursuit of public knowledge, secrecy and competition are the guiding principles of research, whether in the name of national security, corporate profit, or professional prestige.

Even in the research university, where the myth of "value-free science" is most evident, it is the golden rule of patent protection that increasingly decides when public sharing of scientific information ends and when the principle of product development and monopoly control over profits takes over. In the heyday of the Cold War, the national security state benefited directly from the lip service paid to value-free objectivity. (Remember the case, among many others, of the chemist Louis Fieser, who developed napalm, tested it on Harvard's playing fields, and wrote up his research in a book innocently entitled *The Scientific Method*?)[5] Today, it is more likely to be the biotechnology company that lines the pockets of molecular biologists in a field notorious for its two-way corporate-academic traffic. It is no coincidence that this is the same field where the Holy Grail myth of a unified theory of knowledge is thriving, as the search for a genetic "code of codes" in the Human Genome Project feeds the old alchemists' romance of unlocking the secrets of the natural world.

With late-twentieth-century science so compromised, industrialized, and commodified, the militant resurgence of belief in its pristine truth claims is not hard to understand. But the crusaders behind the Science Wars are not about to throw the moneylenders out of the temple. Their wrath is aimed, above all, at those who show how the temple was built and how its rituals are maintained—the constructionist academic left, which is allegedly driven by resentment at seeing its politics devalued at the ballot box and is now apparently engaged in a toy soldier crusade to seize some of the academic authority and power exercised by scientists. These resenters include ethnographers, such as Bruno Latour, who dissect the cultural belief-systems of scientists' communities; sociologists of scientific knowledge, such as the adherents of the Edinburgh school, who expose the interest-driven nature of scientific research; multiculturalists, such as Donna Haraway, Sandra Harding, and Evelyn Fox Keller, who uncover the gender-laden and racist assumptions built into the EuroAmerican scientific method; philosophers who capsize or redefine claims to objectivity; historians who show the relationship between science's empiricial world-view and that of mercantile capitalism; and cultural studies folks who show how the powerful language of science exercises its daily cultural authority in our society. The political aims of these groupings are varied: (a) some simply want to provide an accurate, scientific description of empirical scientific practice; (b) others want, more ambitiously, to see science redeem its tarnished ideals from internal abuse and external impurities; (c) others, more normative, would persuade scientists to be self-critical about the political nature and social origins of their research and to engage in advocacy science to combat the risks and injustices that are side effects of technoscientific development; (d) still others want, more radically, to create new scientific methods that are rooted in the social needs of communities and accountable to social interests other than those of managerial elites in business, government, and the military. The last aim is driven by the principle that people whose lives are greatly affected by the effects of scientific superindustrialism ought to have a role in the decision making that determines research. But it also involves taking seriously the proposition that Western technoscience is a highly local form of knowledge and is therefore unlikely to have a world monopoly on good scientific ideas.

Far from recognizing these critiques as the outcome of scientific self-scrutiny, apocalyptics like Holton only see Spengler's vision of science "falling on its own sword" and society plunging into unreason.[6] And despite the weighty differences between these perspectives, all are collapsed by Gross and Levitt into a caricature of the enemy as boffo nihilists who deny outright the existence of natural phenomena, such as recessive genes or subatomic particles or even the law of gravity. If nothing else, Gross and Levitt have reinforced a weird pre-Copernicanism that views the entire social universe as revolving around scientists and that suspects all bodies with slightly eccentric orbits of displaying antiscience tendencies. Thus they have made bedfellows of the most disparate movements: creationism, holistic medicine, and feminism, to name just three. Creationism is more pro-Christian than it is antiscience; alternative medicine is more prohealth than it is antiscience; and feminism is more pro-democracy than it is antiscience. Nothing connects these movements except Gross and Levitt's conspiratorial fantasy. It remains to be seen whether the science warriors' attempts to summon up an "anti-science movement" specter within the public imagination is successful. If so, their attempts will certainly have been helped along by the hunt for the alleged Unabomber, a media event that has revived the more familiar specter of the leftist Luddite technophobe.

The currency of such stereotypes is particularly unfortunate right now, a time that requires more understanding of the massive tech-noskepticism that has evolved among the populations of advanced industrial societies. If the critique of science has become a very ordi-nary activity, science itself has become more and more exclusive in the age of technical expertise. The telecommunications and informa-tion processing industries that have transformed the workplace into a site of knowledge production have also boosted the power and authority of scientific expertise, establishing cyberculture as an inter-national lingua franca of the new elites that is well nigh unintelligible to the uninitiated (perhaps for the first time since the preliterate, preindustrial age). The rise of a privatized knowledge society does not translate into a scientifically informed citizenry; it creates a hierar-chy of technical expertise and, in particular, releases scientists from public accountability on the grounds that their critics "just don't know enough." But how much do you need to know technically about nuclear fission to conclude that nuclear energy is a socially destruc-tive idea? And what is irrational about questioning the manipulation of genetic material? The narrowness of scientific expertise that poorly qualifies its practitioner for the broad exercise of social reason is the same specialist knowledge that guarantees science its immunity to public criticism.

Science studies has met the charge of technical inexperience before. This complaint met with an empirical response in the "descriptive turn" of the SSK (Sociology of Scientific Knowledge) school, repre-sented in the work of Barry Barnes, David Bloor, Steve Shapin, Michael Mulkay, and Harry Collins, and followed by Nigel Gilbert, Joan Fuji-mura, Ian Hacking, Eric Livingston, Trevor Pinch, Nancy Cartwright, David Gooding, and others.[7] Rather than study the products of sci-ence, or scientists' own representations of science, the strong program of SSK endorsed the study of science in practice—an approach that generated celebrated ethnographic accounts of laboratory procedures, such as Bruno Latour and Steve Woolgar's *Laboratory Life*, Karen Knorr Cetina's *The Manufacture of Knowledge*, Michael Lynch's *Art and Artifact in Laboratory Science*, and, in a different phase, Sharon Traweek's *Beamtimes and Lifetimes*—and generally analyzed the role social interests played in every aspect of research. Such studies demonstrated that scientific knowledge is not given by the natural world but is produced or constructed through social interactions between/among scientists and their instruments, and that these inter-actions are mediated by the conceptual apparatuses created in order to frame and interpret the results.

The result of this kind of attention was a more scientific description of "science in action," one better equipped to demystify scientists' perception that the working environments of their research laboratories and communities are value-free quarantine zones. But the relativism inherent in the descriptive turn was much criticized as a theory that ultimately declared the sociological equivalence of all beliefs, whether true or false, according to internal or external criteria. It did nothing to dislodge the elitist perception that only scientists can do science, and it shied away from all normative or evaluative analysis that might produce change in scientific institutions.[8] In opting for a program of social realism that eschewed value-laden moralistic critique, SSK's passive explanation of science's social construction met with charges of political quietism. Its proof that, contrary to the positivist view, there was nothing distinctive to differentiate science from any other social activity was a crucial accomplishment. But this evidence would not in itself lead to alternative ways of doing science. Nor would it suggest alternative criteria (i.e., criteria other than those framed by the needs of the military, capital, or the state) for the successes and failures of science. Other, more normative critiques, especially from feminists, argued that science is only a partial representation of knowledge and that its methods and procedures not only foster inexperience about social divisions of race and gender but actively perpetuate such divisions.

At a time when the division of labor is increasingly stratified by levels of technical knowledge and when scientific knowledge is systematically whisked out of the orbit of education and converted into private capital, it is all the more important that public critics not be cowed by accusations of technical ignorance. Expertise is not inherently undemocratic, but its deployment in a technocratic society reduces citizen input to zero.

We need to have nonscientist participation in decision making about science's priorities; the concept of "science for the people" is more relevant and more remote than ever. We need to acknowledge that there is more than *one* version of science, because a singular method can always be "purified" of its social contaminants and refortified. We need to have scientific methods that are built around the experience of people in living environments, as opposed to the closed environment of laboratory instruments. In the face of the destructive history of technology transfer in the Green Revolution, we need to encourage the movement for indigenous, Third World technoscience that is sustainable, appropriate, and rooted in local knowledge and local resources. We also need to challenge the dividend system of military science, whereby colossal amounts of funding are justified by technical spinoffs, as an inappropriate way to fund and disseminate scientific benefits to society. And when the Science Warriors and their caricatures command op-ed space and are permitted to bend the ears of academic administrators and government officials, we need to hear from nonconservative scientists who share some of these concerns.

This book is an expanded version of a special issue of *Social Text*. The issue was conceived of as a forum where scholars in science studies could comment directly upon the discussion generated by the Science Wars. Since there had been no concerted, published response to *Higher Superstition*, other than isolated reviews, the editors felt that such a forum would serve a useful purpose. Alternately, contributors were asked to present examples of their own work or to comment upon responses, such as those of Gross and Levitt, to their work within their own fields. The responses are appropriately varied and are drawn from natural scientists, sociologists, anthropologists, historians, and scholars in cultural studies and literary studies, both in the United States and the United Kingdom.

The "Science Wars" issue of *Social Text* also included an unsolicited article by Alan Sokal which attracted a good deal of notice in the public media and in academic discussion when it was revealed by its author to be a hoax. Sokal's declared intention was to present a critique of science studies in the vein of *Higher Superstition*, and so the hoax became something of an event in the Science Wars themselves. In addition, it was used as a vehicle in the Culture Wars, not only by conservatives but also by declared progressives seeking to denounce broad sectors of the "academic left" for straying from the paths of righteousness. While the Sokal affair perpetuated the climate of caricature that Gross and Levitt's book had established, it also brought a much wider audience to the topics addressed by science and technology studies. Unfortunately, Sokal's faux version of science studies may have been taken at face value by many who read no further than his article or the press reports about it. The outcome, once again, was to distract attention from the *bona fide* voices in the field, many of whom were included in that issue of the journal. In presenting a broad spectrum of responses to the claims that initiated the Science Wars, this volume seeks to remedy the neglect.

(Interested readers can consult the Sokal article in *Social Text* 47/48, and the editorial response and exchanges in the May/June 1996 issue of *Lingua Franca*).

Thanks are due to my editorial colleagues at *Social Text*, especially Bruce Robbins, Randy Martin, and Monica Marciczkiewicz, and to Ken Wissoker, Angela Williams, and Jean Brady at Duke University Press.

Notes

1   Ulrich Beck, *Risk Society: Toward a New Modernity*, trans. Mark Ritter (London: Sage, 1992); and *Ecological Enlightenment: Essays on the Politics of the Risk Society*, trans. Mark Ritter (Atlantic Highlands, N.J.: Humanities Press, 1995).

2   See Emil Altvater, *The Future of the Market: An Essay on the Regulation of Money and Nature after the Collapse of "Actually Existing Socialism,"* trans. Patrick Camiller (London: Verso, 1993).

3   Robert Proctor, *Value-Free Science: Purity and Power in Modern Knowledge* (Cambridge, Mass.: Harvard University Press, 1991), 224.

4   Gerald Holton, *Science and Anti-Science* (Cambridge, Mass.: Harvard University Press, 1993), 184.

5   Cited by Hilary Rose and Steven Rose in "The Radicalization of Science," in *The Radicalization of Science: Ideology of/in the Natural Sciences*, ed. Hilary Rose and Steven Rose (London: MacMillan, 1976), 15.

6   Holton, 132.

7   For a representative selection of some of the concerns that have shaped the SSK tradition, see Andrew Pickering, ed., *Science as Practice and Culture* (Chicago: University of Chicago Press, 1992).

8   See Steve Fuller, "Social Epistemology and the Research Agenda of Science Studies," in Pickering, 390–428.

Science Is "Good to Think With"

Sandra Harding

Thinking Science, Thinking Society

I T is ironic that the major criticism of the new social studies of science and technology from the antidemocratic right in fact provides yet more evidence for the value of those science studies. The new science studies shows how the "order of knowledge" has also been the "order of society." When challenges to the social order have arisen, these challenges have also changed the prevailing ways that the production and legitimation of knowledge have been organized, and vice versa: the social order and the structure of a culture's sciences are generated through one and the same social transformations (see, for example, Merchant 1980; Restivo 1988; Shapin 1994; Shapin and Schaffer 1985). This is pretty close to what the antidemocratic right believes: the new science studies, feminism, "deconstructionism," and multiculturalism threaten the downfall of civilization and its standards of reason.[1] The latter criticism does not contest that the order of knowledge and the social order shape and maintain each other, but only the way science studies reveals how such science-society relations have worked in the past and operate today, and the proposal in some of these science studies tendencies for more open, public discussion about the desirability of prevailing science-society relations. It is significant that the Right's objections virtually never get into the nitty-gritty of historical or ethnographic detail to contest the accuracy of social studies of science accounts. Such objections remain at the level of rhetorical flourishes and ridicule.

Democracy-advancing social movements, building on and expanding the earlier class-based analyses, have argued that the natural and social sciences we have are in important respects incapable of producing the kinds of knowledge that are needed for sustainable human life in sustainable environments under democratic conditions. The conventional conceptual frameworks of the natural and social sciences have been designed for quite different projects—in short, for producing the kinds of information useful to the administrators and managers of nation states, multinational corporations, and militaries. These institutions have central interests in continuing European expansionist projects that benefit primarily elites, and they all embody particularly destructive ideals of manliness and womanliness that facilitate such expansion. Important kinds of information and understanding wanted by those who bear disproportionate shares of the costs of how these institutions organize social relations, both in the North and the South, cannot be gathered through the kinds of conceptual frameworks that natural and social sciences have developed for such administrative/managerial projects.

The antidemocratic right often accuses the new science studies of relativism, but it is wrong about just what it is to which the new science studies "relativizes" sciences. Science studies does not claim that sciences are epistemologically relative to each and every culture's beliefs such that all are equally defensible as true. Rather, the point is that they are historically relative to different cultures' *projects*—to cultures' questions about the natural and social orders. Different questions produce different answers containing distinctive, sometimes conflicting, representations of nature and, indeed, of science, and the representations that conflict do not fit together like pieces of a jigsaw puzzle.

For example, the representations of nature, society, and science appearing in the answers to those questions of most interest to nation states, multinational corporations, and militaries often conflict with the representations that appear in the answers to the most pressing questions for those who bear the costs of those institutions' projects. If you want to do modern agribusiness, modern technosciences can help; if you want to maintain a fragile environment and biodiversity, those sciences, so far, have been of little assistance. This is so in part because the representations of nature's regularities and underlying causal tendencies in technosciences and in some forms of environmentalism are conflicting. The representations of nature, society, and maximally effective knowledge production in the sometimes more effective "local knowledge systems" conflict with those in modern technosciences.[2] This is as true of different sciences within the same culture as it is of "the same science" in different cultures, of course. The conceptual frameworks of modern physics, chemistry, and biology on the one hand, and environmental sciences on the other hand do not fit together perfectly. The latter require learning to negotiate between the principles of these modern sciences and of both local and social knowledge of environments, neither of which has a place in the conceptual frameworks of those modern sciences (see Seager 1993). Indeed, the conceptual frameworks of those three modern sciences no longer appear unifiable (Dupre 1993).

Of course, this kind of historicization of scientific projects does indeed generate a different epistemology for science than the standard ones that set out to show how reasonable it is that truth always manages to land on the side of the latest arrival among modern sciences. This epistemology supports thinking with/between more than one knowledge system, from their "borderlands," so to speak (Anzaldúa 1987; Collins 1991). It is exactly in opposition to the claim that there are no rational standards for deciding between them. This kind of epistemology simply openly asserts that whatever their explicit topic, rational standards are always about both the natural and social orders. I turn shortly to suggest how important such an epistemology is for the barely initiated project of developing anti-Eurocentric Northern science studies.

Displacing Anxieties

Another way to state my point here is that science is always "good to think with," as Lévi-Strauss said about sex. For "educated classes" whose own status depends on the same appeals to objectivity, rationality, expertise, and progressiveness on which science's legitimacy depends, science discourses can be mobilized to encourage people to think in politically seductive ways about any and all social issues. And, of course, when it is possible to mobilize both sex and science discourses to reflect on the resources and dangers of social change, one can be assured of generating a great deal of heat, however much light gets produced. The necessity for "society" to control "wild women" and powerful, malevolent "mothers" has again and again been mobilized in appeals for support for modern science's ways of predicting and controlling nature, from Machiavelli's famous concerns for "man's fate" to the antidemocratic right's recent clarion calls for the citizenry to join in stamping out feminism and, in particular, feminist science studies. One of the sins of the latter, we are told, is critically quoting the same kind of earlier appeals by figures such as Machiavelli and Bacon. I guess we are not supposed to ask for whom science is supposed to be kept a safe haven from these phantasmic wild women and malevolent mothers. Clearly, feminist science studies is a threat because it makes visible what is not supposed to exist—therefore, what does not really exist, according to this kind of magical thinking.

Ideals of manliness are also threatened by what are perceived as national economic and political failures in the United States. Purportedly morally and culturally inferior others, femininized in the national discourses of domestic and international relations, are swarming across the U.S. borders, reproducing at higher rates than European Americans and claiming powerful voices in national and international politics (see Enloe 1990; Parker et al. 1992; Tickner 1992). Who is entitled to set the national standards for rationality, objectivity, and the goals of science is no longer so obvious.

However, for many scientists and well-intentioned liberals—and there is a large overlap between these groups—it is not clear that displacing anxieties about social change onto the already most vulnerable segments of the social order will be effective. Liberalism is an appealing political philosophy to scientists, since its basic concepts and ways of thinking are precisely those of modern science's self-image. Liberal judges, lawyers, political figures, and administrators have been in the forefront of public agenda struggles against sexism and racism. It is important to remember that liberal political philosophy is being pulled in both anti- and prodemocratic directions these days, as can be seen in public resistance to the way much originally prodemocratic liberal rhetoric has been appropriated for antidemocratic projects—for example, "right to life," "racial discrimination," and, of course, "objectivity," "rationality," and "science."

This situation highlights the importance of continuing to struggle to keep liberal concepts headed in prodemocratic directions rather than abandoning them. After all, feminist and antiracist science studies have called for more objective natural and social sciences, not less objective ones! They want sciences that are competent at detecting the culture-wide presuppositions that shape the dominant conceptual frameworks of disciplines and public discourse. Such presuppositions, if unexamined, function as evidence, "laundering" sexism or racism or class interests by transporting them from the social order into "the natural order." Women, peoples of color, and poor people want to know how our bodies do work and how to protect them from the effects of sexism, racism, and class exploitation. Women want to know how to serve the health needs of the three generations dependent on them. Just what is in the food we eat and the air we breathe? Concepts such as objectivity, rationality, good method, and science need to be reappropriated, reoccupied, and reinvigorated for democracy-advancing projects (see Harding 1992).

Multicultural and Global, Gendered, Science Studies?

At this point I want to suggest that in some ways the new science studies still has far to go to achieve a maximally objective, accurate, and comprehensive representation of relations between the social order and the order of knowledge—past, present, and future. The Right consistently links science studies, feminism, deconstruction-ism, and multiculturalism, but science studies, both feminist and "pre-feminist," has barely begun to link its projects with those of mul-ticulturalism. Moreover, the links between deconstructionism and sci-ence studies are only beginning to be fully explored, and most of the social studies of science woefully lack an understanding of gender relations as a part of social relations. In the latter case, these fields still tend to conduct their analyses as if feminist analyses were relevant only when women appeared in the historical record or, perhaps, as contemporary lab directors. For them, no gender relations are present when the scientists and/or the science theorists are only men, unless women are complaining.

However, the gap I shall focus on here is the first one; the postcolo-nialism and multiculturalism of current thinking in history, literature, cultural studies, and political economy have hardly begun to inform science studies or its feminist component. How should multicultur-alism or, let us say more generally, anti-Eurocentrism, shift the largely still-Eurocentric focuses of contemporary science studies? The anti-Eurocentric histories of the North and the South have now trans-formed K–12 history texts. How can our understanding of modern sci-ences be located more firmly in such "single-stream," postcolonial histories, and in the related critical studies of so-called development that have in the last decade changed the way many in the North are coming to think about the origins and subsequent history of "Western Civilization"? (See, for example, Blaut 1993; Sachs 1992.)

Development studies themselves have for the most part appeared to the social studies of science, including its feminist components, to lie far off in the conceptual and political distance. They appear as a com-pletely separate discourse, with few obvious points of contact to the histories, sociologies, ethnographies, and philosophies of modern sci-ence that have been centered in science studies. There are many rea-sons for this phenomenon. For one thing, the notions of science cen-tral to many development accounts are largely the older, purportedly value-free, "mirror of nature" ones, not the social and cultural ones developed by the social studies of science.

Moreover, even for the new science studies that insists that science cannot be "pure" of social dimensions, development appears primarily to be an issue of applied sciences and technologies—not of the "high sciences," such as abstract physics, chemistry, and biology that, paradoxically, are still centered in so much of the social studies of science. Thus the new science studies still assumes the liberal concept of development as the transfer of Northern sciences and technologies to the South; not, to imagine another scenario, as the South's development of "its own" sciences and technologies that need not be restricted to what the North has found useful, and that might thereby creatively make important contributions to the storehouse of global scientific knowledge. Thus the institutional structure of U.S. universities tends to locate development studies and their projects mostly outside the science departments, and also outside the humanities and social science departments and colleges that have largely produced science studies. Development studies are more likely to be found in schools of "applied science" such as those of agriculture and public health.

Feminist science studies and technology studies have been largely separate and noncommunicating fields—or, at least, science studies doesn't draw on technology studies. Neither has been much concerned with the issues central to the interactions between technological and scientific knowledge, or to the various forms of women in/and development. Colonial science and its continuation in so-called development is more obvious to historians of science in some European countries where, for example, the archives of national colonial science histories are accessible in Paris, Copenhagen, and other European centers. Such archives are evidently scarce in the United States, with the latter's different pattern of "settler colonialism" and imperialism.

These obstacles are by no means negligible, but neither are the losses to science studies from failing to overcome them. There are two routes into the issues here that will feel at least partially familiar to science studies. One would locate studies of the history of modern science and of the empirical knowledge traditions of other cultures on anti-Eurocentric maps of European expansion, its causes and effects. The other would then construct multicultural and global feminist science studies on the map created by this first project. I sketch out these routes briefly not only for their intrinsic interest, but also because they both strengthen the case for articulating more clearly the advantages of epistemologies that can historicize sciences to determinate social projects, about which a great deal of rational, objective dialogue is possible, if not always easy to organize.[3]

The Growth of Modern Science and European Expansion: A Problematic Co-dependency?

Only occasionally visible in Northern histories, philosophies, sociologies, and cultural studies of modern science, including their feminist components, are (1) anti-Eurocentric representations of the scientific traditions of other cultures; (2) accounts of the effects that European expansion and the growth of modern science had and continue to have on each other, including analyses of the consequent and otherwise existing cultural "localness" of modern sciences; (3) analyses of the losses that continued reduction in cognitive diversity insures for the human future; and (4) anti-Eurocentric philosophic discussions of the resources and challenges of other "local knowledge systems" and, especially, their interactions with modern science in both the North and the South. A still relatively small but vital literature has been produced on this array of topics (see, for example, Crosby 1987; Goonatilake 1984, 1992; Haraway 1989; Harding 1993, 1994, forthcoming a; Hess 1995; Joseph 1991; McClellan 1992; Needham 1969; Petijean et al. 1992; Sabra 1976; Sachs 1992; Sardar 1988; Seager 1993; Shiva 1989; and Traweek 1988, n.d.).

Chinese and Islamic sciences and technologies are the topic of the most developed comparative studies. Careful readings of these accounts also reveal understandings of the causes of the strengths and limitations of modern European sciences that are otherwise hard to come by. For example, Needham discusses how central fruitful scientific concepts, such as the "laws of nature," were indebted to Christian precepts, no less than were the unfortunately long-lasting beliefs in the heavenly crystal spheres. Needham is saying that religious beliefs that constituted central cognitive notions in modern sciences greatly advanced these sciences in some respects and retarded them in others. This kind of discussion stands on the borderlands between Chinese and modern European sciences, looking back and forth critically from one to the other (Needham 1969).

Studying codependency relations between European sciences and European expansion from their beginnings through the present offers another way to explore strengths and limitations to how sciences and societies advance their interests. European expansion turned the world into a laboratory for European sciences, and European sciences made possible that expansion. Modern sciences answered questions about how to improve European land and sea travel; to mine ores; to identify the economically useful minerals, plants, and animals of other parts of the world; to manufacture and farm for the benefit of Europeans living in Europe, the Americas, Africa, and India; to improve their health (and occasionally that of the workers who produced profit for them); to protect settlers in the colonies from settlers of other nationalities; to gain access to the labor of the indigenous residents, and to do all this to benefit only local European citizens—the Spanish versus the Portugese, French, or British.

Most models of the scientific future, including most of those implied in the new science studies, imagine "one true science." They do not imagine as existing or desirable many, different, and in some respects conflicting representations of nature. Yet this vision is beginning to emerge in the new Northern science studies and throughout those positioned on the single-stream historical maps (for Northern, see Dupre 1993). Different cultures are located in, and regularly exposed to, different parts of nature, and such distinctive environmental locations create culturally distinctive scientific resources. Moreover, metaphors, models, and other discursive resources of the sciences are not just the unmitigated threats to knowledge that they are usually figured as being; rather, they enable us to "see" our particular parts of nature in ever more diverse ways. Furthermore, culturally distinctive knowledge traditions have different interests in those parts of nature with which cultures interact, and each can generate both systematic knowledge and systematic ignorance—each has its strengths and limitations. Finally, different cultures have different ways of socially organizing the production of knowledge—European expansion is no less such an organization (today as in the past) than is a more conventional laboratory with its surrounding apparatus of educational feeds, funding agencies, peer-review journals, and so on. Different organizations of knowledge generate different illuminating representations of nature. These four "causes" of all sciences being "local knowledge systems," valuably historicized to culturally local projects, have been obscured by preoccupations with representations of ideal science as an undistorting mirror of a fixed and perfectly coherent nature.

An epistemology for such a world of many different sciences is not a damaging relativist one but, rather, a borderlands epistemology that values the distinctive understandings of nature that different cultures have resources to generate. The goal of such an epistemology is not to try to integrate them all into one maximally ideal knowledge system, for such a process would necessarily lose the advantage of the conflicting cognitive/moral/political interests, discursive resources, ways of organizing the production of knowledge and, thus, conceptual schemes that cultures have developed. Instead, each of us, and our local institutions, would learn the resources and limitations of diverse knowledge systems. Instead of the sciences being "smart" and cultures dumbly following their "directions," cultures and their members would become "smart" and learn when to use one science as a reliable guide and when to use another; what to value in modern sciences and what to value in other knowledge systems. Obviously, we all already do this when we choose which Northern science to go to for a particular kind of information, or when we decide when to listen to biomedical advice and when to go ahead and do our vitamin or other "folk science" therapies. This is an epistemology seeking the most useful knowledge "collage" rather than one perfect representation of the world. Or, to switch metaphors, a "borderlands epistemology" seeks the best set of scientific maps for the different purposes for which one needs sciences, rather than one colossal map that provides a maximally adequate guide to anywhere that anyone might ever want to travel for any purpose whatsoever. Such a perfect "universal map" could only be the world itself, of course.

Science Theory for Multicultural and Global Feminisms?

A second, related site for beginning to transform science studies through postcolonial lenses can be identified by asking this question: how should the gap be bridged between the theoretical approaches of multicultural and global feminisms, on the one hand, and science studies, on the other? These two feminist theories have developed far more accurate, complex, and comprehensive concepts of gender relations than those that have directed the projects and provided resources for most of the existing feminist science studies.

Multicultural feminisms focus on the distinctive political histories and practices shaping women's conditions, interests, and desires in different local, national, and transnational cultures and the necessity for thinking about these distinctive histories in designing social change. And they focus attention on the powerful resources of bi- and multicultural communities in creating "outsiders within" (Collins 1991) and thinkers socially positioned to see the world from the "borderlands" (Anzaldúa 1987). They show the valuable resources to be gained by learning to negotiate on individual, institutional, and discursive levels between conflicting cultures and their conceptual frameworks. Global feminist approaches want to explain the role of gendered institutions, practices, and cultures in the global political economy. They want to show how such institutions depend upon the maintenance of masculine cultures in militaries, governments, and multinational corporations such that the continued appropriation of women's activities and resources in different ways in every part of social relations can be facilitated and legitimated. They challenge the adequacy of Liberal and Marxian analyses of world systems, showing the commitments to male supremacy that have shaped both world systems and such accounts of them (see, for example, Enloe 1990; Seager 1993; Shiva 1989).

Thus feminist science studies, like science studies more generally, needs to conduct its analyses with the more accurate and complex conceptions of gender generated by multicultural and postcolonial feminisms—gender as mutually constructed by historically dynamic formations of class, ethnicity/race, sexuality, and whatever other social relations structure social orders. And, then, questions about the distinctive gender relations of modern and other local knowledge systems' histories and practices can be raised. Women have standpoints on nature and on the production of technosciences, not just on social relations, that are different in different cultures and systematically related across global political economies.

The categories of cultural difference mentioned in the previous section can also provide useful resources for conceptualizing multicultural and global gender differences. Within cultures, women and men have different biologies (though not as different as sexist biology would like to prove) and, to the extent that they are socially assigned different activities, they will tend to have interactions with different aspects of parts of nature's orders. These exposures to nature and assigned social interactions will provide women and men with partially different interests in and desires about nature's regularities and underlying causal tendencies, different relations to local and global cultural discursive traditions about sciences and technologies, and different ways of organizing the production of knowledge. Knowledge *for* women and their activities does not completely coincide with knowledge *for* men and their activities (let alone *for* the managers of nation-states, multinationals, and militaries) though much scientific knowledge can be useful to everyone (see Harding forthcoming b). Thus the history of modern science's development through European expansion, of other culture's sciences and technologies, as well as of their and our prevailing understandings of these histories are gendered in ways scarcely explored in prevailing science studies or its feminist component.

These more complex conceptions of gender, which many of us use to analyze topics other than science and technology, also can be used to locate issues about the gender relations structuring sciences and technologies, and vice versa, on postcolonial maps.

Only a Beginning

Science is "good to think with." There is much more thinking that will be done with it by the antidemocratic right, since the social transformations that are the only mildly occluded object of this thought show no signs either of abating or of escaping protest. However, there is also a good deal more thinking both with and about science that contemporary science studies can do to create more accurate, comprehensive, and useful representations of science/society for prodemocratic sciences and politics. The Science Wars are about the future, as well as about the past and present, and it is important that prodemocratic tendencies take the lead in defining the terms within which science is good for public dialogue. In the North we have only begun to articulate how global social orders and orders of knowledge generate each other past, present, and future.

Notes

1   I use *antidemocratic right* and *democracy-advancing* movements or tendencies in a somewhat simplistic way throughout this discussion. A central focus of dialogue needs to be on the circumstances in which sciences, modern and otherwise, advance and retard democratic social relations. These are extremely complex and ill-understood matters, and the Science Wars are located firmly in the middle of them.

2   Local knowledge systems ("ethnosciences," "indigenous knowledge systems") are by no means always more accurate and effective than modern scientific knowledge, but sometimes they are.

3   One could discuss this as a relativist epistemology, but I think that is a poor strategy, since it is so easily misunderstood. There are very good reasons for claiming that such an epistemology, in fact, generates "strong objectivity," rather than only the weak standards for objectivity that are incapable of identifying the historical conceptual frameworks that contain their sciences' concerns (see Harding 1992).

References

Anzaldúa, Gloria. 1987. *Borderlands/La frontera*. San Francisco: Spinsters/Aunt Lute.

Blaut, J. M. 1993. *The colonizer's model of the world: Geographical diffusionism and Eurocentric history*. New York: Guilford.

Collins, Patricia Hill. 1991. *Black feminist thought: Knowledge, consciousness, and the politics of empowerment*. New York: Routledge.

Crosby, Alfred. 1987. *Ecological imperialism: The biological expansion of Europe*. Cambridge: Cambridge University Press.

Dupre, John. 1993. *The disorder of things: Metaphysical foundations of the disunity of science*. Cambridge, Mass.: Harvard University Press.

Enloe, Cynthia. 1990. *Bananas, beaches, and bases: Making feminist sense of international politics*. Berkeley: University of California Press.

Goonatilake, Susantha. 1984. *Aborted discovery: Science and creativity in the third world*. London: Zed.

———. 1992. The voyages of discovery and the loss and rediscovery of the "other's" knowledge. *Impact of Science on Society*, no. 167: 241–64.

Haraway, Donna. 1989. *Primate visions: Gender, race, and nature in the world of modern science*. New York: Routledge.

Harding, Sandra. 1991. *Whose science? Whose knowledge?* Ithaca, N.Y.: Cornell University Press.

———. 1992. After the neutrality ideal: Science, politics and "strong objectivity." *Social Research* 59: 568–87.

———, ed. 1993. *The "racial" economy of science: Toward a democratic future*. Bloomington: Indiana University Press.

———. 1994. Is science multicultural? Challenges, resources, opportunities, uncertainties. *Configurations* 2: 301–30.

———. Forthcoming a. Is modern science a European knowledge system? In *Science and technology for the south: Sociology of the sciences yearbook 1995*, edited by T. Shinn. Dordrecht: Kluwer.

———. Forthcoming b. What makes possible women's standpoints on nature? *Osiris*.

———. Forthcoming c. Multicultural and global feminist philosophies of science. In *Feminism, science, and the philosophy of science*, edited by Lynn Hankinson Nelson and Jack Nelson. Dordrecht: Kluwer.

Hess, David J. 1995. *Science and technology in a multicultural world: The cultural politics of facts and artifacts*. New York: Columbia University Press.

Joseph, George Sheverghese. 1991. *The crest of the peacock: Non-European roots of mathematics*. New York: I. B. Tauris and Co.

McClellan, James E. 1992. *Colonialism and science: Saint Domingue in the old regime*. Baltimore, Md.: Johns Hopkins University Press.

Merchant, Carolyn. 1980. *The death of nature: Women, ecology, and the scientific revolution*. New York: Harper and Row.

Needham, Joseph. 1969. *The grand titration: Science and society in East and West*. Toronto: University of Toronto Press.

Parker, A., et al., eds. 1992. *Nationalisms and sexualities*. New York: Routledge.

Petitjean, Patrick, Catherine Jami, and Anne Marie Moulin, eds. 1992. *Science and empires: Historical studies about scientific development and European expansion*. Dordrecht: Kluwer.

Restivo, Sal. 1988. Modern science as a social problem. *Social Problems* 35: 25–42.

Sabra, I. A. 1976. The scientific enterprise. In *The World of Islam*, edited by B. Lewis. London: Thames and Hudson.

Sachs, Wolfgang, ed. 1992. *The development dictionary: A guide to knowledge as power.* Atlantic Highlands, New Jersey: Zed Books.

Sardar, Ziauddin, ed. 1988. *The revenge of Athena: Science, exploitation, and the third world.* London: Mansell.

Seager, Joni. 1993. *Earth follies: Coming to feminist terms with the global environmental crisis.* New York: Routledge.

Shapin, Steven. 1994. *A social history of truth.* Chicago: University of Chicago Press.

Shapin, Steven, and Simon Schaffer. 1985. *Leviathon and the air pump.* Princeton, N.J.: Princeton University Press.

Shiva, Vandana. 1989. *Staying alive: Women, ecology, and development.* London: Zed.

Tickner, J. Ann. 1992. *Gender in international relations: Feminist perspectives on achieving global security.* New York: Columbia University Press.

Traweek, Sharon. 1988. *Beamtimes and life times.* Cambridge, Mass.: MIT Press.

———. n.d. *Big science in Japan.* Unpublished manuscript.

Does Science Put an End to History, or History to Science? Or, Why Being Pro-science Is Harder than You Think

Steve Fuller

THE recent revival of the "two cultures" problem between scientists and their cultural critics runs much deeper than the noisy clash of axe-grinding, agenda-mongering egos would suggest. One need only compare the sense of history that runs through the most unassuming textbook in any natural science, with the sense that informs any meticulously documented professional history of the same science. Though neither is polemical by intent, set side by side the two historical sensibilities could not be more opposed. This fact has been an open secret among historians of science for quite some time, though until now few have cared to make much of it, for fear that scientists would start pronouncing on how historians should do their work. However, bolstered by a growing historicism among philosophers and sociologists of science as well as by changes in the global political economy that have not been especially favorable to science, we may no longer want to claim—as Stephen Brush did in 1974—that history of science be rated X for scientists.[1] In what follows, I will present these two clashing sensibilities, in both their public and academic guises, and then show how the Japanese encounter with Western science during the Meiji Restoration enacts a practical resolution of the clash. However, historicizing science is only half the story. Science's cultural critics themselves (ourselves) need to be historically situated. While many scientists have tried to portray us as the latest incarnation of neoromantic irrationalism, I argue that historically truer precedents may be found in the Protestant Reformation and its aftermath, whereby we critics appear as science's secularizers. I end on a cautionary note about how this project of secularization might intersect with the institutionalization of history, philosophy, and sociology of science as the field of "science studies."

The Public's Understanding of Science: The New
Battleground for the Two Cultures

For the past few years, just before the ides of March, surveys have been conducted in Britain to find out what the "average person"—an unsuspecting weekend shopper in a London suburb—thinks about science. The results, usually a source of horror for readers of the national broadsheet newspapers, are then used to kick off the vernal ritual known as "Science Week," a steady stream of public lectures, experimental demonstrations, and museum exhibits designed to promote the cause of Science in barnstorming, nineteenth-century style. The official government reason for holding Science Week is that student enrollment in natural science courses (including teacher training courses) has been dropping steadily over the past decade, a trend that is thought to be somehow related to the perceived decline of British technological innovation and industrial productivity.

Much of the hoopla surrounding Science Week has the feel of a moral panic, a breakdown in social order, the nature of which is neither understood nor controlled by those in charge. The temptation, then, is to scapegoat a relatively defenseless group that can somehow be made to symbolize the crisis. Thus, primary and secondary school science teachers with degrees in education rather than in one of the natural sciences are often scapegoated for not sparking the spirit of inquiry in the classroom.[2] Occasionally, however, someone will observe that the decline in enrollment may not be such a bad idea, given that the unemployment rate for first-degree holders in the natural sciences (20 percent) is roughly twice the national jobless rate.[3] Nevertheless, defenders of Science Week maintain that the malaise runs deeper. Here the survey results are made to prove the point. Basically, they show that people do not have a very clear idea of what "science" or "the scientific method" is supposed to be and cannot fathom what makes such seemingly disparate disciplines as paleontology, high-energy physics, industrial chemistry, and genetic engineering "scientific" and hence worthy of their sustained interest and support. When it comes to specific research projects, the public seems benign toward inexpensive research that is unlikely to interfere with their daily lives but insists on the accountability of any research that is likely to result in a substantial change of lifestyle. In the latter case, the public seems to have a much more vivid sense of what might constitute such research than scientists themselves normally do.[4]

Does the public understanding of science, as characterized above, suggest a genuine state of crisis? I seriously doubt it. Rather, the sense of panic among government and scientific leaders is the result of factors that are largely beyond their direct control. First, the expectation of ready employment for vast numbers of research scientists is an artifact of the Cold War economy that is now being dismantled. (Recall that at the height of the Cold War, 30 percent of state-supported research in the U.S. and U.K. was military-related.) Second, the remaining need to employ scientists in the corporate and academic sectors may well be filled by the swelling ranks of competent scientific personnel in the Third World and former Soviet satellites who are willing to do contract work for lower wages than their First World counterparts. Third, the anticipated turn toward part-time work and midlife career shifts in First World economies cuts very much against the image of science as a lifelong vocation that is justified more in terms of its pure practice than its tangible results.[5]

Moreover, I would go so far as to argue that, if the surveys are to be believed, the average person's understanding of science has the makings of a much more sensible and sophisticated view than the one being promoted by leading scientists and government officials under the rubric of Science Week. Of course, I am not claiming that the public could not benefit from technical instruction in science.[6] But that is quite a separate matter from whether public skepticism about the essential unity and purposefulness of the scientific enterprise is well founded. Here I think a historicist would add her voice to the vox populi. Despite at least a four-hundred-year-long track record of Western philosophers promoting the idea that natural science is a unique form of knowledge whose power transcends the particularities of time and place, these thinkers have profoundly disagreed over what counts as such knowledge and wherein its uniqueness lies. The following question epitomizes the traditional formulation of this dispute: Is science special because of how much it can explain or how much it can control?

Explanation and control need not go together. Notwithstanding wishful thinking by military strategists and racial hygienists, Newton and Darwin explained much more than they could ever control, let alone predict. In fact, only in the last hundred years have scientific theories appreciably extended our control over natural phenomena—as opposed to their simply explaining the control that had already been obtained by strictly nonscientific means. Before the rise of logical positivism, few people thought that a scientific explanation had to predict that an event would occur; rather, an explanation told you why the event had to occur, given that it already has.[7] In other words, the scientific explainer was an expert in the reverse engineering of reality, not in its construction. In this respect, science has been continuous with the Western philosophical project in always coming after the fact (as in Hegel's Owl of Minerva taking flight at dusk). Failure to grasp the significance of this point has been the bane of most social scientific efforts to emulate the natural sciences.[8] To be sure, all along scientists have become better at controlling the phenomena they manufacture in laboratory settings. However, only the combined forces of government and industry, flanked by a compliant public, could institute the background conditions needed for turning these laboratory artifacts—the medicines, machines, and techniques—into social innovations and ultimately weaving them into the fabric of everyday life. This is the feature of science's social contract that the public seems to want renegotiated.

Related to the difficulty of identifying an ultimate "nature" of science is the problem of disciplinary specialization. Where science's boosters see "functional differentiation," both the unlearned public and the very learned historians of science find only an increasingly fragmented division of labor. The problem arises at two levels: within sciences and between sciences.

In the first instance, scientific training and career paths have acquired the labor segmentation patterns of industrial capitalism. The generation of European physical scientists who came of age in the early 1880s—including Hertz, Planck, Ostwald, the Curies—was the last to be trained as "the complete scientist," someone who could construct her own theories, design her own experiments (perhaps even construct some of the instruments), and evaluate the results of those experiments.[9] After that, largely in a conscious effort to scale up the scientific enterprise to match the needs of government and industry, specialized training rapidly set in. It seems nowadays that the only "generalists" left in science are the opportunists who can transfer their adeptness at a certain technique to a variety of research settings. (The luminous qualities alleged for the cluster of mathematical modeling techniques known as "chaos theory" is a good case in point.) A subtler consequence of this increased specialization has been the tendency for recent science popularizations to generalize about the nature of science not merely from the author's discipline but more specifically from the kind of work that the author has done in that discipline. Thus, two prominent popular works that take on science's cultural critics—written by theoretical physicist Steven Weinberg and experimental biologist Lewis Wolpert—diverge wildly on what constitutes the essence of science that the critics have failed to discern. Whereas Weinberg treats experiments as little more than stimulants to the theoretical imagination, Wolpert regards them as the backbone of the entire scientific enterprise.[10] Should we be surprised, then, that an avid lay consumer of science popularizations would be thoroughly confused by a request to define *science*?

Nevertheless, increased specialization within each science has only served to reinforce in scientists' minds the idea that the sciences are functionally differentiated yet joined as one in their search for The Truth. Nineteenth-century images of epigenetic embryology and ortho-genic evolution—movements in vitalist science whose original champions were Goethe and Lamarck—still seem to inform the stories that scientists tell of how their diverse disciplines can ultimately issue a unified picture of reality. While scientists are quick to say that their fields became scientific once the quest for vital forces was exchanged for more mechanistic explanations, this very sense of the history is structured in the vitalist terms that scientists have been taught to reject in their research. Science's cultural critics should not, however, be quick to savor yet another "problem of reflexivity" here, for scientists are able to convert such "meta-vitalism" into an effective public rhetoric.

Physicists, chemists, and biologists can convey an overall sense of purpose in their disparate inquiries by contributing to acts of collective remembering, in which each scientist recalls when her domain branched off the main historical narrative to be pursued in greater depth. This is in contrast to humanists and social scientists, who are so lacking in this sensibility that it is common for disciplinary narratives in these fields to cast aspersions on the legitimacy of neighboring fields, if not one's own.[11] Needless to say, natural scientists are not immune to interdisciplinary turf wars, but, for the most part, they manage to conduct them in the privacy of their own specialist journals and not, say, in the newspapers or before a government panel. Thus, whenever this unwritten rule is violated—as in the publicly aired skirmish between chemists and physicists known as the "Cold Fusion Episode"—it is treated as a profanation.

Historicism's Punches Pulled: The Kuhn-Fukuyama Entente Cordiale

The preceeding considerations suggest that if the public were exposed to a strongly historicist curriculum in science, its skeptical views would probably be reinforced. Of course, at the moment, scientists have little to fear from this outcome because of the limited reach of historicism in the curriculum. Ironically, the main reason for this turns out to be the professionalization of the "history of science" (as well as the other science-studies fields) as an autonomous field of inquiry that operates outside the watchful eye of the science faculty. Based on professional society rosters and departmental location, most natural science historians are identified as "historians of science," whereas social science historians are identified as members of the discipline whose history they study. No doubt the fact that the latter group of historians are well-positioned to remind fledgling social scientists of the contingency surrounding the fortunes of their fields emboldens them to think that a changed world requires new forms of inquiry. There is no reason to believe that a similar injection of historicism would not have the same effect on natural science students. Indeed, an especially artful way of producing this effect would be to highlight the critiques that distinguished scientists have made of the scientific establishment. It would soon become evident that virtually every criticism of science made by a humanist or social scientist today has been anticipated by at least one major figure, usually in reaction to obstacles thrown up against his or her own scientific pursuits.[12]

However, with the exception of feminist scholars, few science-studies practitioners have actually called for the infusion of their perspectives into the science curriculum. Thomas Kuhn's ambivalence on this score is emblematic of the entire field. *The Structure of Scientific Revolutions* begins by claiming that the history of science has the potential for radically transforming our understanding of science, but by the end Kuhn is arguing for the need to distinguish the history of science told by scientists from that told by historians. According to Kuhn, scientists need to tell stories of collective progress in order to motivate labors that may look quite trivial when taken on their own terms or, if not trivial, may never issue in the results being sought. The professional historian is wont to tell a story filled with so many accidents and failures that it would dispirit fledgling natural scientists. In a manner that Kuhn dubbed "Orwellian," every scientific revolution must be followed by a rewrite of the discipline's history in order to make the victorious party appear to be the natural heirs, thereby motivating the specialized work on which they and their students are about to embark.[13]

I have called the tension in the two sorts of historiography identified by Kuhn *overdeterminationist* and *underdeterminationist*.[14] The overdeterminationist holds that history is heading somewhere, though most of the particular events along the way need not have happened as they did. Scientists are generally overdeterminationists in their histories of their fields, and therefore have no problem believing that the same truth can be arrived at by many different cultural paths, and that beyond a certain point, historical detail shades into minutiae. For her part, the underdeterminationist turns this picture on its head, arguing that history acquired whatever direction it has only through a series of contingent "turning points," the outcomes of which narrowed the number of paths that history could subsequently take.[15] When historians stress the need to write history from the standpoint of the past, and not the present, they are making the underdeterminationist point that the original historical agents saw their future as open to many more possibilities than we can easily imagine, since our knowledge is colored by later events. Curiously, although Kuhn recognizes the opposed character of the two historiographies, the historical account provided in *Structure* straddles them to such an extent that Kuhn is the darling of both under- and overdeterminationists (or, in philosophical terms, relativists and realists). Before showing how the Japanese appropriation of Western science effectively enrolled the European peoples in a historicist curriculum that deflated many of their overdeterminationist pretensions, it will be instructive to examine those pretensions, late-twentieth-century style, in the hands of one of Kuhn's biggest recent fans, the Japanese American ex-Sovietologist, Francis Fukuyama.

In *The End of History and the Last Man*, Fukuyama argues that history has revealed all the shapes that the good society can take and that 1989 turned out to be the annus mirabilis when liberal democratic capitalism proved, once and for all, that it was much better than even the strongest socialist society.[16] Thus, in Hegelian fashion, it will not be long before liberalism triumphs across the globe, bringing with it an unprecedented period of peace and prosperity. People who seriously entertain arguments like Fukuyama's typically limit their disagreements to the exact vision of the good society that awaits us at the end of history. They generally agree on the means by which we will have gotten there. The "logic of natural science," in Fukuyama's words, plots an inevitable course that both transcends and transforms even the most historically entrenched of cultural differences.[17] In that sense, science puts an end to history: Once the good society has been hit upon by one country, history then becomes simply a matter of the rest of the world catching up by repeating the steps that the first country took. Well into the 1970s, this was how both capitalists and socialists in the first two "Worlds" thought that the Third World would be "modernized." Socialists pointed to science's role in the creation of labor-saving technologies that eventually undermine the basis for any sharp distinction between the workers and their bosses. Capitalists emphasized the role of science in enhancing people's innovative capacities and hence their ability to compete more effectively in the marketplace. The roles assigned to science are different, but in both cases they are meant to have global application. And to whom does Fukuyama turn for science's crucial role in providing humanity's "mechanism of desire"? None other than Thomas Kuhn.[18]

It may seem odd that Fukuyama would cite Kuhn, and not some full-blown scientific realist, for the idea that science displays an irreversible trajectory that propels the progress of humanity. While Kuhn certainly does not see a unified theory of reality waiting for us at the end of science, he nevertheless holds a crucial piece of Fukuyama's puzzle. Specifically, Kuhn's account of scientific change is both irreversible and Eurocentric. After the Royal Society of London was founded as the first self-determining scientific association, the advancement of science proceeded in terms of greater problem-solving effectiveness over more rigorously defined domains. Persistent failures to resolve differences in theory and method among researchers provided a pretext for functionally differentiating a "paradigm"—a process, as we have seen, not altogether different from what scientists themselves think.[19] And despite the inspiration that Kuhn's work has given to methodological relativism in the human sciences, his theory of scientific revolutions is in the grand style that infers a universal logic of science from about 250 years of the history of the physical sciences in Western Europe (roughly 1660–1920).[20] Admittedly, Kuhn differs from his philosophical rivals in capturing more of what Auguste Comte called the "dynamics" than the "statics" of history. But that only highlights the fact that Kuhn requires an aspiring science to pass through a life cycle that is based *more* closely on this quarter millenium of European history than the philosophical accounts of scientific change proposed by Carnap, Popper, Putnam, or Hacking. All considered, Fukuyama is rather perceptive to regard *The Structure of Scientific Revolutions* as amenable to his special brand of modernization theory.[21]

In fact, I would go further. The ultimate vagueness of Kuhn's evolutionism—his idea that science progresses *from* something but not *toward* anything—makes the theory easier to insert into Fukuyama's teleology than a strongly realist theory of scientific change that implies a clear end in sight. The reason is that if science were portrayed as not only the engine of world history but also an institution with over-arching ends of its own, then that could be (and has been) taken to mean that the ends of science must take precedence over—if not simply overtake—the ends of humanity.[22] We would therefore not be far from a vision of scientists, like the philosopher-kings in Plato's *Republic*, who clarify their own vision in order to superimpose it on everyone else. This would do more than simply inconvenience a classical liberal like Fukuyama. Granting such a prerogative to the scientific community would unwittingly expose the antinomic character of realism. On the one hand, if the ends of science appear too autonomous from the concerns of the rest of society, then scientists look like a special interest group with hegemonic designs. On the other hand, if scientific techniques are shown to advance a wide range of personal and social goals, then perhaps the particular value orientations of the scientific community are dispensable. In the one case, science appears to be a totalizing ideology, in the other a high-grade tool. Thus, unless it is kept at a strategically vague level, scientific realism is likely to devolve into a species of either imperialism or instrumentalism. Kuhn succeeds in maintaining such vagueness, and so the feared devolutions never transpire in his text.

A measure of Kuhn's rhetorical success can be taken by noting the ease with which more explicit forms of scientific realism have been criticized by philosophers who hold fairly traditional views of science.[23] For starters, in plotting the realist's overdetermined historical trajectory, how do we count cases to decide that the natural sciences have indeed been more "successful" than other knowledge practices? Do we count only what happens at research sites, or do we include applications as well? If the latter, does the construction of an operational atomic bomb count as one of the "successes"? How do we compare Western medicine's ability to treat colon cancer with a shaman's inability, given that his society never thought of creating the food additives that cause the cancer? But let us say we take as uncontroversial the fact that science has had at least some genuine success stories. We may still wonder whether all of these successes require a common explanation. In other words, were we not to assume that there is something common to why Newton, Maxwell, and Einstein produced scientifically successful theories, scientific realism would have nothing left to explain. Doubts of this kind are normally raised in the context of dispelling superstitions, beliefs that acquire force by over-investing in the significance of common links between spatiotemporally disparate events. Thus, the scientific realist will stress that, say, Einstein studied closely the works of Newton, while downplaying that Einstein read him unlike the way Newton's original readers did.[24]

Finally, there is the realist's holy grail: A theory that explains all of science's successes by the fewest number of principles. The realization of this quest may end up doing more harm than good for the state of knowledge and its producers. At the level of pure inquiry, it encourages a "reductive" mode of thinking that tends toward a lowest common theoretical denominator (for example, the laws of physics) in terms of which unreduced qualities appear epistemically suspect. At the level of knowledge transfer and utilization, it implies that one has not gained legitimate access to a technology unless the scientific principles on which the technology is based have been mastered. But given that, as was mentioned earlier, this basis typically appears after the technology has already been developed and used, the proposal amounts to systematically disempowering the users and then reacquainting them with their artifacts by means of theory-driven instruction. This is not an idle possibility, but in fact a pattern that has been noted historically in the "scientization" of the professions.[25] It can also be seen in efforts to Westernize Third World curricula for purposes of rendering the natives "governable" by making them epistemically accountable to standards that Western authorities can understand and evaluate.[26] This a good point to turn to the strategy that the Japanese used in the last quarter of the last century, not only to prevent themselves from suffering this fate but also to force the West to rethink the sources of its own epistemic legitimacy.[27]

How Japan Taught the West the Secret of Its Own Success

When people talk about the "Japanese miracle" nowadays, they usually mean the country's rapid ascendency after World War II, understood in terms of its remarkably productive corporate culture. However, Japan has generally been recognized as one of the five or six leading world powers since the first decade of the twentieth century. This was less than a half century after it opened its doors to Western influences, following over two centuries of self-imposed isolation. The real "Japanese miracle," then, occurred in the thirty-five-year period between 1869 and 1904, as the country moved from a decaying feudal order, through national consolidation, and finally to an imperial world power. In this period, Japan accomplished each year what had taken Europe at least a dozen to achieve. For our purposes, the crucial transition starts with the selective appropriation of Western science and technology and culminates in the first military defeat of a Western power (Russia) by a non-Western power (Japan) on the basis of superior technical skill.

However, the conclusion of the 1905 Russo-Japanese War had more far-reaching consequences than simply marking the emergence of a new superpower. More importantly, it demonstrated that Western science and technology could be successfully transferred from the West to a culture that had not first been Westernized. This came as especially shocking news to Westerners, most of whose theories of progress and modernization presupposed that any country must pass through a more or less fixed sequence of stages, which together constitute the logic underlying European history. Here one immediately thinks of the great nineteenth-century philosophers of history, Hegel and Marx, Comte and Spencer (but we may also want to add Kuhn and Fukuyama as latter-day company).

Just based on the West's own history, the Japanese link between military vistory and scientific achievement should not have come as such a surprise. After all, it was by no means clear that Europe was "superior" to Islam until the Ottomans were held off from taking Vienna in 1683. This was one of the first skirmishes in which a "scientific" understanding of ballistics revealed its practical benefits. Over the next century, distinctive features of Muslim culture were gradually reinterpreted as marks of decadence rather than sublimity, though the Muslims themselves were among the last to see things this way. Nevertheless, as if in recognition of science's public relevance, the first Christian and Muslim state institutions of higher education to standardize the natural science curriculum were the military academies, which in postrevolutionary France were called *polytechniques*.[28]

Although the Japanese listened to Westerners, they felt no obligation to follow all the Westerners' advice. The Meiji modernizers were especially sensitive to the historically contingent pattern of science's institutionalization in the West. From this perspective, the migratory tendencies of German students, who rarely did all their courses at one university, suited the Japanese pick-and-choose mentality better than the more campus-bound tendencies of students in Britain, for whom academic study was too often tied to a medieval conception of university life. However, the Japanese did not care for the discursive style of classical German university instruction or the patterns of dependency that it created between professor and fledgling researcher.

Instead, they preferred the routinized and "hands-on" curricular plans of engineering degrees at the German polytechnics. They enabled students to pick up necessary skills and then go on with their lives. Overseas Japanese often remarked on the low status that the Germans themselves ascribed to what struck them as an obviously more efficient and empowering form of instruction.

It is worth recalling that the "liberal arts" basis of the Western university prejudiced it against the inclusion of laboratory-based subjects until shortly before the Japanese themselves became interested in those subjects. The "liberal arts" were traditionally defined as subjects that can be done without using one's hands—the mark of a free person. The university was conceived as a place for aristocrats, clerics, and civil servants to learn how to think and speak well, and those goals dictated a discipline's relative value. The heavy reliance on machinery and the artisanlike character of their work marked the lab-based natural sciences as lower class endeavors. Consequently, the major European scientists before 1900 who spent their careers in universities typically held chairs in mathematics or natural philosophy. The rest worked in polytechnics, hospitals, or in autonomous institutions like the Royal Society where class distinctions were not so crucial. In 1826 Justus von Liebig managed to establish a chemistry lab at a marginal German university, but it was only after a recently united but technically superior Germany trounced France in the 1870 Franco-Prussian War that the European nations came around to the idea that the natural sciences ought to be brought from the polytechnics into the universities and made part of general education.[29] From that standpoint, the Japanese were hardly "catching up," as the Imperial University of Tokyo was founded in 1877 with the natural sciences and engineering as its academic centerpiece.

Here the Japanese ultimately took their cue, not from the Germans, but from the Americans. In 1862 President Lincoln had signed into law the Morrill Act, which established the first "land-grant" universities, which were designed to produce knowledge in the agricultural and industrial arts that could be readily transmitted to inhabitants of the university vicinity. Large laboratories and "research stations" were the focal points of academic activity—often with ready access to farms and factories—while the status of traditional liberal arts subjects were reduced to providing "service" courses in basic literacy and numeracy. However, neither the German nor the American model provided an entirely adequate administrative model. On the one hand, the historical fragmentation of German principalities made it difficult to coordinate the administration of educational policy in the newly unified Germany. The result was a highly competitive but disorganized academic culture which enabled professors to lord over their disciplinary fiefdom with virtual impunity. On the other hand, the U.S. Constitution offered principled resistance to the idea of a unified educational policy by endorsing the rights of local authorities to dictate curricular matters. Here the highly centralized and stratified French educational system seemed the best model for Japan's nation-building purposes.

Most of the European advisors hired by the Japanese cautioned against the promotion of scientific training that was not securely grounded in what Kant had called the "public" use of reason. Kant's original idea was the Enlightenment idea of people participating in the decisions that are taken to apply their knowledge. They would be no mere technicians serving another master's ends, but knowing agents who would share responsibility for the situations in which their expertise plays a role. Thus, once the Kantian ideal was institutionalized in the modern German university, students of the natural sciences had to first be trained in the humanistic subjects of philosophy, history, and the arts to ensure that the appropriate value orientation—one friendly to Western (that is, Prussian-national) sensibilities about democracy, open-mindedness, criticism—is transmitted. Aside from an obvious concern with the Japanese possibly using their newfound technical skills against the West, the European advisors also believed that the spirit of scientific innovation would not be sustained unless Japan acquired the cultural context in which such a spirit had developed in the West. Certainly, the historical schemes of Hegel, Comte, and Spencer made it clear that the genius of Galileo, Newton, and Faraday was due more to the Renaissance, Enlightenment, and Victorian cultures in which they were embedded than to anything that could be explained by their work alone.

For their part, the Japanese were bemused that the Europeans could have such a superstitious sense of "reason in history," in Hegel's words. Thus, when offered advice on what to do about the prospect of students taking offense at Darwinism and other scientific theories that challenged traditional religious (that is, Judeo-Christian) conceptions, the Japanese observed that they had not encountered the West's ideological difficulties with the original "revolutionary" scientific theory, Copernicanism, once it had been introduced as part of Newtonian mechanics in the late eighteenth century. This was mainly because the Confucian basis of Japanese culture does not invest any overriding cosmological significance in the idea of an earth-centered universe.

While Confucianism holds that each thing has its "governing center," it does not hold that everything has the same such center. Consequently, it was possible for the Japanese to harbor a "multiple truth" conception of reality that would not have sat well with either Galileo or his inquisitors, but would nevertheless have avoided the need for any scientific revolution to take place. In that sense, Japan enjoyed a version of what economic historians call "the relative advantage of backwardness," because they did not have to overcome centuries-old Western cultural barriers to the development of the natural sciences. Since at all times the Japanese recognized Western science to be irreducibly alien, they were able to treat, say, Newtonian mechanics as a set of tools for certain purposes, but not others, and thus physics was never perceived as a generalized threat to traditional cultural beliefs and values.

Symbolic of this circumvention of European history was the rewriting of technical scientific terms in ideographic script. Whereas Western science students are forced to confront their Greco-Roman roots every time they decipher the meaning of a technical term, Japanese students can simply read the meaning off the ideogram, which depicts the main properties of the element or process to which the term refers. The person most responsible for this move, Shizuki Tadao, first rendered Newton in Japanese—but not by what was then the usual method of phonetic translation into Chinese sound characters. That would have meant having Japanese students learn neologisms that sound like Newton's original English words but which have no clear place in the semantic universe of the Japanese language. Such a translation strategy would have made Newtonian concepts more alien to Japanese students than Latinate words like *gravity* and *inertia* are to modern European students. Instead, Shizuki selected ideograms that functioned as primitive operational definitions of Newton's concepts. For example, gravity was depicted as "power to create weight." The distinctly instrumental spin to Shizuki's rendition of Newtonian mechanics managed to avoid the protracted discussions of the ontological status of gravity and inertia that were continuing to haunt philosophically oriented Western physicists, such as Ernst Mach and Albert Einstein, even as Japan was mobilizing its scientific forces in the late nineteenth century. Indeed, physics in Japan never came to enjoy the philosophically privileged status that it has had in the West. That place was largely taken up by medicine, a science that the Japanese regard as more directly bearing on the relationship between humanity and the larger world.

A final worry about the importation of Western science came from the Meiji emperor's Confucian advisor who had noticed the historical tendency of science to destabilize traditional forms of authority in the West. Would not the same happen in Japan? Here Prime Minister Ito Hirobumi took an unintended lesson from the Europeans. The Confucian advisor read the history of Western science as Westerners themselves often do, namely, as a story of critical inquiry gradually overcoming the barriers strewn on its path by tradition. If Japanese scientists adopted a similar mindset, and developed a taste for pushing back the frontiers of knowledge rather than simply applying knowledge to national needs, then Japan would soon lose its distinctive cultural identity. It was clear that the European advisors desired just the prospect that the Confucian advisor feared.

However, the prime minister realized that this sense of history presupposed that the social role of the scientist was still that of the heroic individual—epitomized by Galileo—who directly confronted the social order with his revolutionary theories. The nationalization of science had put an end to those days in the West, especially in the most scientifically advanced country, Germany. What one finds instead is a self-policing group of professionals who treat "freedom of inquiry" as a guild right to work on narrowly focused topics of no direct relevance, and hence no direct threat, to the larger social order. At the same time, the demand for research publication enabled applied scientists and policymakers to selectively appropriate the results of these "free" inquiries for their own purposes. Germany's Iron Chancellor Bismarck was thus often credited with having turned the universities into a safety valve for overheated intellectuals clamoring for political reform. In effect, the Prime Minister told his Confucian advisor not to worry: look at what the Westerners do, not what they say in their myths.

It was not long before humanists in the metropolitan powers acknowledged the implications of Japan's selective appropriation of Western science. By 1905 the uniqueness of Western science was conceptualized as a matter of contingency, as if it were only by accident that the natural sciences had emerged in Europe rather than, say, China, India, or Egypt. By posing the question of science as one of historical accident, not of historical necessity, "history of science" as a recognizable field of study came into being. The force of this question was to suggest both that the natural sciences were within any culture's reach *and* that Europe's domination of the globe was by no means guaranteed in perpetuity. In any event, it shifted scholarly interest from science as the reflection of more general European attitudes to science as a relatively autonomous and hence easily exported enterprise.[30] Indeed the "realist core" of science has been nothing more (and nothing less) than those aspects of science that have held up well in cross-cultural translation.

When Japan first took to the world stage in 1869, most Western intellectuals thought that a humanist education was necessary for pursuing knowledge in any field, including the natural sciences. But once Japan defeated Russia without the supposed epistemic prerequisite, Westerners reworked the essence of science so that it no longer required a knowledge of philosophy and the other arts subjects. From this standpoint, the pursuit of a scientific realist agenda amounts to an attempt to arrest the process of essence construction, as an ever larger share of the world's population appropriates the science that is suitable for their needs. If I am correct in this diagnosis, then to discover the essence of tomorrow's science, we should look to the ways in which recently enfranchised citizens of the republic of science—women and especially people of color from all over the world—separate the wheat from the chaff in the West's scientific legacy.

In Search of Historical Precedents for Our Predicament [31]

It was perhaps inevitable that scientists would mount their own historical arguments against science's cultural critics. The most popular ones have involved invidious comparisons. The Harvard physicist and historian of science Gerald Holton has persuaded many scientists that science studies is "antiscientific" based on the alleged resemblance between certain provocative statements by two leading science studies practitioners—constructivist sociologist Bruno Latour and feminist philosopher Sandra Harding—and attacks on the natural sciences that were made during the Weimar Republic (1918–1933).[32] This period is generally regarded as the historical benchmark for widespread antiscience sentiment in the modern era. Nevertheless, there are profound but instructive dissimilarities between that period and today.

Weimar thinking about science was marked by the belief that Germany had lost World War I because of the collusion of scientists and capitalists, whose abstract ("calculative") detachment prevented them from dealing effectively with both the concrete and spiritual dimensions of war. Habits of mind that had been virtues before the war were now stigmatized. For example, scientists' "cosmopolitanism" was reinterpreted as an indifference to the national interest. It was in this context that the idea of a culturally rooted "German science" began to emerge, which would come to fruition during the Nazi regime. Public and private funding shifted toward fields that accented cultural relevance. Not only did astrology and psychotherapy benefit at the expense of astronomy and experimental psychology, but scientists themselves adapted their research to the prevailing irrationalist and subjectivist temper. Weimar engineers were no longer "applied scientists" but "folk practitioners" whose discipline would be better located in the human, rather than the natural, sciences. The most celebrated case of cultural adaptation, however, was made by the physics community, whose financial impoverishment and status degradation prompted its members to close ranks around the quantum indeterminacy principle.[33]

The analogy from this situation to contemporary science studies proceeds by claiming that the deconstructions of scientific practice offered by the likes of Latour and Harding are motivated by their own self-loathing for having to participate in a culture that was responsible for the atomic bomb, the military-industrial complex, and attendant scientifically induced forms of destruction and exploitation. Like the antiscientific practices that flourished in the Weimar period, science studies were committed to preserving, not perverting or destroying, the objects of their inquiry. The analogy breaks down, however, once one realizes that science studies typically portrays itself not as antiscience, but, in certain important respects, as more scientific than science itself. Whether or not one ultimately accepts this self-portrait, its rationale is worth understanding.

In at least two respects, science studies claims to be on the side of science. First, the homely picture of science that emerges from science studies—whereby scientists look more like craftsmen than geniuses—is not primarily meant to debunk (though, admittedly, it seems to have had that effect). On the contrary, it is meant to lead the public to have saner expectations of science.[34] Words like *truth*, *rationality*, and *objectivity* have inspired unrealistic hopes for what science can accomplish. These have often backfired on the scientific community, thereby opening the door to genuinely antiscientific movements, as was the case in Weimar Germany. Second, because science-studies practitioners do not have a vested interest in maintaining the public rhetoric of *truth*, *rationality*, etc., they may be (ironically) in a better position than scientists to approximate the attitudes that the public rhetoric is supposed to capture. A relevant consideration here is that, short of the sheer elimination of academic posts, science studies is

less vulnerable than the natural sciences to changes in funding patterns.[35] Thus, science studies may be less susceptible than, say, physics or biology to Weimar-like cultural adaptation strategies for its
survival.

However, I would argue that a more appropriate historical precedent to the controversies currently embroiling science's cultural critics is secularization, the state's refusal to grant any religion a monopoly over political and economic resources while at the same time
protecting the rights of any religion to profess its creed within state
borders. The immediate cause of secularization was the destabilizing
effect of religious wars on the emerging nation-states of Europe in the
sixteenth and seventeenth centuries. The decoupling of political legitimation from religious affiliation was just as much the product of
Machiavellian survival instincts as of any interest in ensuring maximum freedom of expression. And while the institutional ascendancy
of the natural sciences in the mid-nineteenth century is often regarded as a major vehicle of secularization, we may have reached a
point at the end of the twentieth century—given the concentration of
state resources on scientific research—that calls for the secularization
of science itself.[36] To paraphrase the Enlightenment critic Gotthold
Lessing, the true test of science as a form of knowledge may lie in its
ability to command believers even after its material support has been
removed.

Assuming the aptness of the analogy, we may speak of two "waves"
in the critique of the social dimensions of science and technology.[37]
The first wave is akin to the Protestant Reformation of the sixteenth
and seventeenth centuries, the second to the radical hermeneutics of
the "Higher Criticism" in eighteenth and nineteenth century theology.
The first wave, during the 1960s, was marked by the rise of scientists
who "conscientiously objected" to their colleagues' complicity with
the state in escalating the Cold War, just as Luther, Calvin, and their
associates called for the Church to recover its spiritual roots from
corrupt material involvements.[38] A secularized science would never
have given us the nuclear arms race, just as a Protestantized Christianity in the Middle Ages would not have been able to mobilize the
material and spiritual resources needed to field a series of Crusades
against Islam. In this context, the late epistemological anarchist, Paul
Feyerabend, appears as the purest of Protestants in calling for the
complete divestiture of state support for science as the best way of
retrieving the spirit of critical inquiry from Big Science's inhibiting
financial and institutional arrangements.[39]

The second wave of secularization occurred once the Enlightenment transformed the intellectual orientation of academic theology from the professional training of clerics to a form of critical inquiry conducted independently of religious authorities. The last and leading generation of these theologians constituted the Young Hegelians under whose spell Karl Marx fell during his student years. David Friedrich Strauss's *Life of Jesus* and Ludwig Feuerbach's *The Essence of Christianity* were the texts from this period (the 1830s) that have had the longest impact. Much in the spirit of recent sociologists who have subjected the laboratory to ethnographic scrutiny, these theologians applied the latest techniques of literary archaeology and naturalistic social theory to demystify the Scriptures. Far from blaspheming God, they believed their demystified readings of early church history liberated genuine spirituality from the superstition and idolatry that remained the primary means by which the pastoral clergy kept believers in line. However, the ironic style of these authors put them seriously at odds with both political and religious authorities, causing many of them to lose their professorships and preventing others—such as Marx himself—from ever pursuing academic careers.

Marx wrote *The German Ideology* largely as a series of didactic reflections on how it was possible for the Young Hegelians, despite the attention they paid to the material conditions of Christianity, to be so oblivious to the material conditions of their own times, and hence to be caught off guard by those who accused them of sacrilege. Perhaps a similar book is now in order, given the surprise that science studies practitioners have expressed about the reception that the scientific community has given their work. It would seem that even these critics of science have underestimated the extent to which threatening the transcendental rhetoric of science threatens science itself. As befits Hegel's "cunning of reason" in history, it may turn out that more effective vehicles for the secularization of science will be found among the customized knowledges promoted by such New Age movements as homopathic medicine, parapsychology, dianetics, and (*mirabile dictu!*) Creation science.[40]

It is often forgotten that the clause separating church and state in the U.S. Constitution was originally written to protect the former from the latter, not vice versa. In other words, the point of the separation was to foster spiritual diversity rather than to circumscribe its public relevance. Clearly, this was before the introduction of mandatory public schooling and before religion came to be seen as retarding the spread of democratic values (by no less than John Dewey) in the decades following the publication of Darwin's *Origin of Species*. These two events are not unrelated, since the public school system was designed as a mechanism for converting a disparate populace into modern citizens. On this view, religious sectarianism divided people, inhibiting their ability to share a national identity, whereas the scientific method could unite them.[41]

This development in itself would not be such a cause for concern if science were presented in the classroom simply as a set of techniques for empowering the citizen. However, the situation grows more complicated once scientific theories are presented as "our" best understanding of reality, without which those techniques would not make sense. Here, Creationists argue, science instruction starts to encroach on spheres of spiritual expression traditionally addressed by religion. At a philosophical level, the Creationist can be read as posing the following challenge: Does the scientific research program that historically led to a finding, technique, or artifact have intellectual property rights over how that thing is subsequently deployed and disseminated? To put the question in perspective, recall the classic positivist distinction between the contexts of *discovery* and *justification* in scientific research. This distinction was discredited largely because the suggestion that the former was irrational and the latter rational could not be maintained in light of empirical studies of scientific practice.[42] However, the distinction may be worth drawing in a somewhat different spirit, one that would make a point of detaching expert knowledge from the original expert community, and, in that sense, would separate matters of justification from those of discovery.

Recall the case of Japan's negotiated settlement with the Western scientific tradition. Westerners argued that the Japanese would not be able to match the West's scientific achievement unless they also reproduced the cultural background against which that achievement had occurred. In effect, they held that the Japanese would need to retrace the West's discovery process in order for its scientific knowledge to be fully justified. However, the Japanese essentially constructed alternative means to the desired scientific ends—in some cases capitalizing on their metaphysical and religious differences with the West while in other cases hybridizing Western institutions and practices. Likewise, from a Creationist standpoint, just because some important findings and perspectives in environmental science were originally developed under the rubric of Darwinian evolution, it does not follow that those findings and perspectives cannot be understood or appropriated without the Darwinian framework. In order to protect students' freedom of inquiry, teachers should try, whenever possible, to show that similar results can be reached holding alternative theoretical presuppositions. Admittedly, this injunction may cut against the idea that science instructors should encourage students to frame their understanding of nature as professional scientists do, but then who ever said that "citizen science" was simply a watered-down version of professional science?[43]

Because of their own oppositional stance to the scientific establishment, Creationists often realize that "science" lies more in the (critical) attitude one has toward certain theories than in the content of the theories themselves. Without the presence of competitors, any putatively scientific theory can easily drop into dogma. Thus, Creationists argue that the preclusion of natural theology from physics and biology classes encourages teachers to treat cosmic design as a priori impossible, a judgment that goes considerably beyond the available evidence. Given that two thirds of those who believe in evolution also believe that it reflects a divine intelligence, it would seem that such ex cathedra dismissals fail to engage the average student's intellectual starting point. At the very least, this constitutes bad pedagogy. At most, such an attitude corrupts science itself. None of this implies any explicit endorsement of the Creationist's positive theses. I am arguing only for Creationism's dialectical role in keeping science "honest" in the context of a captive audience. In its determination to block Creationists from the classroom, the scientific community has acquired some of the very characteristics that they claim are objectionable in the Creationists.[44]

Creationism is rather unique in resorting to a political arena to challenge the scientific establishment head-on. Most New Age movements thrive on private funding, which can be considerable. Although half of federal science funding is concentrated in the thirty universities that effectively constitute the scientific establishment, federal sponsorship is itself a declining fraction of overall science funding in the United States. Private sector interest in science is, of course, longstanding and has often reflected industry's attempt to circumvent the peer review process, which was designed to judge projects more as contributions to extant research programs than as prospects for market innovations. In the past, private investment in science has not only been good for business, but it has also helped promote interdisciplinary fields that would otherwise not flourish in rigid, discipline-based academic institutions.[45]

However, there is a new twist to privatization, one that reflects the scientific establishment's inability to satisfy the range of political interests that is nowadays needed to get action on increasingly expensive and specialized research proposals. In such an environment, it is only natural for the federal government to formally "divest" its interests in certain cost-ineffective lines of research. The recent demise of the superconducting supercollider in the U.S. Congress is a good case in point. It is widely believed that if the experiments proposed for the supercollider are as world historic as its proponents claim then they will ultimately be financed by either corporate investors or multinational public investment.

As governments continue to let market demand drive science policy in this fashion, scientific teams in search of funding will need to adapt their research goals to the interests of potential investors. This, in turn, will bring them closer to the kind of customized knowledge production that is characteristic of New Age movements: that is, they will gradually lose the universalist gloss of knowledge per se and become knowledge for specific constituencies. Two clear signposts of this development are the expansion of the domain of intellectual property law and the emergence of "knowledge engineering" professionals who earn a living by translating human expertise into user-friendly computer systems.[46]

Consider the ongoing controversy involving Distinguished Professor of Chemistry at Texas A&M University, John Bockris, who has received massive private funding for studying how base metals may be transmuted into gold, an idea that belongs to the alchemical prehistory of modern chemistry. Bockris, who had previously gained notoriety for supporting Pons and Fleischman's cold fusion experiments, has been roundly condemned by his colleagues, some of whom have called for his resignation. However, Bockris remains unfazed, noting that he runs one of the best-funded labs in his economically strapped institution and that, as a well-regarded senior scientist, he has an obligation—to science no less!—to pursue controversial leads that would cost lesser-placed scientists their jobs. Is Bockris either anti- or pseudoscientific in his approach to inquiry? Or, is he symptomatic of the diffusion of scientific authority that takes place once the state has begun to divest its interests in science funding?

What does it mean to be "pro-science" in these times? Does it mean identifying "free inquiry" with the kind of scientific entrepreneurship championed by Bockris? If so, how can one then fault the New Agers who operate on roughly that market-driven principle? Presumably, in each case, funding will continue only as long as there seems to be some return on investment. Perhaps, instead, *pro-science* implies a consolidation of science-policy decisions into one peer-review–based state funding agency that dwarfs all private competitors. While that would certainly ensure the maintenance of uniform scientific standards, doubts arise about not only its economic feasibility but also its ability to foster the sort of alternative research trajectories that have often led to major scientific advances.

One way to cut the Gordian Knot of science policy is to simply say that a truly "pro-science" attitude would refuse to let financial considerations dictate the logic of inquiry at all. On this view, the mission of science has been perverted, if challenges to the scientific orthodoxy are assessed primarily in terms of the amount of specialized equipment and technical personnel at stake. This was certainly Feyerabend's view—and maybe even Kuhn's.[47] The first order of business, then, would be secularization: to reconstitute research programs so that they are no longer so dependent on vast material resources. For example, theories in high-energy physics could be tested on a computer that simulates the paths of accelerated particles rather than on an actual particle accelerator. After all, particle accelerators are themselves attempts to stimulate the first few milliseconds after the Big Bang that, given a more advanced (that is, economical) computer technology in the 1930s, would probably have been deemed an extravagant way of testing physical hypotheses.[48]

But even if we do not question the amount of funding that is devoted to science-based projects, there are serious questions to be asked about how that money is allocated. As I observed earlier in this essay, the public long ago realized what science's cultural critics have only recent discovered, namely, that the natural sciences range at least as widely, in terms of methods and goals, as the human sciences. In science studies, this disunity originally reflected a turn away from two ideas passed on from logical positivism to the history and philosophy of science: that all sciences are ultimately reducible in method and/or content to physics and that all physics are ultimately reducible to moments in the history of physical *theory* (that is, its development, testing, and application). Undoubtedly, the resulting disunified picture of science will take an indirect route from the scholarly imagination to the corridors of power. Competing scientific agendas have yet to openly vie for a common pool of resources, and politicians still regard entire disciplines as being best represented by their most elite practitioners.

Caveat Scholar

Despite our arch criticism of scientists' sense of history, we science-studies practitioners surprisingly lack much of a sense of our own place in history, and hence we are taken off guard when the scientists strike back. We need to ask: Why did a field like science studies arise when it did, and to what extent has the field remained captive to its origins? Admittedly, there is always a tendentious quality to addressing questions of historical reflexivity while the plot of our story is still unfolding. Nevertheless, certain facts contribute to a framework for thinking about these matters.

In both America and Britain, the impulse to promote academic programs in the history, philosophy, and sociology of science came from senior scientists concerned with the sudden emergence of the natural sciences as a fixture of Cold War public policy. These sciences became the lightning rod for the public's most intense hopes and fears. The remedy that the scientists proposed for this public anxiety was to normalize the role of science in society so that it not seem irretrievably alien. Thus, Harvard's General Education in Science program was focused mainly on teaching future managers without serious scientific training to recognize "good science" when they see it. It was here that physics Ph.D. physicist Thomas Kuhn received his first job, which enabled him to hone the lectures that eventually became the basis for *The Structure of Scientific Revolutions*. In Britain, C. P. Snow's 1959 Rede Lecture, "The Two Cultures and the Scientific Revolution," drove home the need for scientists to enrich their training with an understanding of the broader cultural contexts in which their work will increasingly figure. Once Harold Wilson's Labour Party came to power, several interdisciplinary support teaching units were established to carry out Snow's proposals. The most important of these turned out to be the Science Studies Unit at Edinburgh University.[49]

The point worth emphasizing is that these programs did not begin —as had many movements in the social sciences—in the spirit of social transformation. Quite the contrary. The science critics of the 1960s whom we earlier compared to the Protestant Reformers had little to do with the establishment of the science-studies programs. Indeed, they would have preferred to have the history, philosophy, and sociology of science made part of the core of the science curriculum, not merely enrichment courses taught outside the science departments. The last thing they would have wanted to see were autonomous Science and Technology Studies graduate programs in which doctorates could be awarded for research that would fail to connect with practicing scientists, let alone challenge their professional commitments. Nevertheless, the support-teaching origins of science-studies programs continue to haunt the field. Specifically, the interpenetration of science and society so vaunted by our field rhetorically functions to discourage inquirers from looking far beyond those objects of fascination—the laboratories—to see how science reflects larger societal forces. Instead, science is portrayed as "always-already social," which implies that whatever larger forces need to be taken into account will be "inscribed" in the people and things located in the laboratory. Not surprisingly, science-studies practitioners have endured an uneasy relationship with Marxist and feminist science critics. They have been united more in terms of a common foe—the scientific establishment—than a common methodological and axiological orientation.[50]

Science studies is telling in this regard as it reproduces the value orientation of the sciences it studies by privileging research in contemporary Big Science and continuing the misleading impression that one is a scientist only in research, but not in teaching or administration.[51] Of course, science studies typically tells a rather different story from the one that scientists are inclined to tell about these privileged research sites. Usually, more people and things are incorporated into the science-studies narratives, thereby providing a more complex picture of how science manages to "succeed" as well as it does. But at the same time, this added complexity diffuses responsibility for any of the actions taken in the name of science. On the one hand, this helps redistribute the credit for scientific work from the few "geniuses" who normally receive all the glory; on the other hand, it makes it difficult to hold anyone accountable for anything.[52] If all this sounds familiar, it may be because the "actor network" image of "technoscience" that has recently come to dominate science-studies research looks like a postmodern version of Hannah Arendt's account of modern totalitarian regimes.[53]

According to Arendt, the success of such regimes is due partly to the way they divide the work of tyranny into sufficiently self-contained routines that make it extremely difficult to hold anyone responsible for the overall violence that the regime commits: everyone just seems to be following orders, and the orders themselves are issued without making reference to the violence involved in their implementation. Indeed, the orders may be expressed in such a way that the functionaries identify with their abstract aims without fully realizing their concrete consequences. While the decentered and fragmented character of the postmodern condition would seem to be light-years away from Arendt's image of the Nazi juggernaut, a flexibly organized technoscience suggests disturbing simulations. What, on a sympathetic reading, may appear to be an amorphous network of highly contingent nodes (the so-called strength of weak ties) may be portrayed, less sympathetically, as an all-pervasive system whose general structure cannot be purposefully altered by some strategic intervention, let alone a social movement. In this way, science studies practitioners may be able to continue their steady stream of detailed case studies for both collegial and cliental consumption without offering counsel to those interested in a fundamental renegotiation of science's social contract. It may be, then, that the joke is on us when we fail to recognize that Bruno Latour plays it straight when he says that science studies does not pose any serious threat to the scientific establishment.[54]

Notes

1   Stephen Brush, "Should the History of Science Be Rated X"? *Science* 183 (1975): 1164–83.

2   This became very apparent when I debated with the Minister for Science and Technology and the self-styled Marxist brain scientist Steven Rose during Science Week '95: "Newstalk," BBC Radio 5, 20 March 1995.

3   I raised this point in Steve Fuller, "Death to All Magic Bullets: If We Need More Scientists, What Sort Should They Be?" *New Scientist*, 6 May 1995, 53–54. The unemployment figure was reported for the U.K.'s "old" (i.e., pre-1960) universities in the *Times Higher Education Supplement*, 9 December 1994. A dispute over these figures has recently erupted in the *Financial Times*, Britain's answer to the *Wall Street Journal*. On 19 November 1995, the U.K. government science and technology policy advisor, Robert May, extolled the marketability of science graduates. When I questioned this rosy employment picture (on 21 November), May revealed (on 28 November) that he had in mind the employment figures covering all scientists age 21 to 65 rather than just those who have received their degrees since the end of the Cold War. Because "scientists" have not normally been regarded as an endangered species of employment, their labor statistics have proved elusive to government surveys. Nevertheless, it is now clear that scientists are about as likely as humanists to be employed in a field related to their training (a 35% chance). However it is worth noting that even granting that scientists are eminently marketable in a variety of nonscientific fields, it is highly improbable that these people would have pursued science majors had they known that they would ultimately land in, say, accounting, stock brokering, or computer programming. My conjecture is based on the campaign statements of the four write-in candidates for the 1994 elections to the Council of the American Physical Society. These recent Ph.D.s were calling for a "sustainable physics" that calibrated the production of new physicists to match a shrinking labor

market. Implied in their critique was that physics' main professional spokespeople had lured them with the elementary socio-epistemic confusion of "research that needs to be done" and "jobs that are available." As it turns out, two of these maverick candidates were elected.

4 A good summary of British research in this area is John Ziman's "Not Knowing, Needing to Know, and Wanting to Know," in *When Science Meets the Public*, ed. B. Lewenstein (Washington, D.C.: American Association for the Advancement of Science, 1992), 13–20. See also the quarterly, *Public Understanding of Science* (Bristol, U.K.: Institute of Physics Publications, 1992–).

5 For more, see Stanley Aronowitz and William DeFazio, *The Jobless Future* (Minneapolis: University of Minnesota Press, 1994).

6 However, the ultimate value of such instruction may lie more in simply acquainting the public with the distinctive character of scientific reasoning than in persuading the public to replace its everyday (or "ethnoscientific") modes of reasoning with scientific ones. Given that the scientist's reasons for wanting to understand the natural world are not the same as the nonscientist's, it remains an open question as to whether what a scientist thinks is always needed to mediate, let alone legitimate, what the nonscientist thinks. For a brief survey of the ethnoscience literature, see Brian Wynne, "Public Understanding of Science," in *Handbook of Science and Technology Studies*, ed. S. Jasanoff, et al. (Thousand Oaks, Calif.: Sage, 1994), 370–75.

7 Perhaps the best modern statement of this classic position is in Michael Scriven, "Explanation and Prediction in Evolutionary Theory," *Science* 130 (1959): 480.

8 See Peter Manicas, *A History and Philosophy of the Social Sciences* (Oxford: Blackwell, 1986).

9 For an account of Hertz as a member of this last generation, see Jed Buchwald, "Design for Experimenting," in *World Changes*, ed. P. Horwich (Cambridge, Mass.: MIT Press, 1993), 169–206.

10 Steven Weinberg, *Dreams of a Final Theory* (New York: Pantheon, 1992); Lewis Wolpert, *The Unnatural Nature of Science* (London: Faber and Faber, 1992). I consider the contradictory nature of their attacks in Steve Fuller, "Can Science Studies Be Spoken in a Civil Tongue?" *Social Studies of Science* 24 (1994): 143–68. Replies by Weinberg and Wolpert, as well as my response to them, can be found in *Social Studies of Science* 24 (1994): 745–57.

11 This difference between the group conduct of natural and human scientists was driven home during a conference on "Science's Social Standing" that I organized (2–4 December 1994) at Durham with the express purpose of initiating dialogue between scientists active in the British "public understanding of science" movement and their cultural critics. Although a wider range of natural scientists than human scientists were represented, the former group was much less internally divisive. For a sense of the different rhetorics deployed, see the opening conference papers, published as "Making Sense of Science" (with contributions by I. Velody, P. Atkins, J. Christie, D. Cooper, and S. Fuller) in *History of the Human Sciences* 8 (1995): 91–124. For more theoretical consideration of the fractious character of social science historiography, see Steve Fuller, *Philosophy, Rhetoric, and the End of Knowledge: The Coming of Science and Technology Studies* (Madison: University of Wisconsin Press, 1993), 102–38.

12   At the Durham conference (see note 11), Graeme Gooday, a historian of science at Leeds University, demonstrated this point with a paper on T. H. Huxley's argument that standardized laboratory training in experimental physiology would stifle students' critical and creative faculties. This led to a lively discussion in which some of the scientists admitted to feeling the same way as a result of observing their own students at work.

13   Thomas Kuhn, *The Structure of Scientific Revolutions*, 2d ed. (Chicago: University of Chicago Press, 1970), 167.

14   See Fuller, *Philosophy, Rhetoric, and the End of Knowledge*, 210–13.

15   The locus classicus for this point is Steven Shapin and Simon Schaffer, *Leviathan and the Air Pump* (Princeton, N.J.: Princeton University Press, 1985), which argues that the exclusion of Thomas Hobbes from membership in the Royal Society marks the beginning of experimental science's rise to epistemic dominance in the West.

16   Francis Fukuyama, *The End of History and the Last Man* (New York: Free, 1992).

17   Ibid., 80–81.

18   Ibid., 352–53, n. 2.

19   Kuhn has become more explicit about this point over the years. See Kuhn, *The Trouble with the Historical Philosophy of Science: Rothschild Distinguished Lecture* (Cambridge, Mass.: Harvard University Department of History of Science, 1992), esp. 17.

20   Kuhn's preference for discovering the nature of science by following the development of one exemplary science (physics) rather than by comparing several sciences marks his historiography as "vitalist" in the sense highlighted earlier in this article.

21   I am completing a book (for University of Chicago Press) that locates Kuhn in the political and intellectual climate of the Cold War, which more strongly suggests that Fukuyama's affinity with Kuhn is not arbitrary. For early findings, see Steve Fuller, "Being There with Thomas Kuhn: A Parable for Postmodern Times," *History and Theory* 31 (1992): 241–75; Fuller, "Teaching Thomas Kuhn to Teach the Cold War Vision of Science," *Contention* 4 (1994): 81–106.

22   Steve Fuller, "Towards a Philosophy of Science Accounting: A Critical Rendering of Instrumental Rationality," *Science in Context* 7 (1994): 591–621.

23   A good source for what follows is Jarrett Leplin, ed., *Scientific Realism* (Berkeley: University of California Press, 1984). The articles by Hilary Putnam and Richard Boyd offer a canonical statement of the realist position, while those by Arthur Fine and Larry Laudan provide the most incisive critiques. See also Steve Fuller, *Social Epistemology* (Bloomington: Indiana University Press, 1988), esp. 65–98.

24   An important methodological difference between the "old" (Marxist) and "new" (constructivist) historical sociology of science turns precisely on this point, since the old school accepts the realist premise that the successes of Newton and Einstein demand an explanation that links the two together in a common trajectory, whereas the constructivist is prepared to give quite separate explanations for each (Fuller, *Social Epistemology*, 233–51). Not surprisingly perhaps, Putnam developed the most widely discussed version of scientific realism while passing through his Marxist phase in the early 1970s (see

n. 23). The connection between scientific realism and Marxism is even more evident in the U.K., especially via the work of Roy Bhaskar and his followers.

25   Andrew Abbott, *The System of Professions* (Chicago: University of Chicago Press, 1988).

26   Brian Holmes and Martin McLean, *The Curriculum: A Comparative Perspective* (London: Routledge, 1989).

27   The following account of the Japanese appropriation of Western science is taken from my forthcoming book on multiculturalist epistemology. Its historical detail draws primarily upon two works: James Bartholomew, *The Formation of Science in Japan* (New Haven, Conn.: Yale University Press, 1989); and Scott Montgomery, *The Scientific Voice* (New York: Guilford Press, 1996), 294–359.

28   Bernard Lewis, *The Muslim Discovery of Europe* (London: Phoenix, 1982), esp. 135–170. Corelli Barnett, "The Education of Military Elites," in *Education and Social Structure in the Twentieth Century*, ed. Walter Laqueur and George Mosse (New York: Harper and Row, 1967), 15–36. On a subtler level, the motivation for national improvement of public health in the nineteenth century can be understood in terms of the need to render the citizenry mobilizable in time of war.

29   The introduction of the natural sciences into general education in the 1870s triggered nearly fifty years of debate in Europe about whether science needed more protection from society or vice versa. Max Planck and Ernst Mach were the principals in the German debates (see Steve Fuller, "Retrieving the Point of the Realism-Instrumentalism Debate: Mach vs. Planck on Science Education Policy," *PSA 1994*, vol. 1, ed. D. Hull, et al. (East Lansing, Mich.: Philosophy of Science Association, 1994), 200–207.

30   Lewis Pyenson, "Prerogatives of European Intellect, History of Science and the Promotion of Western Civilization," *History of Science* 31 (1993): 289–315.

31   This section of the paper was originally prepared as an invited piece by *Academic Questions*, the journal of the National Association of Scholars (NAS). Before 1994, the NAS was noted primarily for its well-funded campaigns against so-called academic left critiques in the human sciences. However, the NAS extended its remit to the natural sciences upon the publication of Paul Gross and Norman Levitt, *Higher Superstition: The Academic Left and Its Quarrels with Science* (Baltimore, Md.: Johns Hopkins University Press, 1994). I had hoped to engage in dialogue with these latest opponents in one of the major forums, a principle that motivated me to bring the scientists and sociologists together in Durham at roughly the same time (see n. 11). Unfortunately, after over six months of silence and several promptings on my part, the editor Sanford Pinsker sent a perfunctory rejection letter, showing no signs that the paper had been read by anyone other than himself, a scholar renown for his history of the *schlemiel* as a character in American literature. This is hardly the best editorial practice for an organization that aims to recover and maintain traditional academic values.

32   Gerald Holton, "How to Think about the 'Anti-Science' Phenomenon," *Public Understanding of Science* 1 (1992): 103–28, esp. 108–9. The article has since been expanded into a chapter of his book, *Science and Anti-Science* (Cambridge: Harvard University Press, 1993).

33   Jeffrey Herf, *Reactionary Modernism: Technology, Culture, and Politics in*

Weimar and the Third Reich (Cambridge: Cambridge University Press, 1984); Paul Forman, "Weimar Culture, Causality, and Quantum Theory: 1918–1928," *Historical Studies in the Physical Sciences* 3 (1971): 1–115.

34   See the introduction to a recent attempt to popularize science studies: Harry Collins and Trevor Pinch, *The Golem: What Everyone Needs to Know about Science* (Cambridge: Cambridge University Press, 1993).

35   This point was raised in the *Times Higher Educational Supplement* editorial of 30 September 1994 that launched the recent debate over "science's social standing" in the U.K.

36   This was the theme of my inaugural lecture as Professor of Sociology at Durham, delivered 30 November 1995: "Is Our Faith in Science Superstitious? An Argument for the Secularization of Science Policy."

37   I have called the two waves, "Low Church" (for the one resembling the Protestant Reformation) and "High Church" (for the one resembling the Higher Criticism), in Fuller, *Philosophy, Rhetoric, and the End of Knowledge*, xiii. (See note 11). In the US, low church members tend to affiliate with the National Association for Science and Technology in Society (NASTS), whereas the High Church is associated with the Society for Social Studies of Science (4S).

38   Perhaps the best book in this vein is Hilary Rose and Steven Rose's *Science and Society* (Harmondsworth, U.K.: Penguin, 1969).

39   Paul Feyerabend, *Science in a Free Society* (London: Verso, 1975). An extension of the analogy would have to include the recent charges of scientific misconduct, which have precedent in the personal corruption of Church officials that made the calls for reform most vivid for the average devout Christian.

40   For a science studies review of New Age movements, see David Hess, *Science in the New Age* (Madison: University of Wisconsin Press, 1993). My source for a nonfundamentalist Creation science is Phillip Johnson, *Darwin on Trial*, 2d ed. (Downers Grove, Ill.: Intervarsity Press, 1993). For the U.S. legal background to the Creationist controversy, see Stephen Carter, *The Culture of Disbelief* (New York: Doubleday, 1993), 156–82. It is worth noting that Carter's final verdict on Creationism is much harsher than mine here.

41   Interestingly, this sentiment has been recently revived in David Hollinger, "Science as a Weapon in the *Kulturkaempfe* in the United States during and after World War II," *Isis* 86 (1995): 440–54.

42   A good source for the historical development of this distinction is Larry Laudan, *Science and Hypothesis* (Dordrecht, The Netherlands: Reidel, 1982). The distinction is ultimately traceable to Francis Bacon's refashioning of the old rhetorical distinction between the method by which one invents arguments (cf. discovery) and the method by which one persuades audiences (cf. justification).

43   Here I am updating Ernst Mach's arguments against Max Planck on the connection between theory-free science education and students' academic freedom. See notes 22 and 29 for an analysis of this debate, which occurred in Germany just before World War I. One can begin to see how Mach turned out to be Feyerabend's philosophical hero (see n. 39).

44   One final remark on this score concerns the role that philosophers of science have played as expert witnesses and spin doctors in the various trials concerning the place of Creationism in U.S. public schools. Some philosophers

have strategically suppressed their internal disagreements in order to fend off the perceived Creationist menace. Thus, during the Arkansas trial of 1981, Michael Ruse presented Karl Popper's criterion of falsifiability as the essence of science, knowing full well that the criterion is at best a caricature of actual scientific practice. However, the appeal to falsifiability enabled Ruse to locate Creationism on the wrong side of the science/nonscience divide. What troubles me about Ruse's strategy is not that it invoked a norm that failed to capture scientific practice. After all, norms usually represent ideals that are not always met. Rather, I have a problem with his failure to mention that *both* evolution and creation suffer—albeit differently—when judged by Popper's criterion. It is here that the philosopher slips into ideological special pleading. A frank account of philosophical participation in the Creationist trials (including interviews with Michael Ruse and Philip Kitcher) is provided in Werner Callebaut, *Taking the Naturalistic Turn: How Real Philosophy of Science Is Done* (Chicago: University of Chicago Press, 1993), 194–99.

45   For a discussion of the role of U.S. business (and associated philanthropic foundations) in the creation of interdisciplinary fields in the natural sciences, see Robert Kohler, *Partners in Science* (Chicago: University of Chicago Press, 1991). For the social science equivalent, see Ellen Lagemann, *The Politics of Knowledge* (Chicago: University of Chicago Press, 1989).

46   On the implications of these developments for the future of academic work, see Steve Fuller, "Why Post-Industrial Society Never Came," *Academe* 80, no. 6 (1994): 22–28. For the privatization of knowledge in historical perspective, see Yaron Ezrahi, *The Descent of Icarus* (Cambridge, Mass.: Harvard University Press, 1990), 237–92.

47   Perhaps this is why Kuhn has never referred to any research after the 1920s in his account of scientific revolutions: namely, too many material resources have been involved in the pursuit of science to enable the logic of inquiry to be dictated solely by the "paradigm." An underrated attempt to be less tactful about such matters is the idea of "postnormal" science advanced in Jerome Ravetz, *Scientific Knowledge and Its Social Problems* (Oxford: Oxford University Press, 1971).

48   David Lindley, *The End of Physics* (New York: Basic, 1993).

49   For a fuller account of the history, see Steve Fuller, "On the Motives for the New Sociology of Science," *History of the Human Sciences* 8 (1995): 117–24.

50   A good sense of this ambivalence can be gotten from Stanley Aronowitz's *Science as Power* (Minneapolis: University of Minnesota Press, 1988).

51   Consider the sites of the classic case studies: Bruno Latour and Steve Woolgar, *Laboratory Life* (London: Sage, 1979); Karin Knorr-Cetina, *The Manufacture of Knowledge* (Oxford: Pergamon, 1981); and Harry Collins, *Changing Order* (London: Sage, 1985).

52   This problem is made central to understanding the role of "agency" in social theory in Steve Fuller, "Making Agency Count," *American Behavioral Scientist* 37 (1994), 741–53. Here I argue that intellectual property law may offer some clues for constructing agents with humanly tolerable dimensions.

53   See, for example, Hannah Arendt, *The Origins of Totalitarianism* (New York: Harcourt, Brace, Jovanovich, 1951).

54   See, for example, T. Hugh Crawford, "An Interview with Bruno Latour," *Con-*

figurations 2 (1993): 247–69. It is worth noting that science studies in France has been supported largely by state-industry initiatives that aim to improve the design and diffusion of technoscientific innovation in a country that, while ideologically prone to scientism, has nevertheless witnessed some of the most dramatic cases of innovations failing to engage the public. Among these have been the electric car and a customized commuter rail system, two projects that Latour has turned into classic case studies.

Meeting Polemics with Irenics

in the Science Wars

Emily Martin

ACCORDING to Webster's, a *polemic* is "an aggressive attack on, or the refutation of, others' opinions, doctrines or the like." In today's academy, professors and students often have cause to be polemic, but seldom have cause to remember that *polemic* has an opposite.[1] Webster's defines that opposite, *irenic*, as "fitted or designed to promote peace; pacific, conciliatory, peaceful." Recent skirmishes in the Science Wars have seemed to me so polemically bitter on all sides that rather than sending back another volley intended to hurt and destroy, I want to try moving irenically toward common ground.

I will do this by discussing a few recent occasions in which I have been involved in the Science Wars. The first was an occasion when defenders of natural science directed a polemic in my direction; the second, an occasion when I delivered a polemic at the natural sciences myself; and third, an occasion when I exchanged polemics with other practitioners of the *social* sciences.

Moments from the New York Academy of Sciences Conference, "The Flight from Science and Reason"

At this conference,[2] explicitly designed to awaken unaware scientists to the "clear and present danger" posed to them by social constructionist analyses of science in anthropology, feminism, and cultural studies, I listened quietly in the audience. I had just completed a cultural study in which I treated scientific forms of knowledge as one strand among the many that make up knowledge about the body, health, and the world in the contemporary United States. In that study I did participant observation in a research immunology lab in which there were mutually respectful relations between me and the scientists, technicians, and graduate students. Because I had so recently experienced cordial working relations with natural scientists, I

was discomfited by the barrage of negative sentiments some participants expressed at this conference, sentiments so negative that they
seemed intended to leave no room for cultural studies of science at
all. Some examples follow.

–While discussing means of increasing science literacy among the
  American public, James Trefil revealed his dream that people would
  have film clips of scientific information in their heads. These film
  clips would be played automatically whenever people heard scientific words like *star* or *proton*.[3] Why, I wondered, must he insist
  that knowledge only flows one way, out from science to the rest of
  us, and why must he imagine that this knowledge is best impressed
  on passively waiting minds? Why did Trefil and others at this conference ignore recent findings in anthropology and cultural studies
  that knowledge production occurs at all times and places, inside
  and outside the natural sciences?

–Numerous speakers made plain their contempt for popular culture. Alternative medicine was said to be practiced by "fascists,
  autocrats, and bizarre magicians"; quoting Poe, Gerald Weissman
  remarked that "conventional ideas are foolish"; and Sheldon Goldstein commented that we are living in a "new age of unreason" and
  that, among the public, "logical thought itself is in bad order."
  Most speakers seemed to agree: the public knows nothing. How, in
  the face of this, could I ever get across my own admiration and
  amazement for the complexities of how nonscientists struggle to
  develop knowledge of health and the body, often incorporating
  sophisticated readings of current biomedicine? How could I make
  clear that many alternative practitioners, far from aiding and abet-
  ting a conspiracy against science, are eagerly seeking understanding of their empirical observations through nonreductionistic biomedical disciplines, such as immunology?

–Numerous ominous predictions were made that social constructionism's questioning of the authority of science as the preeminent
  arbiter of truth risks leading us to another Third Reich. As Weiss-
  mann put it, "Once the restraints of reason are cut, the public is prey
  to the wanderings of a debauched brain." Richard Lewontin (1995,
  265) pointed out, in the face of similar warnings from Gertrude
  Himmelfarb, that the Third Reich was surely founded on a passionate belief in an absolute Truth, exactly the opposite of the relativist, comparativist, and situational understanding of values fostered by social constructivism.[4]

A number of responses to the tone and content of this conference have gone through my mind. Can the participants' dedication to a Truth that can be known only by science and that exists completely independently of politics, culture, or history be questioned by pointing out their reliance on links to thoroughly political right-wing organizations and foundations? The list of participants at the New York Academy of Sciences conference overlaps heavily with the list of participants at an earlier conference in a similar vein funded by the National Association of Scholars (NAS).[5] Conveners of the New York Academy conference are frequently represented in the pages of the NAS's journal, *Academic Questions*, and are frequently actively promoted in the weekly NAS science newsletter on the Internet. It is by now well documented that the membership of NAS overlaps with the leadership of major conservative organizations such as the Madison Center for Educational Affairs (Messer-Davidow 1993, 47–50). In addition, the NAS receives major funding from conservative foundations such as Bradley, Coors, Olin, Smith Richardson, and Scaife (Messer-Davidow 1993, 60; Cowan and Massachi 1994, 11).

More important than this particular conference's evident link to powerful right-wing funds, organizations, and agendas is the question of whether the many practicing natural scientists the conference organizers were trying to reach would share their attitudes at all. It is possible that the participants in this conference, determined to present a caricature of cultural studies, feminism, and cultural anthropology, could by doing so elicit a horrified reaction from natural scientists. However, in my experience, most practitioners of science are intrigued by cultural studies of science when it is presented in a subtle way. During my fieldwork, for example, reproductive biologists used my insights about cultural stereotypes that get into accounts of the egg and the sperm to ask new research questions; immunologists used my findings about understandings of the immune system in the wider culture to prepare for their presentations to Congress. In the following, therefore, I direct my remarks not to the organizers of this conference or their allies, but to the practicing scientists they are trying to enlist.

Moments from My Life as a Lecturer on Cultural Studies of Science

Far from always being a silent target of polemics in the Science Wars, I have actually been responsible for hurling many polemics myself. On many occasions as a visiting lecturer on college campuses, in particular as the guest of medical students, I have had as my main purpose the upsetting, and dislodging, of biomedicine's received wisdom about the body.

Once when I was invited to speak to a group of medical students at Johns Hopkins Medical School, I introduced the idea that the fundamental models medicine uses to imagine the body are filled with historically specific cultural content. In the ensuing discussion, a male student from Puerto Rico picked up the polemic I had hurled and threw it again, at one of his white-coated professors. His mother, he told us, was a subject in the first trials of oral contraceptives in Puerto Rico. He detailed her suffering as a consequence of taking high-dose hormones and bitterly castigated U.S. medical research for so ruthlessly using Third World populations as experimental subjects. His professor, one of the principals in the development of the pill, was enraged. He lashed out at the student, telling him he knew nothing, forcing him to acknowledge he had not read the literature in the medical library on the development of the pill, and insisting that he had no right to criticize the science that had saved so many women from death or damage in pregnancy and labor. The student was utterly silenced, at least for the moment.

On another occasion, after I lectured about the cultural presuppositions in biomedicine to the first-year class at Ohio State Medical School, a group of second-year female medical students told me about something that disturbed them deeply. During gross anatomy, their instruction sheets told them to cut off the breasts of the female cadavers and throw them away. They had been bold enough to question the instructor, who told them the breasts were not interesting anatomically and so the time it would take to dissect them was not justified. They had been silenced, but now, armed with fresh ammunition, they began to lay plans for a new confrontation.

At stake in these modest skirmishes is what counts as knowledge and who gets to determine this. Clearly, from the side of the professors, what counts is determined by current professional medical standards, what has been published, and what is proving productive in the ongoing process of research. What counts for the students in these incidents is deeply felt personal experience that is not adequately acknowledged in ongoing science—recollections of a mother's pain from side effects that for her and her son were anything but on the side, and fears of susceptibility to breast disease, fears which are growing, along with rising rates, especially among young women.

To begin to talk irenically instead of polemically, is there any way these accounts can be read across each other? Is there any way to encourage the doctors to acknowledge that the students also "know" something—something the current state of accepted knowledge may have missed? In short, is there any way to begin to encourage the identification of common ground between the practitioners of science and the critics, if for no other reason then for the sake of students like these who are caught in the middle? After all, these young men and women, having gone through college recently, have often been well exposed to feminism, cultural studies, and the like, yet they still want to be practicing scientists.

Even amidst the polemics of the "Flight from Science and Reason" conference, one practicing scientist was able to eloquently delineate common ground between cultural studies of science and science itself. Dudley Herschbach, Nobel laureate in chemistry, toted up a list: as in the humanities, scientific articles certainly use rhetoric to make their points, and this should be celebrated; like a work of art, a scientific experiment can have spiritual value in that it can change the way we see the world; like literary criticism, science could be seen as a form of translating or decoding language. Science tries to decode the many strange and difficult tongues in which nature speaks.

As Herschbach expressed it, science is like a pathway up an unexplored mountain, which could be called truth or understanding. Truth waits there for us scientists until we discover it. Perhaps if we put these remarks of Herschbach's together with a remark by Stephen J. Gould quoted approvingly by one of the conference organizers in the NAS publication *Academic Questions*, we can add to the common ground Herschbach began to identify. Gould is quoted as saying that a scientist "has to be some sort of realist, if only for the sake of one's motivation, 'because 99 percent of the time the scientist's work is so boring, and there are all those mouse cages to be cleaned up again at the end of the day'"(Holton 1995, 15). In other words, within the worldview of its practitioners, belief in the reality of the mountain of knowledge that science discovers is a sine qua non of doing science. Is there a way, in the interest of opening common ground, that we critics of science can stop trying to dissuade scientists from having this worldview? Richard Rorty describes knowledge not as getting reality right but as acquiring habits of action for coping with reality (Rorty 1979; Jenkins 1994). In this vein, the habits of action necessary for most natural scientists to cope with the reality that is the sine qua non of their practice include detailed record keeping; precise measurement of time, quantity, and space; repetition; replication; and reduction. The worldview that supports these habits and makes them necessary is that, when using these tools, you get down to the bones, you find reality.

In a parallel way, the worldview often found among practitioners of another discipline, cultural anthropology, is that what reality is taken to be, down to the bones, depends on ways of seeing that differ. The habits of action that support and are necessitated by this worldview include empathy, shared insights, vivid and effective writing, imagination, poetic interpretation, and understanding things in context. To exercise the effort necessary to participate imaginatively in another world, one must believe that meaning statements are irreducible. So if a psychologist or neuroscientist says that a myth, say, is really caused by repressed anxiety or disrupted brain waves, the anthropologist might well object. Such an account would dissolve the understanding of meaning that was gained by the anthropological account. The anthropological way of knowing is thus made irreducible and irreplaceable; the anthropological epistemology is no less situated in the practice of its discipline than that of a natural scientist.

Moments on the Sidelines in the Science Wars, Social Science Version

The struggle for common ground must go on not just between the natural sciences and the humanities, but also within the social sciences. This became clear to me when I was invited to give a lecture sponsored by the School of Social Sciences at the University of Chicago on feminism and the social sciences.[6] I was astonished at the polemically bitter reaction to my talk. I had made the point that feminism has changed what counts as the production of knowledge by transforming (not just adding to) the very tenets of what counts as knowledge, and by transforming ways in which the academic institutions into which feminists and feminism have been incorporated produce knowledge. As a consequence, feminism would be properly described as a structure of change, or a paradigm of transformation. I argued that, just as the ethnography of science suggests that knowledge production, informed by diverse situations, is going on at all times and places, inside and outside the natural sciences, the same could be true for the domain of social science. The traditional domain about which the social sciences sought knowledge, the public sphere, might be rehabilitated to become one diversified by race, gender, age, and class, in such a way that its members could be seen as participating in the production of knowledge about society. This would depend on realizing that the people social scientists have been studying not only live their own lives but can analyze them, that their "untutored minds" are not necessarily mired in pure experience. If this structural change in the social science view of knowledge could take place with the help of a feminist paradigm of transformation, the ways in which the institutional forms of the social sciences "police" or accommodate knowledge production would change.

To make the prospect of changing what counts as knowledge in the public sphere more concrete, I turned to an example from my recent fieldwork in the neighborhoods of Baltimore. One person, whom Gramsci might have called an organic intellectual (1971, 330), illustrates particularly well how, through ethnography, we can see that "the public" contains theorists who analyze their experience. John Marcellino, a community leader in an integrated, poor working-class area, discussed his experience of health care:

> I got bad teeth, ok? And one of the things all poor people have is bad teeth, because of being poor, ok? . . . [I]t's one of the first things I noticed when I started moving around, because when I talk to Indian people, they got bad teeth. You know what I mean? And I go down south, and they got bad teeth, so it's one of the things that poor people share all over the world in common, we all got bad teeth. You know what I mean? (laughs) And I, what I come to realize it's for a number of reasons. One is, I think that government doesn't want us to have, they don't care about our teeth. . . . I never could figure it out, it just pissed me off, you know what I mean, but I figure it's one of the ways of distinguishing poor people from the rest of the population is we all got bad teeth (laughs) . . . And so you won't get rid of the drug abuse or the prostitution or the crime or stuff, until the people who live here are no longer here. And that to me is the same as the underclass thing, disposable people . . . As soon as they can't figure out a need for us, they'll get rid of us.

> [What's the need for you right now?]

> We still make money for somebody or another. They still need us some, like they needed people to come up out of the south to work in the mills, so they attract them all up. Now there's not as much need for the people to work in the mills, they need some people in the service economy, they try to retrain . . . but if not they're no use, they'll put you in jail . . . they'll choke you off so that you can't make a living doing anything else, so they get rid of

you, or you know, hopefully you'll go back to Virginia or somewhere else, right? You know, you'll crawl in a crack or you won't
have children or something.

Marcellino is pessimistic about the ability of this community to
survive:

> We think it [AIDS] could kill us. It could just kill a lot of people in
> our community, that's what I think about . . . like the way to not get
> AIDS is to not touch people, you know what I mean? . . . we learn
> that it's real specific in terms of how you can catch it. You know
> like being monogamous and only having one partner, . . . that's not
> usual in our community, particularly among young people and,
> you know what I mean. It's not usual, and then you have all this
> interaction between needle users and folks involved in prostitu-
> tion, and a lot of interaction in the community, and so it's not like
> separated, it's not like, well there's a needle community up here
> and people shoot up, and they don't have nothing to do with our
> community. They're part of our community, and having relations
> in our community, and there's people who, you know, are in-
> volved in male prostitution, you know what I mean? And they're
> over here, you know, and there's a gay community over there, you
> know, it's integrated in our community. So our fear was, oh my
> God, you know, when I first heard about it I thought Jesus, we're
> going to be like death . . . then they said it can be seven years be-
> fore you know you have it, and I thought well Jesus Christ then,
> we're really in trouble. Cause now, all these people got it, but no-
> body knows they got it, right? And eventually it's going to be like,
> you know, all of a sudden it's going to be like, you know, like a
> butterfly, you know? [It's a] little thing and all of a sudden . . . one
> day, it's all going to be, you know, all through the community.

This man's account of the fate of his particular local community, in
which he speaks for and with many others, reaches from the speci-
ficities of working-class interdependency to the fear of the community's death through the agency of a microorganism, whose spread is
enhanced by the community's very interdependency. His account is
linked by implication to differential survival in the society: some
will have good jobs, healthy bodies, strong immune systems, and
protected living spaces and will live. Others, with none of these
things, will die. His account speaks about the state, public health, science, medicine, capitalism, and an ethos of neosocial Darwinism, in

which some are seen as "unfit" disposable people and some are seen as fit people of "high quality."[7] Marcellino is doing critical theory, in the sense specified by Marx in 1843: "The self-clarification of the struggles and wishes of the age" (quoted in Fraser 1989, 113).[8] Self-clarification implies that people located in different social circumstances produce different sorts of knowledge. Through specific experiences located in specific times and places, people can reach partial but analytic knowledge about forces, institutions, or powers much larger and stronger than they are.

However compelling the problems Marcellino raises (and the fact that he is able to articulate them), by itself his story would not count as doing "social science." What counts as "science" is in part an effect of institutions (giving credentials, granting authenticity, and so on). By himself Marcellino has none of these things. This is where the institutionalization of feminism in the academy plays a small but crucial role. In this particular example, it is my position in a university, itself made possible by the activities of earlier generations of feminists, that gives me the time and resources to talk to people like Marcellino and to publish books and articles (and give lectures) in which I argue that what he says should begin to participate in what counts as social science.

The attack on this lecture by senior social science faculty and administrators at Chicago was uncompromising. One dean even dropped into his (presumably) childhood cockney accent as he launched an invective at me. The reasons for this hostility are complex. I believe they lie in the fact that the social sciences are themselves deeply implicated historically in the conceptions of the natural sciences and in natural science notions of truth. When feminism entered the scene in the social sciences in the 1960s and 1970s, most canons of discovery and proof were based as closely as possible on the natural sciences. Objectivity was taken to be desirable and achievable; quantitative measurement was thought to yield the most significant results. A properly value-free discipline, it was taken for granted, could reveal reality (Zalk and Gordon-Kelter 1992, 5).

This "scientism" of the social sciences was fed, just as it was in the natural sciences, by fears of feminization. As Dorothy Ross explains in her recent history of the social sciences, the aggressively masculine language of penetrating, knowing, and controlling nature (and now society) helped set the "masculine boundaries of sociology against the feminine precincts of social work and reform" (Ross 1991, 394–95).

One way feminist scholarship upset this enterprise was to complicate the universal "man" who was the object of study. In studies of kinship, psychology, work, politics, and so on, a new object of study was created: "woman." "Woman" often did not behave according to the generalizations discovered for "man." Creating a new universal "woman" to combat the old universal "man" would eventually turn out to have its difficulties, but at the time it served the important function of raising these questions: How value-free was the science of the social sciences? (Or did the social sciences actually reproduce ideological discourses that justified and recreated the social relations of gender? [Zalk and Gordon-Kelter 1992, 9].) How limiting was the strong preference of the social sciences for the study of public over private matters and for matters of the mind over matters of the body? And what could be learned by reversing these preferences, making women the foremost objects of study, women's personal experiences the source of data, and the body rather than the mind the subject of analysis?

These were not the only positions feminist approaches turned on their heads. From the early days of American social science, the belief was that only social science could show the way toward the good, the rational, society. "The true social standard . . . lay in the fullest development of the organic society, and only science could discover and enforce its conditions" (Ross 1991, 368). Individual feelings were an inadequate measure of the social good. There was a total distrust of "untutored" human nature and a desire for complete control of "wayward human experience" (Ross 1991, 368–69). Science provided a model of "disinterested" discourse in contrast to the "self-interested" discourse of the citizenry outside science. "In the name of science one thus could (and can) still treat public opinion as mere opinion. . . . A specialized non-public science [was] deployed in the service of administrative rationality and in competition with the public sphere" (Calhoun 1992, 36).

By paying attention to that quintessentially "wayward and self-interested" subject, woman, and by using her "untutored" experiences as a way of questioning the rationality of social scientific conclusions about the public good, feminism in the 1960s and 1970s questioned the tenets of knowledge that had been fundamental in the social sciences. In so doing, and because of the unmistakable contribution to knowledge made by this work, feminism and feminists slowly began to be included in the academy, in its curriculum, faculty positions, and publications.

Following the dislodging of "universal man" came the dissolution of the category "universal woman." During the late 1980s and 1990s this category, which had served earlier decades of feminism very well, dissolved. In one social science discipline after another, "woman," created as an object to dislodge "man," fell apart under the impact of awareness of diversity, whether based on race, ethnicity, sexuality, age, or postcolonial status. Rayna Rapp (1992, 85) explains:

> All . . . unified, universalist, notions of womanhood are suspect, for they are built to the measure of everywoman, who too often turns out to be a white, Anglophone, feminist scholar in disguise.

An enormous proliferation of studies from differently situated points of view made clear that socialization, kinship, marriage, politics, and everything else were experienced very differently (even within the confines of the United States) depending on whether a woman was working class (Stack 1990), Jewish (Prell 1990), fundamentalist Christian (Harding 1990), African American (Carothers 1990), Asian American (AWUC 1989), Italian American (Di Leonardo 1984), Appalachian (Stewart 1990), Hispanic (Fernandez-Kelly 1990), lesbian (Newton 1993), disabled (Hillyer 1993), or transsexual or post-transsexual (Stone 1991), to name but a few.

In the face of all this, it became clear that the unitary term *feminism* masked a diversity of feminisms; that even if all women wanted improvement in their lives, they might not agree on the meaning of "improvement" (Fox-Genovese 1993, 235). Just as earlier the "universal intellectual" who spoke for the universal and abstract "male" was no longer acceptable in the face of the category "woman," so now the universal feminist intellectual became unacceptable. Feminism had once again—this time from a position more solidly within the academy—challenged the tenets of knowledge, tenets ("everywoman" and the "universal feminist intellectual") it was responsible for creating.

At this point we can see that developments in feminism closely paralleled developments in science studies. At the same time as the category "woman" was coming apart, so was the category "science." This happened by means of interdisciplinary work in the social studies of science, in which historians, sociologists, and, more recently, anthropologists have joined hands (Hess 1992). The new wave of feminists in this crowd (unlike their predecessors) bypassed the study of the social sciences per se and went right for the jugular of the natural sciences. To this group, working in the 1980s, there seemed little point in bothering with social science, since its methods were derived from the natural sciences and since at this time social science had little power in Euro-American culture generally compared to natural science.

This new research has dramatically revised our understanding of natural science. It is as if we once thought of science as an isolated medieval walled citadel, and this walled citadel turns out to be more like a bustling center of nineteenth-century commerce, porous and open in every direction. While science likes to be thought of as a luminous citadel, alone on a hill, in actuality many powerful collectives and interested individuals dot the surrounding hillsides. Not only are they nearby, but they interact with the world inside the citadel of science frequently and in powerful ways. Ethnographic research has shown that the strict, fixed borders between the pure realm of knowledge production and the "untutored" public do not hold up to scrutiny. The walls of the citadel are porous and leaky; inside is not pure knowledge, outside is not pure ignorance. This means the way is opened for a more complex, less flatly antagonistic attitude toward science than prevailed among feminists earlier. Scientific knowledge is being made by all of us; we all move in and out of the bustling city of knowledge production. Of course, the Chicago social scientists and the organizers of the "Flight from Science and Reason" conference would probably still find this view flatly antagonistic: others might yet be intrigued.

Toward a Little Common Ground

In social constructionist science studies, whether via cultural studies, feminism, or anthropology, a common goal is to understand the particular form and content of core natural science concepts in their relevant historical contexts. It was apparent at the conference that some scholars have interpreted this effort as an attack intended to obliterate the natural sciences. More moderate responses also occur: in my research, scientists would often find my questions and analyses strange and sometimes even uncomfortable, but usually more interesting than threatening. I want to continue opening up common ground with these scientists by pointing out that social constructionist accounts of the natural sciences are part of a larger enterprise which *includes* producing social constructionist accounts of the social sciences and humanities. In what follows, I will describe the emergence of social constructionist accounts of some central concepts in the intellectual home territories of science studies. If the intent here is to better understand the nature of human cultural activity—not to obliterate disciplines but to go beyond their limitations—perhaps the same could be imagined to be true for the natural sciences.

First, some background is necessary. As we have seen, after the fall of "universal man" we were left with "man and universal woman." What was left after the fall of "universal woman" and "universal science"? This was the context for the burgeoning of what is known as identity politics. With an emphasis on "inventing a new language . . . and defining new bodies of knowledge . . . to replace the institutional forms of knowledge that oppress certain communities or social groups" (Escoffier 1993, 32, 40), an intense need arose for specific knowledges possessed by particular communities. In identity politics, "'experience' emerges as the essential truth of the individual subject, and personal 'identity' metamorphoses into knowledge. Who we are becomes what we know; ontology shades into epistemology" (Fuss 1989, 113).[9]

It follows that we can only know what we experience and that singular experiences cannot be lumped with other singular experiences without producing violent distortion. As Judith Butler puts it, "The very category of the universal has only begun to be exposed for its own highly ethnocentric biases" (Butler 1992, 7). Any time an example or paradigm is made to stand for a whole, the example or paradigm "subordinates and erases" that which it seeks to explain. The whole appears to be produced by the example, made to stand for it, and this is a "gesture of conceptual mastery" (5).

Insofar as identity politics is based on a notion of an inner essence, which some people have and some people do not and could never have, it is an essentializing politics. The inner essence, the identity, operates as a given (Alcoff 1994). Those who doubt the value of identity politics argue that what may be gained through its ability to mobilize political action is lost in its tendency to remove the inner essence from examination. As Joan Scott puts it, "Making visible the experience of a different group exposes the existence of repressive mechanisms, but not their inner workings or logics; we know that difference exists, but we don't understand it as constituted relationally" (Scott 1992, 25). Understanding it relationally would include understanding how the group came to be repressed (in relation to forces, attitudes, or beliefs), where the repression came from (in relation to other groups, family structures, state policies, and so on), and when it occurred (in relation to other historical processes).

To avoid the essentialism of identity politics, some feminists are advocating more fluid concepts of the person. Linda Alcoff (1994, 117) stresses that identity can change over time:

The concept of positionality allows for a determinate though fluid identity of woman that does not fall into essentialism: woman is a position from which a feminist politics can emerge rather than a set of attributes that are "objectively identifiable." Seen in this way, being a "woman" is to take up a position within a moving historical context and to be able to choose what we make of this position and how we alter this context. From the perspective of that fairly determinate though fluid and mutable position, women can themselves articulate a set of interests and ground a feminist politics.

Similarly, Teresa De Lauretis (1986, 14–15) stresses that multiple identities can coexist in the same person:

The female subject is a site of differences; differences that are not only sexual or only racial, economic, or (sub)cultural, but all of these together and often enough at odds with one another. . . . Once it is understood . . . that these differences not only constitute each woman's consciousness and subjective limits but all together define the female subject of feminism in its very specificity . . . these differences . . . cannot be again collapsed into a fixed identity, a sameness of all women as Woman, or a representation of Feminism as a coherent and available image.

But these moves do not solve the problem for other feminists, who fear that making the self temporally changeable and internally diverse risks losing an effective acting self altogether. We would end up with a "subject [with] no internal coherence, [no] . . . means for self-knowledge; instead the subject [would be] seen as dispersed in (multiple) texts, discursive formations, fragmentary readings and signifying practices, endlessly constructing and dislodging the conceit of the self" (Dirks et al. 1994, 12).

Some worry that this fragmented view of the subject merely reflects the kind of person and society created by the condition of postmodernity. This worry is the more acute because notions of a fluid, changeable, and flexible self are found very widely in popular culture (Martin 1994). There is a tendency to "celebrate and transcendentalize the decentered and fragmented subject" which is actually an effect of late capitalism. It may allow escape from the essentialized subject, but only by means of losing any kind of acting subject at all (Harvey 1993; Dirks et al. 1994, 14).

The important point is that we in cultural studies scrutinize our own concepts, just as we do for concepts in the natural sciences. To the extent that concepts of the flexible, fluid, multiple self are part of the general cultural context of late capitalism, feminist or cultural theories based on some version of this kind of self *participate in* rather than critically analyze contemporary culture. The discomfort of this realization begins to open up a form of common ground with the natural sciences: it is akin to the discomfort scientists express when we try to put *their* concepts in historical context.

If we can be adequately analytic about contemporary views of the subject, we may avoid one serious consequence: benefiting the new global order driven by the interests of multinational corporations, because "its extraordinary emphasis on personal expression in effect drives the logic of modernist individualism to its ultimate conclusion" (Fox-Genovese 1993, 246, 253). As Jody Berland argues, "To embrace fragmentation uncritically runs the risk of duplicating the move to a market-driven consumerist model of human populations in which the fragmentation of conventional identities is a fine art" (Berland 1992). Paula Treichler agrees: "To champion postmodernist fragmentation and dispersion does sometimes deflect attention from realities that *should be* brutally (rather than strategically) essentialized" (Treichler 1992, 62).

My own view is that the way forward lies in attending to and incorporating in our explanations what Treichler calls "realities that *should be* brutally . . . essentialized." I would take these realities to mean large-scale political–economic forces abroad on the earth, structural forces that can be universal in their scope and that are often damaging in their effects. They would include the increasing concentration and mobility of capital, which often lead to emiseration of the poor; and the concomitant restructuring of the organization of work, both inside corporations and factories and in the spread of "homework." They would also include vast alterations in how information is stored and retrieved, and in the extent to which biological research focuses on genetics. "Brutally essentializing" these forces might mean naming them, identifying their core features, or describing their various effects. In the process we would be using many of the tools common in the natural sciences: detailed record keeping, precise measurement of time, quantity, and space, repetition, replication, and reduction.[10] For example, counting, measuring, and making controlled comparisons allow us to understand what it means to say that in the United States the economy is expanding and more jobs are created every year. The economy is expanding because the wealthiest

40 percent get 68 percent of the income, creating enough consumer power to keep companies in business but leaving 60 percent of the population unable to participate (Peterson 1994). This helps explain why many people feel as if they are living through a depression. As an example, there is the family with three children who between them hold four jobs but make only $18,000 a year (Johnson 1994). "When it was noted that two million new jobs were created last year, the husband quickly put that statistic in perspective. 'Sure, we've got four of them. So what?'" (Herbert 1994). Embedded in this account are techniques of knowing we share with natural scientists: acknowledging that these techniques are necessary to shed light on processes of large-scale oppression opens up another small amount of common ground between us.

Conclusion

Toward the beginning of this essay, I argued that a focus on the habits of action necessary for disciplines like the natural sciences or cultural anthropology to gather knowledge as defined by their worldview would begin to open some common ground and avoid fruitless polemics. Feminism and cultural studies could also come to be understood as ongoing projects, involving a "motivated and stylized" frame of mind that one actively works at in order to be located in a position to both analyze and see how effective action could be taken (Gupta and Ferguson 1994). For feminism and cultural studies, the goal would not be a pure "positionality" that engages in constant shifting and flexible positioning for its own sake, but a *motivated* positionality collectively sought, in which diverse kinds of alliances could be forged with others (even natural scientists) over a variety of (changing) common interests. In the same spirit, the wont of natural scientists to believe their techniques of knowing can get down to the bones of reality could be regarded as a motivated positionality necessary to support the more arduous aspects of the discipline their way of knowing requires. A benefit would be that the natural sciences could actively use, for particular purposes, the insights of cultural studies of sciences without having to betray the entire basis of their discipline. From my experiences with a variety of practicing scientists—including the young medical students I described earlier who are confronting their professors about broadening the social responsibilities of medical researchers and increasing the responsiveness of medical research design to social problems—I know this can happen.

Notes

I thank Richard Cone, Rayna Rapp, and Mary Poovey for giving me helpful and illuminating responses to portions of this essay.

1   The term was called to my attention by Stefan Collini (1993).

2   Held at the New York Academy of Sciences, 31 May–2 June 1995. Among the main organizers were Paul Gross and Norman Levitt, the authors of *Higher Superstition:The Academic Left and Its Quarrels with Science*. My earlier work was briefly discussed and dismissed in *Higher Superstition* (Baltimore, Md.: Johns Hopkins University Press, 1994), 125–26.

3   In the absence of a tape recording or published proceedings, paraphrases and brief quotes from speakers at the New York Academy of Sciences conference are based on my own notes.

4   For a published account that links multiculturalism's questioning of authority to fascism see Brasor 1995.

5   Some of the papers presented at this conference, "What Do the Natural Sciences Know and How Do They Know It?" are published in *Academic Questions* 8 (1995).

6   The conference, held 31 May–2 June 1994, was called "Boundaries and Trespass: Conventions and Creative Revisions in the Social Sciences."

7   As in a recent newspaper article about universal standards of beauty (Angier 1994).

8   Letter to A. Ruge in *Karl Marx: Early Writings*, ed. T. B. Bottomore (London: Watts, 1975).

9   Fuss 1989 and Spelman 1988 are useful sources on the complex implications of identity politics.

10 For an account of the social power of statistics in recent history see Asad 1994.

References

Asian Women United of California (AWUC). 1989. *Making waves: An anthology of writing by and about Asian American women*. Boston: Beacon.

Alcoff, Linda. 1994."Cultural feminism versus post-structuralism: The identity crisis in feminist theory. In *Culture/power/history: A reader in contemporary social theory*, edited by Nicholas B. Dirks, Geoff Eley, and Sherry B. Ortner. Princeton, N.J.: Princeton University Press.

Angier, Natalie. 1994. Why birds and bees, too, like good looks. *New York Times*, 8 February, C1:1, 12.

Asad, Talal. 1994. Ethnographic representation, statistics, and modern power. *Social Research* 61: 55–88.

Berland, Jody. 1992. Angels dancing: Cultural technologies and the production of space. In *Cultural Studies*, edited by Lawrence Grossberg, Cary Nelson, and Paula Treichler. New York: Routledge.

Brasor, Gary Crosby. 1995. Weimar in Amherst. *Academic Questions* 8: 69–86.

Butler, Judith. 1992. Contingent foundations: Feminism and the question of "postmodernism." In *Feminists theorize the political*, edited by Judith Butler and Joan W. Scott. New York: Routledge.

Calhoun, Craig. 1992. Introduction: Habermas and the public sphere. In *Habermas and the public sphere*, edited by Craig Calhoun. Cambridge, Mass.: MIT Press.

Carothers, Suzanne. 1990. Catching sense: Learning from our mothers to be black and female. In *Uncertain terms: Renegotiating gender in American culture*, edited by Faye Ginsburg and Anna Tsing. Boston: Beacon.

Collini, Stefan. 1993. Introduction to *The Two Cultures*, by C. P. Snow. Cambridge: Cambridge University Press.

Cowan, Rich, and Dalya Massachi. 1994. *Guide to uncovering the right on campus*, vol. 1. Cambridge, Mass.: University Conversion Project.

De Lauretis, Teresa. 1986. *Feminist studies/critical studies*. Bloomington: Indiana University Press.

Di Leonardo, Micaela. 1984. *The varieties of ethnic experience: Kinship, class, and gender among California Italian-Americans*. Ithaca, N.Y.: Cornell University Press.

Dirks, Nicholas B., Geoff Eley, and Sherry B. Ortner. 1994. Introduction to *Culture/power/history: A reader in contemporary social theory*, edited by Nicholas B. Dirks, Geoff Eley, and Sherry B. Ortner. Princeton, N.J.: Princeton University Press.

Escoffier, Jeffrey. 1993. Intellectuals, identity politics, and the contest for cultural authority. *Found Object* 2: 31–44.

Fernandez-Kelly, M. Patricia. 1990. Delicate transactions: Gender, home, and employment among Hispanic women. In *Uncertain terms: Negotiating gender in American culture*, edited by Faye Ginsberg and Rayna Rapp. Boston: Beacon.

Fox-Genovese, Elizabeth. 1993. From separate spheres to dangerous streets: Postmodernist feminism and the problem of order. *Social Research* 60: 235–54.

Fraser, Nancy. 1989. *Unruly practices: Power, discourse, and gender in contemporary social theory*. Minneapolis: University of Minneapolis Press.

Fuss, Diana. 1989. *Essentially speaking: Feminism, nature, and difference*. New York: Routledge.

Gramsci, Antonio. 1971. *Selections from the prison notebooks of Antonio Gramsci*, edited by Q. Hoare and G. N. Smith. New York: International Publishers.

Gupta, Akhil, and James Ferguson. 1994. Anthropology and "the field": Boundaries, areas, and grounds in the constitution of a discipline. Unpublished manuscript.

Harding, Susan. 1990. If I should die before I wake: Jerry Falwell's pro-life gospel. In *Uncertain terms: Renegotiating gender in American culture*, edited by Faye Ginsburg and Anna Tsing. Boston: Beacon.

Harvey, David. 1993. Class relations, social justice, and the politics of difference. In *Principled positions: Postmodernism and the rediscovery of value*, edited by Judith Squires. London: Lawrence and Wishart.

Herbert, Bob. 1994. What recovery? *New York Times*, 13 March, D17.

Hess, David J. 1992. Introduction: The new ethnography and the anthropology of science and technology. In *Knowledge and society: The anthropology of science and technology*, vol. 9, edited by David J. Hess and Linda L. Layne. Greenwich, Conn.: JAI.

Hillyer, Barbara. 1993. *Feminism and disability*. Norman: University of Oklahoma Press.

Holton, Gerald. 1995. "Lumpers," "splitters," and scientific progress. *Academic Questions* 8: 14–19.

Jenkins, Timothy. 1994. Fieldwork and the perception of everyday life. *Man* 29: 433–55.

Johnson, Dirk. 1994. Family struggles to make do after fall from middle class. *New York Times*, 11 March, A1, 14.

Lewontin, Richard C. 1995. Essay Review: À la recherche du temps perdu. *Configurations* 2: 257–65.

Martin, Emily. 1994. *Flexible bodies: Tracking immunity in America from the days of polio to the age of AIDS*. Boston: Beacon.

Messer-Davidow, Ellen. 1993. Manufacturing the attack on liberalized higher education. *Social Text*, no. 36: 40–80.

Newton, Esther. 1993. *Cherry Grove, Fire Island: Sixty years in America's first gay and lesbian town*. Boston: Beacon.

Peterson, Wallace C. 1994. *Silent depression: The fate of the American dream*. New York: Norton.

Prell, Riv-Ellen. 1990. Rage and representation: Jewish gender stereotypes in American culture. In *Uncertain terms: Renegotiating gender in American culture*, edited by Faye Ginsburg and Anna Tsing. Boston: Beacon.

Rapp, Rayna. 1992. Anthropology: Feminist methodologies for the science of man? In *Revolutions in knowledge: Feminism in the social sciences*, edited by S. R. Zalk and J. Gordon-Kelter. Boulder, Colo.: Westview.

Rorty, Richard. 1979. *Philosophy and the mirror of nature*. Princeton, N.J.: Princeton University Press.

Ross, Dorothy. 1991. *The origins of American social science*. Cambridge: Cambridge University Press.

Scott, Joan W. 1992. Experience. In *Feminists theorize the political*, edited by Judith Butler and Joan W. Scott. New York: Routledge.

Spelman, Elizabeth V. 1988. *Inessential woman: Problems of exclusion in feminist thought*. Boston: Beacon.

Stack, Carol. 1990. Different voices, different visions: Gender, culture, and moral reasoning. In *Uncertain terms: Renegotiating gender in American culture*, edited by Faye Ginsburg and Anna Tsing. Boston: Beacon.

Stewart, Kathleen Claire. 1990. Backtalking the wilderness: "Appalachian" en-genderings. In *Uncertain terms: Renegotiating gender in American culture*, edited by Faye Ginsburg and Anna Tsing. Boston: Beacon.

Stone, Sandy. 1991. The *Empire* strikes back: A posttranssexual manifesto. In *Body guards: The cultural politics of gender ambiguity*, edited by Julia Epstein and Kristina Straub. New York: Routledge.

Treichler, Paula A. 1992. Beyond *Cosmo*: AIDS, identity, and inscriptions of gender. *Camera Obscura* 29: 21-76.

Zalk, Sue Rosenberg, and Janice Gordon-Kelter. 1992. Feminism, revolution, and knowledge. In *Revolutions in knowledge: Feminism in the social sciences*, edited by Sue Rosenberg Zalk and Janice Gordon-Kelter. Boulder, Colo.: Westview.

My Enemy's Enemy Is—Only Perhaps

—My Friend

Hilary Rose

ULTURAL struggles in Britain have been privileged to date, in that Gross and Levitt's *Higher Superstition: The Academic Left and Its Quarrels with Science* has, except in specialist circles, received rather little attention. It has, to borrow Margaret Thatcher's useful metaphor, been denied the "oxygen of publicity." The British version of the Culture Wars has been less a generalized attack than a series of assaults—two of the most conspicuous being on psychoanalysis and on the sociology of scientific knowledge (known to its practitioners as SSK). Perhaps this transatlantic difference derives from the scale of the lurch to the right in the United States, by comparison with the slow implosion of the Right in Britain. The attack on both psychoanalysis and SSK claims to come from something that its protagonists speak of as "Science." From the standpoint of this single entity—which I shall describe with a capital *S* to distinguish it from the heterogeneity of the sciences and their methods—psychoanalysis is criticized for not being Scientific; SSK is criticized for taking a constructivist, and repudiating a realist, theory of scientific knowledge.

In these wars, the self-appointed defenders of Science are seeking to police the boundaries of knowledge and to resurrect canonical knowledge of nature, against the attempts of the Others (including feminists, antiracists, psychoanalysts, postcolonialists, leftists, multiculturalists, relativists, postmodernists, etc., etc., in all our bewildering diversity) to extend, transform, or maybe even dissolve the boundaries between the privileged truth claims of science and other knowledges. But first, just because any of us may find ourselves among the Others under attack, I must emphasize that this commonality may not automatically generate bonds of solidarity between this "us." My enemy's enemy is—only perhaps—my friend.

For example, some (usually postmodernists) have claimed such an alliance between themselves and feminists, while numbers of feminists have claimed that postmodernism depoliticizes and weakens feminism. Jane Flax (1993) writes of her distress as a feminist psychotherapist and theorist at a bruising attack on her postmodernism by a feminist audience. Conversely, from a critical realist standpoint, A. N. Sivanandan, the editor of *Race and Class*, passionately criticizes postmodernism's attempt both to rewrite a new nonracialized subjectivity for white working-class youth and to clear ground for this new approach by deploring "moralizing antiracists" (1995). In his judgment the literary turn may diminish discrimination but cannot meet the challenge of racist violence on the streets, which then spreads out, offering a bloody legitimacy to increasing State racism.

In such times alliances between "us" are likely to be provisional and built; the innocent appeals of an old ungendered and "unraced" solidarity of class are no longer available; gone too is the "innocence" of universal sisterhood. Instead, while not abandoning commonality, recognition of complexity and the need to pay meticulous attention to context, not least our own, are the name of the game. Positioning myself within a particular reading of the feminist critiques of science—and they are plural and diverse—I want to contrast the very different attacks on psychoanalysis and SSK. Who is speaking is as important as what is being said and in what arena the debates take place.

The End of Psychoanalysis?

English as an increasingly global language produces some real problems, as well as possibilities, for an offshore island. It assists key weeklies in fanning cultural conflicts by importing cultural conservatives from the U.S. scene—where they seem to grow with distressing abundance—into Britain, as well as by enthusiastically reporting our own indigenous talent. Thus not having been particularly attentive to feminism, the media receives the anti- and postfeminist backlash, in which women claim notoriety if not fame by trashing the work of feminists with tremendous enthusiasm (as typified by the space given to Camille Paglia). The attacks on psychoanalysis have been increasing. The *New York Review of Books* (NYRB) published Frederick Crews's initial hostile piece,[1] but to make sure that a British readership should not miss this attack, the *Times Higher Education Supplement* (THES) followed with a further extraordinarily self-referential article by Crews.[2]

Crews's waspish critique draws extensively on two early 1980s kindred texts, which were clearly supposed to work like the twin blades of a pair of scissors. One blade was fashioned by philosopher of science Adolf Grünbaum (1984), who threw into methodological question "free association," the other by anthropologist and philosopher Ernest Gellner (1984). Also shaped by the Popperian mold, Gellner shares and endorses Grünbaum's methodological critique and goes on both to criticize the claims of the analyzed to provide psychological truth for their patients and to criticize the recruitment of psychotherapists. The intensity of this latter criticism I find quite puzzling, as there is an everyday sense in which those who control access to any profession, discipline, or trade—from the medieval guilds, to the contemporary rituals of admission to the fireservice, to the ceremonial rite of passage of the Ph.D.—all behave rather similarly. Indeed, Lisa Appignanesi and John Forrester (1992) suggest that so far as recruitment mechanisms of professions went, psychoanalysis was more friendly to women than most of science. But I digress; the point I want to underline is the certainty that Grünbaum and Gellner share with their mentor Popper as to what is Science. Hence the most tartly written and denunciatory section of Grünbaum's book is the attack on Habermas and hermeneutics as a radically different way of producing reliable knowledge. The destruction of hermeneutics is a crucial preliminary to using Science to defeat psychoanalysis.

However, this largely literary and philosophical encounter between the critics and supporters of psychoanalysis sets aside the growth of biological psychiatry and the immense infrastructure for both diagnostics and therapies,[3] which has been developed during what in the United States has been designated the *Decade of the Brain*. Thus, while the idea of Science is central to the attack on psychoanalysis, there is little critical attention to the claims of its chief rival, the newly power-charged discourses of biological psychiatry. Technologies we almost only recognize by their acronyms, MRI, PET, and CAT scans, recombinant DNA probes, to the latest generation of psychotropics offer a huge apparatus claiming to diagnose and manage madness. But although few of us would propose throwing away these technologies, not least on a pragmatic basis, the well-financed pursuit of the biological basis of madness has been conspicuously unsuccessful (Breggin 1993; Russell 1995). Claims for the genetic transmission of manic depression amongst the Amish, of schizophrenia and alcoholism are routinely made and as routinely broken. After the latest failure to replicate the claim of the genetic transmission of schizophrenia, John Maddox, the editor of *Nature*, wearily suggested that maybe it was time to give up the attempt. But psychopharmacology listens to Prozac and marches on regardless.

As Denise Russell (1995) observes, even if within the language of the philosophy of science biological explanations of madness are failing and degenerating research programs, there are few signs of them stopping. Commercial interests encourage the pharmaceutical industry to develop new drugs for which medicine is required to find conditions to cure. Despite its scientific weakness, the new biological psychiatry draws cultural strength from its alliance with the basic neurosciences. In the United States it has gradually excluded analytically oriented psychiatrists from the control of the psychiatric departments in the medical schools, where hitherto they had considerable influence. It is a not unreasonable speculation that, when the history of the last two decades is eventually written, biological psychiatry will have been shown to have driven out the psychoanalyst much as the medical profession drove out the midwives, and on similarly modest scientific grounds (Donnison 1977; Ehrenreich and English 1978; Oakley 1984). Meanwhile, psychoanalytically informed approaches in medicine have been both feminized and downgraded, allocated primarily to women counselors, social workers, and nursing staff. Having lost the medical schools largely without the help of the philosophers and literary critics, U.S. psychoanalysis has to join that of Britain in standing up without its medical crutch. But the accounts of the growth of science by the philosophers of science rarely consider the political economy of research.

As a cuckoo in the nest, biological psychiatry has successfully occupied the discourses of madness, less by engaging with the talking cures than by simply pushing them out. The issue of scientificity cedes to a gendered political economy of biomedical research, which sustains the relentless search both for new technological artifacts and for new technorepresentations of the mind/brain. The question of "The End of Psychoanalyis?" (to cite the title of a recent U.K. conference) lies at least as much outside psychoanalysis as within it. It is evident that the philosophical attack makes numbers of practicing psychoanalysts distinctly uncomfortable, not least because, like many laboratory scientists, they have little familiarity with either past or current debates in the social studies of science. Yet as I have been suggesting here, the concept of science is itself under cultural siege, and it is these besieging forces that look to be potential allies.

Apologists for a radically reconstructed psychoanalysis (I find it impossible to accept the collusion of classical analysis with sexual abuse, and so I would exclude this from possible friendship) manifestly have a choice of lively discursive strategies open to them. They are likely to have in common a resistance to the one-Science thesis and an awareness of the limits of biological psychiatry not least as judged by its own canon. Hermeneutics, despite Grünbaum's strictures, is far from dead; the thesis of plurality of the sciences has many supporters. Like Flax (1993) or Forrester (1989), such apologists can adopt a thoroughgoing postmodernist attack on the privileged truth claims of science, or they can follow Irigaray (1985) in her rejection of the possibility of doing theory without reproducing phallocentric culture and so propose the most radical move of all, jamming the theoretical mechanism itself. Such friendlier discourses are well known to the theoreticians, albeit the political problem for analysts as a community is that many practitioners are too busy listening to their analysands, and perhaps too deafened by the culturally loud voice of Science, to hear the still squabbling but quieter voices of possible allies.

The Eye of the Cultural Storm?

While debates about pyschoanalysis sit somewhere at the edge of British cultural life, those about science are central. The most recent debate about the nature of scientific inquiry, which has been grumbling away in the background for some time, is in many ways an extension of the old assumption within the Anglo-Saxon tradition that science equals the natural sciences. The concept of *Wissenschaft*, which holds all of systemic inquiry together in other cultural traditions, is simply not available in an Anglo-Saxon context. A year ago this background grumble took center stage with a public debate, held at the 1994 meeting of the British Association for the Advancement of Science (B.A.), between Harry Collins, a leading sociologist of science, and developmental biologist Lewis Wolpert, Chairman of the Royal Society's Committee on the Public Understanding of Science (COPUS).[4] Collins, together with fellow sociologist Trevor Pinch, *The Golem: What Everyone Should Know about Science*, and Wolpert, *The Unnatural Nature of Science: Science Does Not Make (Common) Sense*. These rival and polarized texts claim to extend the public understanding of science, the former by claiming that science is socially constructed, the latter by its insistence that scientific knowledge is unitary, with unique truth claims derived from its capacity to hold a mirror to nature visible to its illuminati.

The confrontational debate made a considerable impression on those who witnessed it—as an extraordinarily vituperative and unpleasant event. To the extent that considerable sections of the academy have given up those highly adversarial and masculinist exchanges, long criticized in feminist circles as fostering the nonmeeting of minds (Moulton 1983), the debate was a return to a past from which we are not yet free.

The quarrel was then transferred to the pages of the *Times Higher Education Supplement* (THES). The options for those sociologists who disagreed with Collins or those biologists who disagreed with Wolpert (to say nothing of those who as feminists thought both were unreconstructedly macho in thought and style) were shriveled down to the binary choice of either Collins or Wolpert. Within a few months the meeting was reenacted at Durham (financially supported by THES, which saw it was onto a fine controversy), and although Collins would not participate in a repetition of the B.A. savagery, preferring to expound his views on television, Wolpert was very much present. Despite there being many more contributors, so that the adversarial structure was less acute, the meeting still had many unpleasant exchanges.

PUS: The Contested Turf

The ground under contest is the Public Understanding of Science (or to give it its unlovely acronym, PUS), which formed the title of a 1985 report produced by a Committee of the Royal Society chaired by Walter Bodmer. The problem, as perceived by this elite group, was that science was losing its popularity. This sensitivity to loss of trust and thence anticipated loss of support is being played out not only in Britain, but also in the United States as the center of the global research system, not least in terms of the one-third cut to the science budget proposed by the Republicans in 1995. The report sought to overcome this by proposing that a more scientifically literate public would be more supportive. The Royal Society's Committee for the Public Understanding of Science was set up under Bodmer's chairmanship, with the mission of encouraging scientists to assist the public in understanding science. A whole battery of science festivals, hands-on experiences for children, prizes for authors of books that enhance PUS, Science, Engineering, and Technology weeks for schoolchildren, and research awareness programs for industry were brought into being. However, the Bodmer credo that the increased public understanding of science breeds trust and thereby support was indeed a matter of faith. A number of both quantitative (Durrant et al. 1989) and qualitative (Irwin and Wynne 1996) social science studies have pointed to the association between increased scientific literacy and increased skepticism about science. These findings have had no impact on the COPUS credo; this year COPUS launched itself into its second decade with a continuing commitment to one-way communication. The job of the public is to listen and learn.

Communicating science to the public through the media is a long and honorable tradition, pioneered in the 1940s by the socialist, geneticist, and brilliant essayist J. B. S. Haldane with his regular column in the *Daily Worker*. Before that the lecture hall was the chief form. In the second half of the twentieth century this task of communication has become largely the task of professional science writers, who have extended it into popular science magazines (*New Scientist, Focus*), the broadsheets, and television.

As an officially sanctioned activity, recruiting scientists themselves into public understanding of science is relatively new in Britain. Radical and feminist scientists have of course long been at this game, but from a critical standpoint. What remains peculiarly difficult for the British is to move beyond communication as monologue. Unlike some other countries, such as Denmark, which has a strongly democratic approach to technical decision making and takes for granted dialogue between the producers of, for example, new biotechnology and the public as end users, Britain, apart from one cautious experiment at a biotech consensus forum, remains incredibly apprehensive that a nontechnical expert should have an opinion and start talking as well as listening. Such anxiety has two origins: the first is the pathological commitment to a culture of secrecy endemic to British society; the second is specific to scientists and reflects a deep anxiety about letting others talk about scientific matters. Science is one of the few cultural activities where the practitioners have always sought (indeed, rather successfully) to stay in charge of the story about science. Despite very few practicing scientists in the Anglo-Saxon context having any training in the history and philosophy of science, there is more than a whiff of an ideology of the authority of the ultimate expert who is alone qualified to say what is and what is not science. Historians and philosophers, it goes, are all very well in their place, but the ill-concealed subtext is that place lies in being deferential to the natural scientists. The remembered happy marriage is of that relation of mutual admiration between the distinguished immunologist Sir Peter Medawar and the philosopher of science Sir Karl Popper.

Twin groupings born in the late 1960s, one outside the academy (the radical science movement) and one within (the new post-Kuhnian social studies of science) broke with this deference. Science, in this analysis, was not outside culture, independent, uncontaminated by the social, and "pure," but was itself an integral part of culture. The social studies of science, whether radical, feminist, or mainstream, understood itself as having a more or less critical and no longer deferential relationship to science.

The radical science movement was the child of the antiwar movement and the student movement. Born from moral outrage against the Tabus of our science in their dirty war against Third World peasants, the radical science movement gradually learned that science was not outside culture and history, to be used well or ill, but rather was integral to both. From rage against imperialist powers misusing science and technology, the movement slowly struggled into learning that science as a global system of the production of knowledge was intimately connected to political, economic, cultural, and military powers.

To construct this new understanding, the movement had to make its own analyses forged out of an international and shared experience of campaigns and to recover and remake the marxist analyses of a former generation silenced in the brutal years of the Cold War. It learned from campaigns against specific genocidic technologies, such as the herbicides, the gases, and the fragmentation bombs and mines and, particularly in the United States, from campaigns to force military research off the campuses. It learned how technologies were socially shaped and what were the combinations of capital, expertise, and military ambition that brought them into fruition. It learned that the life sciences play a powerful role in representing human nature, but one that also could be contested from both within and without the science itself. As resources were drawn away from the war on poverty in the U.S. cities to support the Vietnam War, the movement challenged the newly popular racist representations of psychometrics that conveniently rose to claim that black people were genetically intellectually inferior, and thus were constitutionally unable to benefit from the educational improvements of the antipoverty programs. Within Europe, and particularly Britain, similar racist representations of natural inferiority were offered to political regimes anxious to limit black immigration and educational systems anxious to justify failing their black children. For a few years around the end of the 1960s through the 1970s the network of the radical science movement extended internationally mapping the global production system of science itself. The victory of the Indochinese peoples diminished the intense focus of the struggle, and different groups within the movement spread their concerns into the environmental and feminist movements.

Struggles over the social standing of science are not confined to visible debates; the strategic positioning of key actors in the committees which bring together "science" and "society" is far from accidental. Within the life sciences one of the fiercest debates, which has mobilized particularly but not only feminists, concerns the potent link between the new reproductive technologies and the new genetics with their claims to diagnostics and therapy. Together these threaten to determine not only which fetus shall survive but who shall mother. As the disciplines of embryology and genetics have come under intense public scrutiny, the elite embryologists and geneticists have mobilized as very visible players within the PUS discourse.[5] Thus, following the "test tube baby" debates of the 1980s, the Warnock Committee was established by the government to consider the ethical issues raised by human embryology and to suggest any necessary regulation. Incidentally, even the Thatcher administration recognized that this was a woman's issue and in addition to the predictable clutch of theological ethicists, scientists, etc., actually allocated half the membership to women; a first for any such committee of what are in Britain called the Great and the Good. (Of course, these were carefully chosen women unlikely to seriously rock any patriarchal boat.) As an embryologist, Fellow of Royal Society, and experienced committee woman, Anne McLaren was appointed as a member.

McLaren, perhaps because of her radical political background but also because of her gender, was rather different from her male elite counterparts in that she was able to engage in dialogue with various publics, including feminists. She energetically entered the public arena advocating a solution that both permitted research to proceed and simultaneously allayed mainstream anxieties. In large measure it was this position that was eventually adopted by the majority of the committee. In recognition of her work, McLaren was awarded the Royal Society's Faraday Medal in the Public Understanding of Science.

That the first chair of COPUS was the geneticist and leading figure within the International Human Genome Organization, Walter Bodmer (also awarded a Faraday medal for his work in promoting the understanding of the new genetics), and that the second was Lewis Wolpert clearly expresses this desire of scientists from disciplines under criticism to stay in charge of the science story (or at least strategically located). For example, Wolpert, with his very public and hostile views about the sociology of science, is also chair of the Medical Research Council's (MRC) grant-giving committee on the Ethical, Legal, and Social Aspects (ELSA) of the new genetics.

In my judgment, this very British strategy of natural scientists not only staying in detailed charge of their own discourse about PUS but seeking to control the discourses of other disciplines will for good reason increase distrust. Indeed, the recent House of Commons Select Committee report on Human Genetics has accepted that ELSA should not be under the exclusive control of the MRC and has recommended that it be jointly managed with the Economic and Social Research Council. The struggles around the public understanding of science go on at a number of levels and places and are not limited to head-on confrontation, as in *The Golem* versus *Unnatural Science.*

Golem Science

Drawing on the mythical Jewish figure of the Golem, a clumsy but powerful fool with "emeth," or truth, written on its (and sometimes his) forehead, Collins and Pinch (hereafter C&P) (1993) set out to explain the political role of science and technology. For them science is a Golem, wobbling in their text between gender neutrality and masculinity. I have many sympathies with C&P's political project, which is to increase the public understanding "about" science rather than merely to increase the public understanding "of" science, as in the COPUS model. While I might use different language, we share a common sociological impulse that people are expert in their own lives and that a desirable cultural and political objective is to move to a dialogue between the several publics and the sciences.

Their book discusses eight case studies of scientific controversies, from the worm runners (a 1960s claim that tissue from the brain of trained worms could be injected into the bodies of untrained worms and that memory was thus transferred), through experiments to prove relativity, to cold fusion theorists. The C&P thesis is that "the scientific community transmutes the clumsy antics of Golem Science into a neat and tidy myth" (151).

For me the central theoretical problem is the lack of reflexivity in C&P's sociological stance. Thus while they show us the scientists actively socially constructing their "neat and tidy myths," we are invited to believe that C&P's own sociological accounts of science are real. They tell us their subjects construct science, while they offer one true sociological story which everyone should know. Arguably, C&P thus reproduce for sociology the authoritarian scientific voice they criticize. Worse, their tactless use of the word *myth* to describe the slow patient work of laboratory scientists pretty much forecloses the possibility of dialogue with them. For *myth*, used in what appears to be a vague everyday sense here, includes any narrative having fictitious or imaginery elements.

Sociobiologist Richard Dawkins is quick to pick up the term *myth* and to attack relentlessly.

> It is often thought clever to say that science is no more than a modern origin myth. The Jews had their Adam and Eve etc. . . . What is evolution some smart people say, but our modern equivalent of gods and epic heroes, neither better nor worse, neither truer nor falser. . . . There is a fashionable saloon philosophy called cultural relativism which holds in its extreme forms that science has no more claim to truth than tribal myth: science is just a mythology favoured by our modern Western tribe. (Dawkins 1995, 31)

Collins (and for that matter Steve Fuller, who claims that SSK is only methodologically relativist) does not quite match up to the ontological slippage that takes place in the C&P school of SSK, rendering it vulnerable to Dawkins's attack. A more sophisticated—and politically more sensitive—account of science as stories about nature is also used by Donna Haraway in *Primate Visions* (1989), but as a fully reflexive scholar she has the grace to acknowledge that her accounts are also stories. Further, like most feminists working in this area, she is sensitive to the need for natural scientists to be realists in order to construct their accounts of nature. Contemporary SSK in all its variants recognizes that a deconstructionist would have a hard time in a lab, but they handle this with varying degrees of sensitivity. But the more general point I want to make by contrasting C&P with Haraway, whose alleged postmodernism has precipitated a conference of mainly environmentalist scientists to discuss its implications for the protection of nature,[6] is that there is a tremendous range of positions available within the social studies of science.

What separates the feminists from the mainstream professionals is that feminists are committed to building alliances, not least with other feminists, and consequently are very sensitive to the delicacy of the relationship between the feminist critics of science and feminists in science. While C&P deliberately rule out the possibility of their scientist subjects entering into negotiation with their social realist account of science, feminists are willing to listen to those voices, and some want to include Nature herself as part of the actor network. Apparent oxymorons bind the feminist discourse: feminist science, feminist empiricism, feminist objectivity, feminist rationality; these mark out a normative discourse and delineate it from the deliberately nonnormative discourse of mainstream SSK.[7] While sensitivity is not synonymous with success, there is much more attention among feminists to choosing language which is likely to foster dialogue as against that which is likely to foreclose it. By contrast, C&P's sociology of science shares more than a little of the same clumsy Golem-like qualities of the scientific Golem they want us all to know about.

There are always problems with popularizing research and in using case studies to resolve truth issues. Thus I read the original case study on, for example, the worm runners rather differently and think that C&P as popularizers go beyond the claims being made. The worm runners' thesis had hilarious and very obvious social possibilities, not least for educational practice—should we eat or mainline the professors' brains? As such it produced lots of media discussion and some wonderful cartoons. However, its reception within the biological community, as against its presence in the popular scientific weeklies always looking for a controversial story to maintain readerships, scarcely matches the establishment notion of "a neat and tidy scientific myth." Indeed, a conversation with sociologist David Travis, who carried out the original research, confimed that practically all the neuroscientists he interviewed were deeply skeptical about the worm runners' claims, both for theoretical reasons within biological discourse and because they did not trust the accounts of the experimental procedures.

Collins and Pinch assert that both the psychologist McConnell's worm and the pharmacologist Ungar's later rat-based claims for memory transfer have not been disproved on what they speak of as "decisive technical evidence" (25). But the very notion of decisive technical evidence begs the question, for it sets aside the possibility of theoretical biological criticism of the kind that I have made here of C&P's sociological stance. Although the unrepeatable experiment plays a part in resisting the establishment of a scientific fact, not least when carried out with the theater of multi-authorship or multilocation, the production of a fact is not convertible to one decisive moment but requires an accumulation of evidence. Hence it is entirely possible for me to take considerable pleasure in their own meticulous empirical work showing the processes of constructing scientific facts at the micro level while still finding myself uncompelled by their theoretical conclusions.

The case studies of C&P set to one side the larger context in which scientific claims are made. Dorothy Nelkin's *Selling Science* (1987) by contrast reminds us of the lure of press releases in a grant-funded research system, especially where there might be potential commercial developments. Today's version of the memory transfer hopefuls may be the current hunt for "smart drugs" and associated big bucks. While some have already gone down the "smart drug" claim drain, the more modest likelihood increasingly under discussion among the neuroscientists is whether it might be possible to intervene chemically to slow down the terrible neural degeneration associated with Alzheimer's. My point—which C&P seem to allude to in some sentences where they acknowledge expertise but exclude in crucial others, as when they sum up Golem science as "myth making," is that science is concerned with the test of performativity. As a cultural project, modern Western science has never been content only to represent nature but desires also and always to act upon it; hence performativity is not a criterion that can be lightly set aside. It is this criterion that enables Richard Dawkins (1995, 32) to explode with, "Show me a cultural relativist at thirty thousand feet and I will show you a hypocrite. Airplanes built according to scientific principles work."

The understanding that science is socially shaped finds increasingly wide cultural acceptance,[8] but it does not follow that because scientific claims are socially shaped they are interchangeable with myths or even stories.[9] What is called the good science/bad science debate remains, just as it does in sociology, or for that matter plumbing and dressmaking. The trouble with Collins is that, where fringe science is concerned, he wants to make heroes (never heroines) and insists that because they have the tools of the trade and appear to follow the procedures, no one is allowed to say that the plumbing leaks at every joint. Ian Hacking suggests that the social constructionists focus on the early stages in the construction of scientific facts, but that they leave the scene too early, and so we are "left with a feeling of absolute contingency. They give us little sense of what holds the constructions together beyond the networks of the moment, abetted by human complacency" (1992, 131).

Indeed, Collins has appointed himself as the defender of fringe science. Thus, when the biologist Jacques Benveniste made his homeopathic claims and received the full *Nature* treatment—the editor John Maddox, plus a magician and a scientific fraud buster, visited Benveniste's Paris laboratory to witness a replication experiment—Collins claimed that this was epistemology in action and that replication could not prove or disprove the claim. But while most biologists were as skeptical as *Nature* about the claim, many thought that *Nature*'s style was bullying and offensive. The combination of what was referred to as the editor, his magician, and his rabbit was seen as unpleasant overkill and robustly criticized in the ensuing correspondence.

Collins and Pinch insist that they are concerned with the political role of science and technology; well, so am I. But their refusal to acknowledge the now substantial body of feminist scholarship is also political. This scholarship has explored the sexualized and racialized representations of nature constructed by an androcentric and Eurocentric science, articulating the social processes through which women have been excluded from science and how, even when they enter, their contributions are erased. C&P's construction of the political is pretty much synonymous with Sandra Harding's concept of "weak reflexivity"; they persistently restrict their analyses to the "micro processes of the laboratory explicitly excluding race, gender and class relations" (Harding 1991, 162). For genuinely smart fieldworkers, their inability to see such social relations at work is quite an achievement. More mischievously, were I a feminist sociobiologist, I might ask whether there was something on the Y chromosome.

Inability to see feminist research is not exclusive to C&P but has been a general weakness of British mainstream sociology of science, which would include the work of Michael Mulkay, David Bloor, Barry Barnes, Steve Woolgar, and Malcolm Ashmore. Nor did Bruno Latour, as the most authoritative voice across the Channel, do any better until his enthusiastic appreciation of Haraway (Latour 1993). In his work on reproductive technology Mulkay (1995) begins to cite a small number of feminist texts, and Woolgar too is beginning to acknowledge feminism in relationship to constructivism (Grint and Woolgar 1995).

Until these hints of conversion, this highly professionalized group has been singularly hostile to normative critics of science, dealing with them by erasure and silence. By contrast, *Science*, the powerful voice of U.S. science, has for some years had an annual issue devoted to feminist debates in science, and even *Nature*, its British counterpart, reviewed Evelyn Fox Keller's brilliant biography of Barbara McClintock, whereas the lead British SSK journal, *Social Studies of Science*, did not. Instead, their sociology, despite their claim to be interested in the political role of science, has chosen to mirror science's claim of being a gender-free culture. C&P's radical impulse concerned with the political role of science and technology is so hedged in by professionalized and pale-male constructions of the political that its capacity to build alliances with other critics, whether within or without the sciences, is severely restricted. Arguably, the unquestioned success of this highly professionalized British approach to SSK (quite apart from the flak it is drawing from some natural scientists) is beginning to run out (Knorr Cetina 1993); there are hints that more normative approaches such as those of feminism look to be more fruitful.

More optimistically, this exclusivity of mainstream SSK does not imply that the others cannot borrow their intellectual tools. These will unquestionably need adapting, as Audre Lorde's epigrammatic question—"Can the master's tools tear down the master's house?"—does not go away. There are also encouraging cracks in the masculine culture of science studies, as feminist research students wishing to work on the social studies of science and technology pressure the departments from below. Gradually the departments are beginning to hire feminists, and the possibility of whole new conversations comes into existence. My reading of *The Golem* is that it contributes to these new conversations almost despite itself.

Unnatural Science

The Unnatural Nature of Science is, for anyone with a more than cursory familiarity with current philosophy of science, an astonishing essay in glassy mirror ideology (GMI). It sees no gap between the word and the thing. Practitioners of GMI such as Wolpert simply do not appreciate the lethal criticism of the mirror theory offered by the Picasso joke. A man troubled by Picasso's portraits with eyes facing both frontward and sideways asks the painter why he does not paint realist pictures. To make his point clear the man takes out a photograph of his wife and says: "Like this." The artist looks at the photo and mildly observes: "Small, isn't she?"

Scientists, unlike postmodernists and other ontological relativists, believe that there is something "out there" and that by following the practices of science they can represent that "thingyness" faithfully. This realism, while a crucial belief for everyday laboratory practice, is not transferrable into a theory of representation with the simplicity of the man with the photograph. Such unreconstructed mirror theory would have a hard time within the philosophy of science and an even harder time within the new ethnographic accounts of science classically represented by Latour and Woolgar's *Laboratory Life*. Many laboratory scientists report their pleasure in this book as a meticulous mapping of the sociotechnical process through which scientists take the inscriptions emanating from their equipment and gradually turn them into scientific facts. They acknowledge that persuading other scientists working in the area that these are the only possible interpretations of the inscriptions—that these are indeed the new facts—is at once both technical and deeply political. But Latour and these scientists both know that while science is always social, it is not only the social which writes the science (Latour 1993).

Many of the scientists in both the THES and at the Durham discussions were entirely unalarmed by the concept of science as a social construct and showed no special discomfort with the idea that science is not independent of the culture in which it is produced. Robert Banks, a biologist, observed that "biology is grossly biased towards those organisms which closely resemble ourselves, or which irritate us, like *E. coli.*" (Just so, I thought, given biomedicine's historical preoccupation with the male body and the female reproductive system and brain, but did Banks have that in mind?) But equally, natural scientists insist that science is also not independent of observation and experiment on the natural world. By contrast, the GMIs seem to need their science to act with a more godlike cultural certainty; thus perhaps it is not by chance that two of the more influential voices to support Wolpert are committed to militant atheism with a positively nineteenth-century fervor. The elision between science and social progress possible in that century was to dissolve during the next. C. P. Snow, in his essay *The Two Cultures* (1964), saw scientists as "the men [*sic*] with the future in their bones." Not for the first time, the heroic masculinist claim was made as Minerva's owl was leaving. That widespread certainty about either scientists or the future was lost at Nagasaki and Hiroshima, which came, and not only for Robert Oppenheimer, to represent the scientists knowing sin. GMIs may try to dismiss that learned cultural distrust of science, but dismissal and heroic nostalgia for an innocent past cannot help in our present cultural uncertainties.

Wolpert takes the localized belief of every scientist in the laboratory that there is one right answer for every phenomenon and turns it into a universal precept. He makes no allowance for the different discourses of the different sciences, for the possibility that an explanation of the same phenomenon within biophysics is likely to be very different from that within biochemistry, to say nothing of the physicists' long-standing capacity to accept both wave and particle theory. For that matter, the account of say "money" will be very different in the discourses of anthropology or economics (think of Marx's wonderful analysis) as against that of the physical sciences.

Wolpert's central argument is that modern science has very strongly defined boundaries; that it is unique and unitary as a way of knowing nature; that its roots lie entirely in ancient Greece; and that it is radically different from something he speaks of as common sense. He thus sets aside, or is unaware of, historical accounts such as that of Martin Bernal (1988), who is concerned with the black African roots of Athenian science, and dismisses the ethnosciences, whether those of the Chinese (monumentally documented by Joseph Needham and his colleagues), the Egyptian, Indian, Islam, or Mayan, to name but a few of the many seeking attention, as technology or trivial. At Durham, sociologist of science Mammo Muchie sought to bring Third World critiques of science into the discussion, arguing that because modern Western science is hegemonic, it appears as natural and universal. Its achievement is to appear as a culture of no culture. In a similar vein to the feminist arguments I was trying to make, Muchie spoke of the exclusion of emotion and ethics from the construction of rationality. Against such arguments for the possibility of other more localized and more environmentally and socially responsible sciences, Wolpert insists that there is only Monoscience.

Science, he claims, is unnatural knowledge in that many of its truths run counter to everyday beliefs—for instance, that the sun goes round the earth or that heavy bodies fall faster than lighter ones. However, he never tries to define what he means by common sense, or to consider whether it too is a culturally relativist concept—that today's common sense is merely yesterday's "good science." Or by common sense does he want to invoke the counterintuitive, which is surely the stance of every systematic approach to knowledge from the arts to the sciences and by no means the unique property of any single one?

As a sociologist I would want to argue that not least because we live in a deeply scientific and technological culture, "lay" people (and outside our narrow expertises we are all lay people) pick up particular areas of science, typically those which are important or have some special interest for them. Often people do this without claiming that their knowledge is "science" but instead speak modestly of "hard facts" or "reliable knowledge." Thus in my own PUS research on people with a genetic cholesterol disorder, most of them knew more about saturated, poly, and unsaturated fats than was asked in a parallel quantitative study ascertaining public levels of scientific literacy. In the same series of sociological studies of PUS, Wynne's Cumbrian sheep farmers rapidly acquired a richer appreciation of radiation in the food chain than the Ministry of Food and Agriculture scientists (Irwin and Wynne 1996). For that matter, effective natural scientist

PUS practitioners perceive their publics as having tacit knowledge of, say, probability theory, or of biomedicine, but also recognize that as nonscientists they often do not equate their knowlege with mathematics or science. The practitioners use this perception as a building block for their pedagogy. Wolpert's insistence on the sharp line between science and common sense is hard to reconcile with his commitment to COPUS.

To return to the debate that followed the original Collins/Wolpert clash, other leading scientific figures strongly supporting Wolpert included Richard Dawkins (recently appointed to a new, privately funded Chair in the Public Understanding of Science at Oxford) and the physical chemist Peter Atkins. The three were as one in their contempt for SSK; as Dawkins put it, it is just "chic drivel." Wolpert's view of the sociology of science is matchingly intolerant: "I've never come across anything that wasn't either obvious, trivial, or wrong. I think they have made zero contribution." Atkins resists the very idea of science as a social construct: "The universal character of science, by which I mean its independence of lasting national, racial, and religous and political influences, must argue strongly against science as a social construct."[10]

At one level I cannot see why C&P and the mainstream SSK in Britain are so under attack, for they never question, as feminists and radicals do, the larger political role of scientific knowledge such as sociobiology, or for that matter the new genetics. Sociobiology's endorsement of rape, polygamy, male violence, male dominance, etc. over the last two decades surely merits criticism both from within its own canon and from without.

For example, after a passage about gene reproduction, Dawkins continues: "The world is full of organisms that have what it takes to become ancestors" (1995, 2). As a sociologist interested in biology I might be skeptical, but I would have to leave effective critical analysis to a biologist. For good reason I would look to biologists willing to enter normative debate, such as Anne Fausto Sterling, Steven Jay Gould, Ruth Hubbard, Richard Lewontin, and Steven Rose. But social scientists need to engage when Dawkins goes on to make what appears to him to be a self-evident claim about the social: "A body that actively works as if it is trying to become an ancestor . . . that is why we love life and love sex and love children" (2). If this purports to be a realist account of the social, we have surely entered the sociological counterpart of the *Hello* magazine. Does such crude biological reductionism in its cheerful sentimental universality purport to explain everyday life in Oxford, let alone mass rape and genocide in Bosnia?

But it seems the business of the cultural weeklies on this occasion was to promote adversarial debate about science as a monolithic entity, which minimized the possibility of any complexity or any interchange of views. Positions merely became more entrenched. The psychological need of these GMIs to speak for all of science with absolute ontological certainty and not to talk in detail about the messier, more provisional discourse of any particular science points to their unease. Indeed, their resistance to talking about their own experiments and observations, instead issuing grandiloquent claims about science, is a hallmark of their style.

Occasionally such adversarial displays eased, but only with a significant effort such as that made by sociologist Michael Lynch. At the Durham meeting he made a close textual analysis of *The Unnatural Nature* which simply left its author excusing his assertions, explaining that he was only an amateur in the history and philosophy of science. Lynch gently prodded, "Why don't you talk to us about embryology then? That could be really interesting." For what Lynch had shown was that Wolpert made assertions without evidence, invoked a commonsense concept of realism, and had fatally confused a scientific grasp of the issues with a scientist's grasp of the issues.

But adversariness was built into the meeting. Even this more scholarly critical exchange did not stop Wolpert the following day from denouncing the sociology of science lock, stock, and barrel. My (optimistic) hunch is that outside such adversarial arenas, the full-blown GMIs are relatively isolated and that there is a widespread cultural understanding that science is a human activity and as such is socially shaped. In consequence, both the boundaries and the nature of science are continuously subject to change over time and place. Nonetheless, the GMIs do command a considerable amount of media and other cultural space and cannot be ignored. The question is how to cope with them. To me it seemed that Lynch's approach of patient public critical opposition was exemplary, even if it was transient in its influence on the author.

The media itself is not a bystander in this; it promotes and feeds off these vicious and unproductive soundbite exchanges and makes little space for thoughtful analysis. Instead it has valorized intolerance, ignorance, and a plurality of authoritarian Moseses each coming down from the mountain with holy writ. Worst of all, the stagey Culture Wars get in the way of quieter and more serious arguments as to whether postmodernism is depoliticizing, whether we can give up the truth claims of science, whether there is still the possibility of a limited conception of objectivity, and whether and how the abstract rationality of science can be replaced by a new socially and environmentally responsible concept of rationality. Should the gently squabbling "we," which includes only a handful of natural scientists, be endeavoring to engage in dialogue with many more, admitting the scientist as collaborator rather than as object (Labinger 1995)?

Or should we set aside these debates and try to examine the new ideas about the changing production system of knowledge (Gibbons et al. 1994) as it moves toward "post academic science" (Ziman 1995)? And if so, what does it mean for our cultural practices? At my most optimistic, I read these changes as opening the possibility of many new actors entering the production system of science, which could indeed include those others historically excluded by modern Western science, not least Nature herself. If the old science and even the sciences have lost public trust, and if there is in consequence a clear political danger of the Newts of this world moving against the public support of systematic knowledge, then the restoration project of the GMIs is both a mistaken and futile strategy. The only effective and creative response is to try to reshape the sciences by bringing the other Others in. There is nothing mechanical or guaranteed about these possibilities; the new production system could, as in the emergent Conservative British model, seek to exclude everyone except industry and the technoscientists. But this new system is developing in a context where dreams of localized, embodied, responsible knowledges press from multiple currents in both the South (Shiva 1989) and the North. Dreams come into existence not, I think, through binary confrontations but through multiple conversations and complex alliances. For my part, anyone who is prepared to help make space and enter seriously—and pleasurably— into such conversations and alliances is my friend.

Notes

1   Frederick Crews. *Times Higher Education Supplement*, 3 March 1995, 17.
2   *New York Review of Books*, 18 November 1993, 21.
3   An overtly political attack has been mounted on therapy by the culturally conservative Social Affairs Unit. Currently, this hostility finds little support in Conservative Government policy.
4   The British Association debate gave equal space to the contesting standpoints. In contrast, the meeting organized by Levitt and Gross at the New York Academy of Science earlier this summer was primarily a platform for their views. Reported in *Nature*, 8 June 1995, 30.

5   The natural scientific elite in Britain corresponds to Fellowship of the Royal Society.

6   The conference is published as *Reinventing Nature: Responses to Postmodernism*. See Soule and Lease 1995.

7   Key figures in the production of these oxymorons would include Patricia Hill Collins, Donna Haraway, Sandra Harding, Nancy Hartsock, Helen Longino, Elizabeth Potter, Hilary Rose, and Vandana Shiva.

8   The European Union Framework Programme for Targeted Social and Economic Research speaks of the "social shaping of science and technology." Many European social scientists claim to be responsible for the language, so writing this into the language is also a matter of the "social standing of science."

9   This is not a new conflict in Britain. Within the radical science movement in the early 1970s there was a sharp precursor in the "science is social relations debate" between historian R. M. Young, sociologist Hilary Rose, and biologist Steven Rose. See Rose 1994 and Pustilnik 1995.

10   Peter Atkins, *Times Higher Education Supplement*, 30 September 1994, 18.

References

Appignanesi, Lisa, and John Forrester. 1992. *Freud's women*. London: Weidenfeld and Nicolson.

Bernal, Martin. 1988. *Black Athena*. London: Free Association Books.

Breggin, Peter. 1993. *Toxic psychiatry*. London: Fontana.

Collins, Harry, and Trevor Pinch. 1993 *The golem: What everyone should know about science*. Cambridge: Cambridge University Press.

Dawkins, Richard. 1995. *River out of Eden*. London: Weidenfeld and Nicolson.

Donnison, Jean. 1977. *Midwives and medical men: A history of interprofessional rivalries*. London: Heinemann.

Durrant, John, Geoffrey Evans, and Geoffrey Thomas. 1989. The public understanding of science. *Nature*, 8 July, 11–14.

Ehrenreich, Barbara, and Deirdre English. 1978. *For her own good: 150 years of the experts' advice to women*. London: Pluto.

Flax, Jane. 1993. *Disputed subjects: Essays on psychoanalysis, politics, and philosophy*. New York: Routledge.

Forrester, John. 1989. Lying on the couch. In *Dismantling truth: Reality in the postmodern world*, edited by Hilary Lawson and Lisa Appignanesi. London: Weidenfeld and Nicolson.

Gellner, Ernest. 1993. (1984). *The psychoanalytic movement: The coming of unreason*. London: Fontana.

Gibbons, Michael, Camille Limoges, Helga Novotny, and Simon Swartzman. 1994. *The new production of knowledge: The dynamics of science and research in contemporary societies*. London: Sage.

Grint, Keith, and Steve Woolgar. 1995. On some failures of nerve in constructivist and feminist analyses of technology. *Science Technology and Human Values* 20: 286–310.

Grünbaum, Adolf. 1984. *The foundations of psychoanalysis: A philosophical critique*. Berkeley: University of California Press.

Hacking, Ian. 1992. Statistical language, statistical truth, and statistical reason: The self identification of a state of scientific reasoning. In *The social dimensions of science*, edited by Eran McMullin. Notre Dame, Ind.: University of Notre Dame Press.

Haraway, Donna. 1989. *Primate visions: Gender, race, and nature in the world of modern science*. New York: Routledge.

Harding, Sandra. 1991. *Whose science? Whose knowledge? Thinking from women's lives*. Milton Keynes: Open University Press.

Irigaray, Luce. 1985. *This sex which is not one*, translated by C. Porter. Ithaca, N.Y.: Cornell University Press.

Irwin, Alan, and Brian Wynne, eds. 1996. *Misunderstanding science*. Cambridge: Cambridge University Press.

Knorr Cetina, Karin. 1993. Strong constructivism—from a sociologist's point of view. *Social Studies of Science* 23: 555–63.

Labinger, Jay. 1995. Science as culture: A view from the petri dish. *Social Studies of Science* 25: 285–306.

Latour, Bruno, and Steve Woolgar. 1983. *Laboratory life: The social construction of scientific facts*. Beverly Hills, Calif.: Sage.

Latour, Bruno 1993. *We have never been modern*. Cambridge, Mass.: Harvard University Press.

Moulton, Janice. 1983. Against the adversarial method in philosophy. In *Discovering reality: Feminist perspectives on epistemology, metaphysics, methodology, and philosophy of science*, edited by Sandra Harding and Merrill Hintikka. Dordrecht: Reidel.

Mulkay, Michael. 1995. Galileo and the embryos. *Social Studies of Science* 25: 499–532.

Nelkin, Dorothy. 1987. *Selling science*. New York: Freeman.

Oakley, Ann. 1984. *The captured womb: A history of the medical care of pregnant women*. Oxford: Blackwell.

Pustilnik, Amanda. 1995. *Looking through "not in our genes": Debunking, epistemology, and practice between biological determinism and the dialectics of liberation*. B.A. thesis, Harvard University.

Rose, Hilary. 1994. *Love, power, and knowledge: Towards a feminist transformation of the sciences*. Cambridge: Polity.

Russell, Denise. 1995. *Women, madness, and medicine*. Cambridge: Polity.

Shiva, Vandana. 1989. *Staying alive: Women, ecology, and development*. London: Pluto.

Sivanandan, A. N. 1995. La traison de clercs. *New Statesman*, 14 July, 20–21.

Snow, C. P. 1964. *The two cultures: And a second look*. New York: Mentor.

Soule, M. E., and G. Lease, eds. 1995. *Reinventing nature: Responses to postmodernism*. Washington, D.C.: Island.

Wolpert, Lewis. 1992. *The unnatural nature of science: Why science does not make (common) sense*. London: Faber and Faber.

Ziman, John. 1995. "Postacademic science": Constructing knowledge with networks and norms. *Royal Society Medawar Lecture*, 29 June.

The Gloves Come Off: Shattered Alliances

in Science and Technology Studies

Langdon Winner

T HE acrimonious disputes surrounding social studies of science today reflect long-standing disagreements about the character and purpose of inquiry in this field. The publication of *Higher Superstition* underscores how nasty these quarrels can be, perhaps foreshadowing explosive clashes between the two cultures in years to come.[1] One might have hoped spirits less malicious than Gross and Levitt's would have been the ones to bring these conflicts to light. But for those who have followed the development of science and technology studies (STS) over the years, it has been obvious that eventually the other shoe would drop, that someday it would occur to scientists and technologists to ask: Why do the descriptions of our enterprise offered by social scientists and humanists differ so greatly from ones we ourselves prefer? How much longer should we put up with this?

In my experience, four basic projects have inspired research and thinking in STS during the past several decades. While these projects are by no means mutually exclusive, they do serve as foci for fairly distinct groups of interests. Within the sprawling, interdisciplinary community of STS in America and Europe one finds widely different intellectual approaches, different expectations about the value of results, and different understandings about the audiences that will ultimately judge and sponsor such work. Tensions between these projects and shifting alliances between them account for much of the vitality of STS, as well as for the flare-ups we see at present.[2]

In the first project, the great challenge is simply to understand how modern science and technology work, how their various practices, institutions, and tangible products have developed. Success in this project comes in providing satisfactory explanations. Thus, one might ask: Exactly how did molecular biology arise? What contexts and influences contributed to its rise? How did laboratories take shape?

How did career patterns in biology change? What was the role of government and of business in the rise of this research? Inquiries of this kind have flourished largely because it has never been common for working scientists and technical professionals to write their own histories or explore the social dynamics of their work. Filling this gap, scholarship in the mode of historical and social explanation has expanded steadily in the United States and Europe during the past thirty years. What was once a dearth of writing in this area has been replaced by a flood of books and journal articles.

Many of those who have embraced STS in its purely explanatory mode have come to the subject from backgrounds in the natural sciences and engineering. Having once made professional commitments to the fields they now study as research objects, it is not surprising that STS scholars of this stripe often have extremely sanguine views of what science and technology are all about. Contrary to what Gross and Levitt assert, it is fairly common among historians in STS to believe that the results of their research should pass muster with colleagues in their former professions and that findings should cast a favorable light on the professions under study. By the same token, it is by no means uncommon for scholars of this stripe to view criticism of science and technology with disdain. The conclusion that STS is a hotbed of critical views simply disregards the extremely conventional but solid scholarship that comprises much historical and sociological research in this area.

For thinkers in the second project, science and technology studies presents a convenient domain in which to develop conceptual models and approaches taken from the home disciplines in the humanities and social sciences. For a number of reasons, none of the fields of human studies organized during the academic reforms of the late nineteenth century chose to focus upon science and technology as such. Intellectual boundaries that seemed sensible at the time meant that the tools of history, sociology, anthropology, political science, literary criticism, and others would not be used to study the inner workings of the natural sciences or engineering. It was widely assumed that these pursuits were not amenable to the same methods used to study politics, world history, kinship relations, or modern fiction. In the decades after World War II, however, it became obvious that because many of the transformations in modern life were crucially connected to developments in science and technology, the social sciences would need to probe their origins and effects or risk being completely out of touch. Hence the rise of a host of new disciplinary and interdisciplinary efforts that by the late 1960s were asking: Why not create a true sociology of science? Why not employ the methods of anthropology to study high-tech professionals? Why not apply the tools of literary criticism to scientific communication?

Research in this vein has spawned a new cottage industry, applying a wide range of concepts and approaches from the humanities and social sciences to the study of science and technology. Much of this work has drawn upon the rise of poststructuralist, postmodernist theoretical strategies, insisting that the discursive practices of scientific and technical fields were ripe for deconstruction. Unlike scholars coming from backgrounds in science and engineering, however, those who have arrived from the humanities and social sciences are little worried about how their findings will be received by the people and institutions they study. Characteristic of such work has been the spirit of imaginative interpretation and unfettered speculation that has enlivened other areas in which the new cultural studies have taken root. If the results appear critical of contemporary science and technology, so much the better, advocates of the new scholarship are inclined to think; it is simply an indication that the probes have begun to touch raw nerves.

For those engaged in yet a third project, the attraction of STS involves the need to respond to a host of practical problems that have arisen in the world as a consequence of changing scientific knowledge and various technological applications. Here the focus is not so much that of explaining the development of science and technology or showing how new intellectual tools can be applied, but rather analyzing specific troublesome issues, for example, the safety of nuclear power plants, environmental devastation caused by overdevelopment, hazards to public health and safety caused by the dumping of toxic wastes, and so on. While research of this kind aspires to both intellectual richness and policy relevance, what matters in the end are proposals to remedy the problems described. For that reason, advocates of this project are closely oriented toward decisionmakers in Washington, D.C., and social movements that might effect positive change.

Concerns of this sort may have little connection to the two projects mentioned above. Those at work on problem-centered research may find little value in writings about closely related topics in the history, sociology, or cultural studies of science. In fact, it is fairly common for them to lament that STS is dissolving in a swamp of arcana—studies of laboratory life and elaborate interpretations of material artifacts, for example—while the real troubles of the world go begging. After all, what good is all this novel theorizing if it does not generate practical remedies?

A fourth collection of concerns in STS attracts philosophers and social theorists. Here the focus turns to what many thinkers have argued is a profound crisis in the underlying conditions of modern life and thought. The development of modernity has gone badly wrong, not only at the level of specific, vexing social problems but in its fundamental core of ideas and institutions, especially those that involve science and technology. While attempts to fathom the nature of the crisis vary from writer to writer—from Marx to Mumford, from Heidegger to Ellul, from Habermas to Foucault—the point of inquiry is to locate philosophical, historical, and cultural origins of phenomena closer to hand.

In its very nature, research of this kind is both radical and critical; it seeks to "look deeper," to probe what may be highly general sources of contemporary disorientation and to suggest change of the most fundamental kind. This does not mean, as opponents sometimes claim, that thinking in this key is essentially "antiscience" or "antitechnology." But it certainly does mean that prevailing ideas and practices in science and technology must be subjected to close scrutiny with no a priori declarations of endorsement. The fact that STS in a philosophical vein resists collegial appeals to join the chorus affirming science as a grand and glorious enterprise has often attracted the wrath of science boosters. No less a man than Lewis Mumford became the focus of such ire when Gerald Holton, his erstwhile friend and one of today's most Quixote-like crusaders against "antiscience," concluded that Mumford's *Pentagon of Power* was simply an attack on scientific rationality, the thesis of Holton's bitter review in the *New York Times*.[3]

Roughly speaking, then, one can locate the purposes of much STS research and teaching within the four projects I have summarized. My own work, for example, flows primarily from two of these: expressing a desire to confront what I perceive to be a systematic disorder in modern life, a disorder manifest in technology-centered ways of living that I regard as unfriendly to any sane aspiration for human being; and applying concepts and approaches of a particular discipline, political theory, to questions about the significance of technology for political life.

Whatever one's personal project, however, all who enter STS must navigate an intellectual terrain characterized by diverse intentions and uneasy alliances. And always just offstage are sources of support and opposition presented by those who have a strong stake in where the discussion moves. As I entered STS in the middle 1970s, the desire to explain the inner workings of science and technology was clearly the one most prominent among my colleagues. It was also the purpose most resonant with university scientists, engineers, and administrators who clearly hoped that better biographies of scientists and inventors, and better histories of the various fields of research and development, would shine a favorable light on their professions. They looked to STS to contribute to the celebration of science and technology as crucial to the progress of civilization, describing in detail how the great wealth of knowledge and social benefit had been achieved.

At the time it appeared that a second important STS project, that of extending the conceptual equipment of the humanities and social sciences into this new domain, would fit the program of hagiography and celebration very nicely. Support for the new research would demonstrate that the long-neglected "human dimensions" were finally receiving emphasis. When the results came in, they would reveal that what sometimes appears to be hard-nosed, even soulless work by scientists and engineers is, in fact, morally complex, aesthetically deep, and culturally refined. Several of the STS programs founded at colleges and universities during this period upheld science appreciation as their central educational aim, one that many faculty members enthusiastically endorsed. Such initiatives, in my view, could well have been named HSTS: Hooray for Science, Technology, and Society.

Other themes in STS research, however, were far more difficult for scientists, engineers, and university bureaucrats to stomach. A growing list of studies on social and environmental ills accompanied by a hefty stack of philosophical critiques of technological society sometimes made it seem that the scientific and technical professions were being unjustly attacked, indiscriminately blamed for all the ills of modern society. In that mood, those who studied particular signs of social stress and ecological decay were often accused of having "negative" and "unbalanced" views of science and technology. Those who studied the philosophical roots of systems analysis, artificial intelligence, and the like were quickly labeled antitechnology malefactors. Thus a senior scientist well-known for his accomplishments in both computing and social criticism was repeatedly rebuffed in his attempts to join the STS program at MIT, his home institution. Persisting in his lively exposé of the follies of instrumental rationality, the man became persona non grata among the timid souls who ran the program and who sought to please the institute's faculty by singing the praises of science and engineering.

Conflicting views about the proper aims of STS occasionally erupted into nasty struggles. Some notable tenure battles, especially David F. Noble's lawsuit against MIT, gave adequate evidence that the field was not one big happy family. Indeed, when scholars with critical views have joined university programs geared to celebrating the progress of science and technology, they have often been headed for trouble, not the least of which is opposition from colleagues in the "hard" disciplines. As if to suggest a bold departure, Gross and Levitt propose that scientists respond to antiscience bias by taking control of the tenure and promotion of their colleagues in the humanities. Those who have been around this track before might ask: so what else is new?

Until *Higher Superstition* broke the silence, the opinions of mainstream scientists about the appropriate boundaries for science and technology studies were seldom openly discussed. Humanists and social scientists were supposed to read subtle messages that informed them of their subservience and act prudently. That is why I was astonished by the frank confession offered by one authoritative spokesman in the early 1980s. The occasion was the meeting of a review panel for the program on Ethics and Values in Science and Technology (EVIST) of the National Science Foundation. Our task was to sift through several dozen research proposals and decide which ones deserved funding. As we began the day's work, an upper-level NSF administrator came into the room to explain the real reason we were there. He noted that EVIST had gained support within the foundation because it helped the scientific community respond to political pressures from Congress and the general public. As problems in ethics and values of science arose in matters of, say, nuclear power or molecular biology, scientists could respond that qualified experts in the humanities and social sciences were looking into the matter. The public's misgivings about problems associated with science and technology could be answered by pointing to research programs in STS where all the social and ethical perplexities, all those knotty "values" questions, were being rigorously investigated.

The contrast in purposes had never been so vividly apparent to me. What some of us in STS saw as a bold new intellectual enterprise aimed at challenging prevailing notions about the workings of science and technology was understood by science administrators as a way to blunt or co-opt signs of discontent that had recently erupted in American society. The role of STS scholarship was, in effect, to help defuse public anger by transforming disruptive protests into tame academic pursuits. As the administrator droned on about this strategy, I remember thinking to myself, "Well, it's nice to know what business you're in, at least in the eyes of the people paying the bills."

As background for understanding today's worries about the character of STS, the musings of the NSF administrator must be seen in historical context. As David Dickson has noted, the late 1970s and early 1980s (the years of the Carter and Reagan presidencies) were years that witnessed a shift in relationships between science and society.[4] During the previous decade, scientists found themselves subject to pressures to orient research toward national priorities in health care, environmental clean-up, and energy research. Many scientists came to believe that the public's influence on R&D had grown too large, that the direction of science by political policymakers had gotten out of hand. Dickson argues that scientists, galled by what they regarded as excessive democratic control of research agendas, were more than willing to form alliances with other sources of social control. Hence, during the Reagan era scientists supported a turn away from research agendas shaped by a sense of social need toward R&D geared to the ongoing military buildup and quest for "national competitiveness" expressed in the priorities of business firms.

In that setting, anything that could be done to diminish public concerns about social and environmental issues that had surfaced in the 1960s and early 1970s was regarded as a positive step. In Dickson's account, the continuing support for programs like EVIST and the Office of Technology Assessment were among the means chosen by leaders in the scientific and technical community to deflect the tendency of a restless public to assert its role in setting research priorities. Many in STS at the time recognized these pressures and fought them by trying to strengthen ties to social movements and by finding ways to address political issues in STS through the popular media. On the whole, however, the development of science and technology studies in the 1980s flowed with the depoliticizing tides of that decade. As reflected in the yearly meetings of the Society for Social Studies of Science, many scholars were content to discuss such things as the social construction of the bicycle during the late nineteenth century rather than, say, the rapid dismantling of industrial workplaces shattering the lives of so many of their contemporaries.

During the past decade the second project on my list has matured, engaging new voices from a wide variety of backgrounds. Once counted on to lend an air of polite refinement to literary and cultural discussions about science, the development of methods in the humanities and social sciences has instead produced a direct challenge to understandings about scientific knowledge and technical application that have been dominant for the better part of two centuries. Feminists, postmodernists, and critical theorists of many persuasions have converged on the idea that, yes, the approaches of the revitalized humanities can well be directed to interpreting what science and technology are all about. In this way of seeing, the descriptions and accounts of scientists and engineers provide useful data, but they are by no means the last word on the matter. This is the feature of the new scholarship that drives Gross and Levitt nuts, because it seems to undermine the authority of science. It is also a feature that has now begun to attract the attention of right-wing militants worried about politically correct speech, multiculturalism, and other supposed dangers lurking in the academy. Indeed, it is not unthinkable that an attack on antiscience could be incorporated into later goose steps of the Contract with America.

The surprise is, of course, that the turn in the 1980s toward abstract, politically disengaged research focused on projects of explanation and interpretation eventually produced results that now seem threatening. As researchers in the social sciences and humanities devise increasingly sophisticated ways of asking, "What's really going on in the development of scientific knowledge? What actually occurs in the construction of technology-centered practices?" their answers are sometimes unsettling to those with a supposedly infallible "insider's view." A consistent finding in recent STS writings has been that the phenomenon of power infuses science and technology through and through. As scientists and engineers read the books and articles that argue this point, some recoil in utter horror. Arguments of this kind do not match their belief that science is a sure-handed method for obtaining beautiful, objective truths. Neither do they match the central ideal of engineering that technological development is a neutral process that simply pursues the best instrumental results. As they puzzle over the new scholarship, people in the "hard" disciplines sometimes conclude that all the power relations said to exist at the heart of science and technology must have been imported from outside, probably from the social scientists' own mistaken models.

It is too soon to tell whether this irate response will be the one that prevails. What is clear is that the gentlemanly social contract that previously linked STS research to allies in the scientific and technical communities is now under fire. Scientists and engineers have been prepared to abide some degree of criticism coming from problem-oriented and philosophical wings of STS, as long as what they assumed to be the central purpose—production of histories and social analyses portraying science and technology as grand and glorious endeavors—would still be the heart of the enterprise. Will they still be prepared to support STS when it appears to them that the entire field of inquiry has taken off in perverse directions? Perhaps not. Rather than come to terms with the new ideas and enter dialogue with those who advance them, leaders in the scientific and technical communities may simply decide it's best to shut the whole thing down. That is, of course, the recommendation of *Higher Superstition*, one that ironically confirms much of what STS scholars have been saying. You claim that science is mainly about power? We'll show you! We'll smash this heresy by taking away your jobs and funding. Quod erat demonstrandum.

As before, it is important to notice the context in which these tensions arise. In the 1990s the basic relations between the scientific and technical professions and the rest of society are once again being renegotiated. Concerns of the previous decade about productivity, national competitiveness, and their promised link to the well-being of the whole American populace have been supplanted by a more bracing recognition of what the global economy demands. The idea that the advance of science and technology must contribute to the prosperity of any particular nation has given way to a recognition that entrepreneurs will create high-tech enterprises wherever in the world it is most advantageous to do so. Thus, scientists and engineers, like everyone else, must be swift enough and flexible enough to satisfy rapidly changing corporate demand. This means that specific needs of national economies and populations must be de-emphasized in favor of priorities established by "the global market." Responding to this situation, educators in science, engineering, and business now commonly warn their students that they must regard themselves as transnational actors and shape their careers accordingly. In a similar vein, government R&D programs formerly aimed at fostering new products for American industries and jobs for American workers are dismantled in favor of letting "the market decide." Preferred now is "basic research" without reference to any well-defined social need, research that will, in the fullness of time, contribute to economic growth—somewhere in the world.

In this new atmosphere, any variety of discourse that helps remind people of the social character of science and technology may seem counterproductive to those who control the flow of resources. Even once-cherished themes of "productivity" and "competitiveness" must be de-emphasized because they may suggest to everyday people that improvements of that kind might actually benefit them. In fact, the scientific advances, technological innovations, and productivity gains in the United States of the past twenty years have gone hand in hand with a rapid erosion of the real wages of working-class and middle-class Americans. To an increasing extent, the science-based, high-tech economy has brought a widening gap between the extremely well-to-do and the rest of society. By almost any conventional measure, progress, seen as the contribution of advancing science and technology to the living conditions of most people, has been stalled for a long while. For those seeking new roles in the corridors of transnational capital, it may seem undesirable to ask the general populace to ponder how science and technology work nowadays, or whether changes taking place in that sphere are truly beneficial to them. Under the circumstances it is no surprise that we see the resurgence of the myth of pure, objective science along with the myth of unfettered technological progress, the STS equivalent of "family values."

Disturbing evidence on this score comes from recent episodes of ideological cleansing at the Smithsonian Institution. To mark the fiftieth anniversary of the dropping of the atomic bomb on Hiroshima, the Air and Space Museum planned to show the Enola Gay along with photos and historical descriptions of related events before and after the blast. But the exhibit was canceled when the American Legion and Air Force Association complained that it would show, of all things, the effects of the explosion on its human victims. According to the veterans, such evidence would soil the patriotic message the commemoration ought to convey, a message of strength, victory, and American lives saved by the bomb. Responding to these pressures, museum officials canceled the original exhibit and replaced it with a vastly scaled-down, inoffensive version.[5]

In a similar episode, the broad-ranging $5.5 million exhibition *Science in American Life*, which opened in April 1994, drew fire from physicists who alleged that it showed too many negative aspects of science and too few positive ones. A letter of complaint sent to the Smithsonian by Burton Richter, president of the American Physical Society, charged that the exhibit gave "a portrayal of science that trivializes its accomplishments and exaggerates any negative consequences. We are concerned that the presentation is seriously misleading, and will inhibit the American public's ability to make informed decisions on the future uses of science and technology."[6] Among the features of the exhibit that the physicists found offensive were those that emphasized links between scientific research and commercial interests (how unseemly!) and those pointing to the underrepresentation of women and minorities in science (telltale signs of political correctness). The physicists also took umbrage at the barbed-wire fences included in depictions of the Manhattan Project, which built the first atomic bomb.

Eventually, the American Chemical Society (ACS), financial backer of the exhibit, joined the physicists in demanding that the exhibit be cleansed of its distressing messages. Paul Walter, chairman of the ACS, wrote to Smithsonian officials demanding that the exhibit be changed because parts of it "demonstrate a strong built-in tendency to revise and rewrite history in a 'politically correct' fashion."[7] Marvin Lang, chemistry professor at the University of Wisconsin and member of the original advisory board for Science in American Life, charged that planning for the show was too greatly influenced by "social scientists and pseudo-scientists who had no idea of how science worked."[8] Yielding to these pressures, the Smithsonian secretary Michael Heyman agreed to set up procedures to rid the exhibit of its supposedly antiscience bias.

The beleaguered Smithsonian curators who have come a cropper of angry patrons are, of course, members of the STS community. As represented in the halls of the Smithsonian, their work is by no means on the outer edge of science theory but instead expresses some straightforward lessons about the social nature of science and technology, lessons that STS scholars have documented so thoroughly they now seem common sense. That even these mundane themes now seem reprehensible to authority figures who oversee the reputation of science and technology bodes ill for the future relationship between STS and its previous sources of institutional support.

The modus vivendi between humanists and social scientists and scientific and technical professionals has fostered the development of challenging new perspectives on both the inner workings and contexts in which science and technology operate. Apart from occasional stresses, strains, and squabbles, it long seemed that a community of interests would cohere. Now, however, it seems possible that frictions within this coalition, intensified by political upheavals in society at large, may tear long-standing alliances to shreds. Rather than abide (much less sponsor) challenging, advanced work in STS, some in the scientific community yearn to return to the good old days, the days of Carnap, Bronowski, and "our friend the atom."

Members of the broad, diverse networks of STS, including a good many sympathetic scientists and engineers, have yet to respond effectively to this disturbing turn of events. The time for organizing that response may be short. Self-anointed "pro-science" forces are now forging powerful alliances with radical reactionaries, ones marching in the forefront of every backward step.

Notes

1   Paul R. Gross and Norman Levitt, *Higher Superstition: The Academic Left and Its Quarrels with Science* (Baltimore, Md.: Johns Hopkins University Press, 1994).

2   For portraits of topics and approaches in STS see Sheila Jasanoff et al., eds., *Handbook of Science and Technology Studies* (Thousand Oaks, Calif.: Sage, 1995). For an overview from fifteen years earlier see Paul T. Durbin, ed., *A Guide to the Culture of Science, Technology, and Medicine* (New York: Free Press, 1980).

3   For details of this sad story see Donald L. Miller, *Lewis Mumford: A Life* (New York: Weidenfeld and Nicolson, 1989), 534–37.

4   David Dickson, *The New Politics of Science* (New York: Pantheon, 1984).

5   For a report on the Enola Gay controversy see Helen Gavaghan, "Smithsonian to Study Museum's Role after Dropping A-Bomb Exhibition," *Nature*, 2 February 1995, 371. Historian of science Stanley Goldberg, a member of the Enola Gay Exhibit Advisory Board who resigned in protest, reflects on the significance of the debacle in "Smithsonian Suffers Legionnaires' Disease," *Bulletin of the Atomic Scientists* 5 (May/June 1995): 28–33. Looking back upon successful efforts by veterans' groups and Congress "to censor the conclusions of sound historical scholarship," Goldberg warns, "That kind of thought control should have no place in a government committed to democracy. I believed that issue had been settled in the 1950s, when McCarthyism was laid to rest. Apparently I was wrong" (33).

6   Quoted by Kurt Kleiner in "Fear and Loathing at the Smithsonian," *New Scientist*, 8 April 1995, 42.

7   Quoted by Colin Macilwain, "Now Chemists Hit at Smithsonian's 'Anti-science' Exhibit," *Nature*, 27 April 1995, 752.

8   Quoted by Colin Macilwain, "Smithsonian Heeds Physicists' Complaints," *Nature*, 16 March 1995, 307.

The Science Wars: Responses to

a Marriage Failed

Dorothy Nelkin

THE Science Wars, the defensive attacks against "the flight from science and reason," are directed against recent studies that treat science as an activity influenced by social and political forces. Similar in their polarization and vituperation to the Culture Wars, their targets are scholars identified with a field of study called, variously, science, technology and society, or science studies, or cultural studies of science. Also targeted are historians and philosophers of science and various literary intellectuals who comment on contemporary scientific affairs.

Intellectual interest in science among "outsiders," that is, nonscientists, has been growing for at least two decades, partly as a consequence of the growing visibility of science in the larger culture, partly as awareness of its significant social implications grows. The success of science has given the once-private domain of the scientific laboratory a high public profile. Contemporary studies of science extend beyond examination of its social impact to include the development of research priorities and the methods and organization of research. Anthropologists look at the relationship among scientists in their laboratories. Cultural analysts explore scientific modes of discourse. And philosophers and sociologists question the nature of objectivity, the construction of facts, and the biases and values that shape scientific interpretations of nature. Such investigations treat science as a profoundly human endeavor, a product not of disembodied minds but of actual people in social interaction. To some scientists this social constructivist approach appears to be a hostile attack on science, and they are responding aggressively. Indeed, their counterattack is remarkable for its emotionalism, hostility, moral outrage, and polemical tone.

These scientists portray science studies scholars as science bashers, ignorant alarmists, self-deluded ideologues, dogmatic feminists, or at best foolish, faddish, muddleheaded, murky, radical, or left-wing. Their extravagant language is carefully selected to attract the attention of a media ever alert to colorful disputes. Science warriors debunk critiques of science as pretentious nonsense, miasmatic New Age drivel, romantic antagonism, unrelenting ecobabble, sophomoric ideology, damaging quackery, and even "hermeneutic hootchy-koo."[1] The social constructivists, we read, are adding to the outright superstition and ignorance of the American public. Their work will lead to the end of science; we are warned that the very "soul of science" is at stake. In the respected journal *Physics Today*, physicist Arthur Kantrowitz compares the power of today's literary and social science critics to that of the Chinese intellectual bureaucracy in the fifteenth century, when "intellectuals inflicted four centuries of insularity and scientific stasis on China."[2]

The theme is loud and clear. Science is being defiled and debased by uninformed outsiders—including radical feminists, postmodernists, humanists, literary intellectuals, socialists, anthropologists, environmentalists, creationists, animal-rights activists, and the president of the Czech Republic, Vaclav Havel. Havel, in a public speech in Philadelphia in 1994, had suggested there was "a crisis in science as the basis of the modern conception of the world"; that the premise of science, its unconditional faith in objective reality, had failed to grasp "the spirit, purpose, and meaning of the system." He and other intellectuals who suggest the limits of science have been defined as enemies bent on destroying science and undermining the scientific worldview. Science is bedeviled, besmirched, and besieged. This is the sort of protectionist language that anthropologist Mary Douglas calls "pollution rhetoric"—the typically defensive response of endangered institutions that seek to seal their doors in an effort to preserve their purity and security in the face of external intrusion.[3]

Normally, scientists are slow to respond to political pressures unless they are immediately and directly affected. Political scientist Joseph Haberer, in a history of the politics of science, describes scientists' relationship to the state as one of "prudential acquiescence."[4] In the 1970s, efforts to mobilize scientists to combat the influence of "scientific creationists" were largely unsuccessful, and scientists virtually ignored this significant threat to science education. Then, in the 1980s, only those immediately affected by direct opposition to animal experiments were willing to mobilize against the growing animal-rights movement. Similarly, most scientists slept through the successful antiabortionist efforts to block fetal research. But in the 1990s, many scientists seem willing and eager to attack social constructivist theories, even though their effect on science is far from obvious. The intensity of their response is curious. Why should scientists be so defensive these days? Why are they so troubled by approaches that demythologize science? Why is it so problematic to explore the gap between truth and knowledge, to ask "how do we know?" What are these Science Wars really about?

It is surely not about the real power of humanists, who are hardly about to topple the scientific enterprise. There is no evidence that science studies scholars are responsible for congressional cutbacks in science funding. A Congress that is planning to eliminate the National Endowment for the Humanities and National Science Foundation funding for the social sciences could not really place so much value on humanistic and sociological studies that it would be led around by historians and sociologists. Nor is there evidence that science studies scholars, even those with wide popular appeal, have ever had much influence on public policy. Philosopher Thomas Kuhn's important and widely read 1962 book *The Structure of Scientific Revolutions* challenged fundamental assumptions about the disinterestedness of science and its receptivity to new ideas, but it was hardly followed by a decline in science funding in the 1960s. One is hard put to find any correlation between historical or sociological analysis of science and changes in science policy.

The pollution rhetoric of scientists, their defensive language and moral outrage, may be best understood in the context of changes within science itself, and especially in the social contract between science and the state, which had developed after World War II. This contract included a set of both tacit and open agreements about the autonomy of science. The government would provide research support, relatively unfettered by requirements for accountability, if scientists would work in the interest of public progress and conscientiously administer and regulate themselves. These conditions helped science to flourish, and scientists have taken them for granted as their due. In the 1990s, however, the terms of the contract appear increasingly obsolete and, in various ways, both sides have failed to meet their side of the bargain. Government is decreasing its funding of science and requiring greater accountability, and scientists, often working in the interest of private profit, are facing increased difficulties of self-regulation. In this context, "outsiders" who study science are convenient scapegoats, and waging war is an easy way for scientists to avoid critical self-inquiry, to deflect responsibility and blame.

Following World War II, scientists developed their own Contract with America that was to shape the future of science for nearly forty years. The social contract was built on the premise that scientific information was a public resource, that what was good for science was good for the state. This relationship was often described as a marriage, implying shared assumptions and mutual trust. The unusual degree of autonomy granted to science reflected its apolitical image, the reputation of scientists as unbiased and "disinterested" and therefore reliable as a source of truth. It also reflected the widespread public trust in the ability of the scientific community to control its internal affairs. Thus science, as described by policy analyst Don Price in 1964, was "the only set of institutions for which tax funds are appropriated almost on faith and under concordats which protect the autonomy if not the cloistered calm of the laboratory."[5]

The marriage had been forged at a time of extraordinary faith in science as the basis of technological progress. It was consummated after Sputnik in 1957, when economic growth and Cold War competition favored expansion of the scientific effort. But conjugal bliss was short-lived. The tensions that were to erode the relationship began in the late 1960s when the antiwar and environmental movements raised questions about many central institutions, science included. Then, over the following years, science became increasingly controversial. The animal rights movement obstructed the use of long-accepted research methods. Creationists challenged the teaching of evolution theory in the schools. Antiabortionists blocked fetal and embryo research. Gay-rights activists challenged the scientific procedures that delayed the availability of AIDS therapies. Religious groups with moral reservations joined farmers with economic concerns to question biotechnology applications. In widely publicized critiques, these groups have continued to question the image of science as the basis of social progress and to challenge the role of scientists as a source of neutral authority and unbiased expertise. And cumulatively, they have contributed to a growing skepticism about the relationship of science and technology to social progress.

By the 1990s, the harmony that had long marked the partnership between science and the state had also deteriorated; the relationship has become less a marriage than a negotiated treaty or perhaps an affair of convenience based less on trust than on mutually recognized self-interest. And, as in all such tenuous relationships, tensions are high.

The government, for its part, has not been able to fulfill its side of the contract—the provision of unfettered support. Just when the costs of cutting-edge research have escalated, economic support for research has declined. This is, in large part, a consequence of world events, especially the end of the Cold War, the cutback in defense-related research, and the growing national deficit. Many scientists had depended on the Pentagon which, under the rubric of national security, had been able to provide significant funds. Even the "health war"—which became, like the Cold War, a source of science funding—is no longer a priority. At the same time, the mood of deficit reduction and the abhorrence of taxation has affected the budget of science funding agencies. It is more and more difficult for scientists to win grant support from government agencies, and those who are grant-funded face greater oversight.

As economic competition overshadows military goals, many scientists are shifting their priorities to commercially relevant research devoted to the solution of short-term problems. Not surprisingly, their corporate supporters demand research in the interest of profit. Thus, the image of science as driven by scientific curiosity, the vision of science as the endless frontier and the driving force of progress, has become increasingly clouded. The extraordinary optimism about the future of science and technology that had maintained the social contract has dissipated, and scientists, like other institutions and most people these days, must cope with fewer resources.

The changes in government policies are mainly driven by economic conditions, but they also reflect public disaffection with "big science." A series of debacles during the 1980s, widely covered in the press, fueled doubts about the wisdom of funding costly big science projects. The Challenger accident was only one of several failed megaprojects; others included the $2 billion Hubble telescope and the Mars Orbiter. Then, in 1993, that colossus of colliders, the superconducting supercollider, became a model for the inefficiency of megascience and an example of the difficulty of managing large-scale scientific projects. In a decision that seemed to many a symbol of the political future of science, Congress withdrew its funds.

The scientists' side of the contract, the promise of self-regulation, has also deteriorated. It has become increasingly difficult to maintain control over the very large number of scientists working in highly specialized fields, especially in the context of increasing competition. The most visible and widely publicized problems have been incidents of scientific misconduct and fraud. Scientists have had mixed responses to reports about abuses: some regard fraud as simply an aberration of a few "bad apples"; others regard it as revealing basic structural flaws in the organization of science. But there is general agreement that fraud in science strikes at the moral roots of the scientific enterprise and that it presents a serious challenge to the ability of the scientific community to regulate itself.

And fraud is only one of the issues raising doubts about the feasibility of self-regulation, the mechanism of which is the peer-review system. But some scientists are bypassing this process, going directly to the media with research findings that they feel are newsworthy and should be publicly available without the time-consuming process of peer review. The cold fusion experiments at the University of Utah and the behavioral studies of identical twins at the University of Minnesota are just two of many examples in which scientists, either rebuffed by professional journals or too impatient for peer review, have gone directly to the press.

Scientists are also struggling to come to terms with repeated revelations about ethical abuses, especially the disclosures of problematic human experiments conducted without adequate precautions or voluntary informed consent. Hazel O'Leary's investigation of the history of government funding of unethical human subject research has revealed not only flawed government policy but also widespread scientific complicity in dubious research. The "autonomy" of the postwar contract always had an element of wishful thinking, as scientists actively participated in the Cold War culture that made it seem necessary to give mentally retarded children plutonium in their milk. In the 1950s, public knowledge of such actions was limited; in the 1990s, it is widespread. So too in the 1990s, incidents involving misuse of federal research funds and conflicts of interest arising from entrepreneurial activities inevitably become matters of public debate.[6]

The visibility of these issues results in part from media attention, but there are real changes in science. Corporate influence on research is significant, especially in fields such as biotechnology. The growing importance of industry-university collaborations has left a public impression that science is for hire, that some scientists are simply indentured scholars to a corporate entity, and that scientific information is less a public resource—the basis after all of the original contract—than a private commodity.

In all, in the 1990s, science has become a very different enterprise than it was when its contract was forged with the state. It is larger, more costly, more controversial, and more difficult to control. More and more funding seems essential, but budget projections, below the rate of inflation, are regarded by scientists as "grim." And the scientific community is faced with internal dilemmas that have undermined its part of the bargain. For scientists, the changes are genuinely problematic; some established and productive projects have been blocked and some young and well-trained scientists are without jobs. Dislocations are always difficult and scientists have reason to be troubled. But who is to blame?

Institutions under siege, in the words of Mary Douglas, seek to "channel perceptions into forms compatible with the relations they authorize."[7] To protect its boundaries, an "instituted community blocks personal curiosity, organizes public memory, and heroically imposes certainty on uncertainty." This is what the Science Wars are about. As journalist Barbara Culliton describes it, the scientific profession is "hell bent on reviving the values that sustained science in the good old days."[8] Thus, in defending their disciplines against critical public attitudes, scientists are arguing with extraordinary passion to support their own dispassionate objectivity. They want once again to be perceived as pure, unsullied seekers after truth, and they want to define their own history and contemporary practice in just such terms. Rather than organizing to confront the politics of the corporate state or the growing influence of religious fundamentalists, science advocates have proposed the creation of "truth squads" to confront "junk philosophy of science." And they want to train working scientists to take over the teaching of the history of science. Such proposals would return to the days when stories of science were of heroes, studies of science denied the influence of social exigencies or cultural beliefs, and analyses of science policy were little more than advocacy to increase science funds.

One target of scientific wrath was a Smithsonian Institution show called "Science in American Life." The exhibit, supported by the American Chemical Society, was not intended to simply popularize science but to examine its role in society. Thus, it portrayed the costs as well as the benefits of scientific progress. Scientists, especially physicists who were feeling especially threatened by budgetary changes, reacted vehemently, sounding much like the veterans who opposed the Enola Gay exhibit at the National Air and Space Museum. They wanted the exhibit to celebrate science, not to dwell on its negative aspects.

It is not really surprising that World War II veterans want to control the interpretations of their history—especially in a cultural climate that, as writer Tom Engelhart put it, has turned away from the "victory culture."[9] Sensitive to critical interpretations of the troubling events that once gave meaning to their careers, the veterans defended their traditional understanding of these events and resented the skepticism of professional historians who based their analysis on archival research. But it seems curious, if not bizarre, that scientists who are themselves in the business of seeking new knowledge should be upholding tradition, that they should seek to control the history of science with such emotional zeal. Surely they realize that participants in events will view them through the narrow lenses of self-interest and that skepticism—disinterestedness—is a part of research integrity for the social as well as physical sciences.

Despite the budget cutbacks, the scientific culture is not about to disappear. While certain areas of science, especially big-science projects such as NASA and fusion research, are challenged, the role of science as a model of rationality in human affairs is not really in question. In particular, the historians and sociologists who study science have always had to justify their work to scientists and to validate its credibility in terms of scientific standards.

Moreover, American society has by no means abandoned science. Unlike welfare mothers, the chronically ill, artists, and other needy groups, many scientists can fall back on industry and private foundations when their government funds are reduced. And the budget plans for the National Science Foundation call for increasing support for basic research while wiping out most support for the social sciences. Perhaps on this front the scientists have won their war.

In sum, the Science Wars seem misplaced. There are many threats to scientific rationality these days—from religious fundamentalists, right-wing politicians, nativists, and other antiliberal forces. Attacking fellow academics is, of course, easier, but it is grossly misdirected. It is strategically misguided as well. As in any other area, historical and sociological studies vary greatly in their quality and their insight; but war does not discriminate among its targets. Equating efforts to demystify science with attempts to destroy research, conflating skeptical scholarly inquiry into scientific behavior with populist trashing of professionals, and engaging in confrontational polemics against all who question science, scientists are fostering polarization and discouraging reasoned discussion. Thoughtful scholars, even those outside the debate, are forced to take sides as the Science Wars challenge assumptions about their mode of research and their freedom to articulate ideas. At a time when academic institutions are generally under siege, dividing the academy into warring factions in this way is extraordinarily counterproductive.

By defending themselves so bitterly against outside critiques, scientists can only reinforce the public image of their profession as arrogant, indifferent to public needs, and answerable to no one. By sealing doors and closing ranks, they appear as simply another self-protective institution looking out for interests and careers. And by making vociferous claims to absolute authority over the definition of truth, they are themselves behaving like fundamentalists.

Finally, through their self-righteous efforts to discredit critics, scientists are diverting attention away from important questions. What are the moral boundaries to scientific investigation? At a time when health care, welfare, education, and even school lunch programs are facing serious cutbacks, should there not also be limits to the continued expansion of research? Given the importance of science in society, what is wrong with a sociologically informed account of the workings of science, the nature of scientific communities, and the debates over the production and interpretation of knowledge? Is science only to be popularized, praised, and promoted? If, as scientists claim, science literacy is an important goal, demystifying science, exploring how it works, and assessing it as a social institution may usefully contribute to public understanding. And it may also contribute to scientists, enhancing their understanding of the changing social and political realities that will inevitably affect their future.

Notes

1   Paul R. Gross and Norman Levitt, *Higher Superstition: The Academic Left and Its Quarrels with Science* (Baltimore, Md.: Johns Hopkins University Press, 1994).

2   Arthur Kantrowitz, "Review of Higher Superstition," *Physics Today* 48, no. 1 (1995): 55.

3   Mary Douglas, *How Institutions Think* (Syracuse, N.Y.: Syracuse University Press, 1986).

4   Joseph Haberer, *Politics and the Community of Science* (New York: Van Nostrand Reinhold, 1969).

5   Don Price, "The Scientific Establishment," in *Science and National Policy-Making*, ed. Robert Gilpin and Christopher Wright (New York: Columbia University Press, 1964), 20.

6   See the discussion in David H. Guston and Kenneth Keniston, eds., *The Fragile Contract* (Cambridge, Mass.: MIT Press, 1994).

7   Douglas, *How Institutions Think*, 92.

8   Barbara J. Culliton, "A Conundrum of Ethics," *Nature Medicine* 1, no. 2 (1995): 97.

9   Tom Engelhardt, *The End of Victory Culture* (New York: Basic Books, 1995).

What Is Science Studies for

and Who Cares?

George Levine

W E ARE, self-evidently, at a moment of aggressive public attack (some call it debate) on science studies. Just as the wars against political correctness managed to deflect attention from the real problems as an entrenched community appealed to democratic virtues while defending privilege, so now well-funded and powerful forces are claiming to be oppressed by the likes of Bruno Latour, Andrew Ross, and Katherine Hayles and are attempting to trash—in the name of reason, truth, objectivity, justice, and the first amendment—people who raise awkward questions.

Conferences of the National Association of Scholars (NAS) (in November 1994) and of the New York Academy of Sciences (NYAS) (May 1995) are symptomatic of an aggression that touches—perhaps dangerously, perhaps only with a rather sad and silly paranoia—on some of the most important issues of our time. While support for all serious intellectual enterprises is being more than threatened by newly empowered anti-intellectual political forces, these organizations are behaving as though the threat to science is really coming from some of the few intellectuals who have taken the trouble to think seriously about it. The excess and silliness of the response may seem the mere pettishness of spoiled researchers used to big-time funding and might well induce in us both complacency and a tendency to enjoy tweaking for its own sake; but complacency or teasing are the last things we need at the moment. Questions about the relations between society and science are among the most important of our time. We need, rather, to be thinking about how our healthy instincts to be oppositional might be channeled in more productive directions.

The language of "holy war" should remain the property of the egregious Norman Levitt and Paul Gross, who find an antiscience leftist lurking behind every paragraph of science studies. For them, irrationality and irresponsibility are pitted against the god of reason, and they characterize scientists who join our enterprise as apostates.[1] There's not much profit for us in adopting a rhetoric that pitches the interested nonscientific public against science. Recent developments in the budget-cutting attempts to scale back government support for work in both science and the arts and humanities ought to be making it clear that our fates and our interests are entangled. We need to be exploring the mutual interests of science and its critics and, to make science studies genuinely effective, we need to persuade a lot of scientists that what we are doing is not only in society's best interest but in theirs. Popular skepticism about science derives not from the work of Latour, for example, but from the distanced arrogance with which science operates its magic, its ostensible contempt for lay questions and worries, its very power. The best thing that could happen to science, if it wants to convince society as a whole that it deserves support, would be to reject the arrogant language of holy war and humanize itself. And in our efforts to assist in understanding how science is involved in culture, how science *is* culture, it might profit us in the long run and for the necessary battles to come to think more about our own strategies and, indeed, about the degree of our own culpability in the obvious failures of relations between practicing science and cultural critique.

Science studies, as we know, was born in controversy and sustains itself controversially, largely against the scientific community that is the subject of so much of its work. It has flourished, certainly, because the work of science has in many ways been socially and politically determining in our time—because, whether we like it or not, science, understood or not understood, shapes our consciousness as it shapes the material conditions of our society. We may be enthusiasts for science or frightened observers of its power, but we know that we can't ignore it and that not to be aware of its power is not to be educated.

Science studies has its own singular genealogies, growing from historical, philosophical, sociological, as well as literary, concerns, but it is often rather amusingly (if its implications weren't ultimately so serious) lumped under the heading of postmodernism, where a lot of kooky, anti-intellectual, politically correct, and subversive types have been thought to hang out. Even the history and philosophy of science, which grew up as a field in the 1950s and 1960s with a strong positivist bent and a deep commitment to science, is now being accused of the sins of postmodernism. The strong positivism of the early century was almost by definition skeptical about the degree to which science's knowledge claims could be said to describe "reality." And Bas van Fraasen, whose contemporary mode of empiricism—which denies the necessity of any claims about a reality out there for the validity of scientific knowledge—would seem also to be "postmodern" in this wide and unacceptable sense so often used as denunciation. But both the positivists and van Fraasen have not worried scientists because, clearly, their projects were not antiscience but attempts to establish the validity of scientific knowledge. Since Thomas Kuhn's decisive *Structure of Scientific Revolutions* early in the 1960s, however, scientists have been particularly wary of historians of science and sociologists.

The squabbles that followed have now been inflated to holy wars because there is so much at stake: intellectual authority, educational direction, disciplinary turf, the allocation of big money. Many more people than Levitt and Gross are taking cultural criticism of science seriously because scientists are themselves feeling vulnerable. *Their* funding is getting cut, too. While many of us linked to the humanities may feel especially ineffectual in the world of politics, big money, and the public sphere, people are getting worried that we are not ineffectual enough. The frightened announcement of the NAS conference notes that our attacks are dangerous because, among other things, they "alter directions of research" and "affect funding." It is hard to believe that my consideration, for example, of the relation between Darwin's views and Trollope's novels has contributed to the death of the supercollider, but that is what the NAS conference is arguing implicitly. Several years ago, when Steven Weinberg turned down an invitation to a conference I was running, he told me that he knew what we were up to. I didn't understand at the time, but clearly Weinberg's affiliation with the NAS conference and his recent stern defenses of scientific rationality have most to do with the supercollider, which should have been in Texas, where Weinberg teaches. I want such serious thinkers on our side, but somehow he and others have adopted the hysteria of Gross and Levitt and believe with the NAS and the NYAS that there must be an important connection between the work of science studies and Congress's decision that the supercollider was just too expensive. As philosopher of science Paul A. Komesaroff put it, oh so delicately, about eight years ago, "The two great bodies of thought, represented by natural science and contemporary philosophy [in which he intended to include sociology of science], have so far failed to make proper contact."[2]

How to do something about that contact? A continuing and even naturalized consideration of science in the context of culture is one of the urgent priorities of research and education at our fin de siècle, but I don't have to convince you of that. From our perspective, what's necessary is some clarification of what we are doing. And so I want to ask some questions and gesture toward some answers.

Are we clear about whom we are addressing, and for what reason? Are we clear about why even our most professionally innocuous stabs at connecting science and literature are often taken as assaults on science? It is certainly true that as an area of study, "science and literature" has some large-scale objectives that extend beyond the Arnoldian virtue of "Curiosity" and the disciplinary injunction that we make our contributions to knowledge. Does our agenda entail reform of the practices of the sciences? Are we involved in a sustained critique of science that might have practical implications for how science is funded by our society? What might it mean for us, in literature or other nonscience disciplines, to suggest to scientists how they should do their work? What do we think about scientists taking our discourse seriously enough to suggest to us how we should do ours? Are we committed to showing how important science is or has been to the way we write, think, feel, act? Or how pernicious it has been?

Let's look at some of the fundamental assumptions of our work. We always talk about science as though it is a discourse. For example, in a brilliant recent book, *Fact and Feeling*, Jonathan Smith begins innocently enough "with the assumption that science is a form of cultural discourse: like literature or history or music or art or religion, it both shapes and is shaped by the culture of which it is a part."[3] Recognizing, as he puts it, that the [nineteenth-century] "elevation of scientific discourse . . . does not also isolate it from, or make it necessarily antithetical to, other forms of discourse," Smith works hard, without denigrating either, to break down the opposition between science and literature. It is difficult for us to realize, because this is so much like mother's milk to us, that the move is not only culturally counterintuitive, it is politically fraught. But such a view is viper's poison to almost everybody else. That is, much that we take for granted is very likely to cause us trouble outside our disciplinary or interdisciplinary area—and all too little within it. Smith is healthily aware that to talk of science as "merely" another discourse demeans science, and demeans literature as well. To what degree, when we argue that science is a discourse, and not an epistemologically decisive one, are we out to diminish science and to elevate ourselves? That, by the way, is the Gross-Levitt line.

Since the current battles so easily reduce to turf wars, we need to answer such questions as clearly as we can. Certainly, many of us turned our attention to science studies because we were fascinated by science and admired it enormously. But science itself is not, as Smith reminds us, a "monolithic" entity; it is, rather, a complex and continuing process that transforms and often supplies the materials for its own critique as readily and as significantly as literary studies do. Insofar as we imagine science as an institution, formally embodied in the academy and in a hierarchy of material practice all over the world, discussion of science always has strong practical implications. We need to consider whether, as we investigate relations between scientific and cultural discourse, we are in fact implicitly attacking the institution. Looking around the university, literary faculty find that science faculty teach less and earn more and have access to funding beyond the humanist's fiscal imagination; science faculty find that literary study is full of ambivalences and ambiguities and, alas, moralisms, that would be totally discredited in their own fields. These practical, cultural differences don't help. But they don't inevitably entail the for-or-against rhetoric that is now dominating public discussion.

Gerald Holton, a speaker at the November NAS conference, but one whose contributions to the cultural study of science have been very valuable, devotes the long final chapter of his most recent book, *Science and Anti-Science*, to the "Anti-Science Phenomenon." Holton's argument is neither so intemperate nor so homogenizing as Gross and Levitt's, but he is distinctly and indeed reasonably concerned about a problem that bothers them, too — the terrible ignorance of science that pervades our society. Holton makes a strong and moving case for the social and political importance of sound scientific knowledge in the tradition of Karl Popper's *Open Society and Its Enemies* (1945), and traces a sometimes convincing connection between the rejection of science and a dangerous right-wing descent into authoritarianism. But that old and valuable liberal critique of "closed" societies really has nothing to do with the present case. We need to make that clear. Why is it that studies of science and culture must be taken as opposed to "sound scientific knowledge"? Bruno Latour is inevitably disparaged as wanting to abolish the distinction between science and fiction.[4] We might argue with Holton and show him not only that Latour is not anti-science (indeed, he has presented himself recently as quite literally in love with at least some aspects of science and technology and he works for the government precisely on matters of technology), but that the abolition of the distinction between science and fiction of which

Latour talks need not be dangerous for science—depending, of course, on what you mean by "science."

Holton's position should remind us that resistance to critiques of science has a long and honorable history, as well as a shady and grubby one. Our job is to show that the Popperian critique doesn't work in relation to the central activities of science studies. Nobody is trying to close down science, and the defensiveness that assumes that anyone who isn't for it is somehow sacrilegiously against it needs to be overcome. Most science studies work I know would never have been written if its authors were not fascinated by science. Substantively, we continue our exploration of the complex relations between science and literature, science and culture—the way they support, reveal, test, and inform each other—most of the time without direct attention to our contemporary disciplinary battles. But how does all this work relate to the practice of scientists themselves—which is, after all, what they are worrying about? One might well ask about our work, who cares? That is, who cares beyond our own professional institutions?

The surprise for me at my home institution was that the first person outside the humanities and social sciences who cared, and the person who to this day cares most, is Norm Levitt. *Higher Superstition* may have been largely provoked by a year the Center for the Critical Analysis of Contemporary Culture at Rutgers devoted to issues of science and culture. We invited people like Latour, Helen Longino, and Simon Schaffer; during this time Levitt sneeringly borrowed my copy of Katherine Hayles's *Chaos Bound*. He cares with a passion, which might in fact be better than the unselfconscious indifference with which the work of science studies is treated by most of the scientific community.

What started in my career as (what I imagined) a not very daring curiosity about the way science was affecting the Victorian writers has suddenly become controversial. This may be because people who otherwise wouldn't have cared at all about, say, the relations between Darwin and Trollope, or thermodynamics and Pynchon (of which I solemnly and academically wrote many years ago), are suddenly finding themselves vulnerable sharers in a skimpily funded academic world. But it has been intellectually and practically invigorating to know that my audience extends beyond my discipline and that what I think and argue might actually get nonspecialists upset. I thought I was working out of a deep, perhaps even mystified, respect for science; I thought science was powerful, interesting, and difficult. Science studies in literature, insofar as it thinly existed, rarely implied disrespect for science.

It's disingenuous not to recognize, however, that many of us are in the humanities because we were bad at or turned off by science; the ethos we joined tended, conventionally, toward antiscience. That was certainly true in the pre–science studies days of the new criticism. The romantic critique of science, peeping and botanizing on its mother's grave and implicit, for example, in Carlyle's rejection of "mechanism," has continued to play itself out in both high and low culture. Its recent manifestations in New Age mysticism and environmentalism is part of what disturbs Holton, as it also disturbs me. When Levitt first attacked me and my center, he used Whitman's "When I Heard the Learned Astronomer" as an example of what's wrong with our attitudes toward science.

However sympathetic we may have become to the astonishing activities and achievements of science, we bring with us the ethos of our discipline, a set of attitudes and skepticisms that are in effect institutionally opposed to the professional practices of science. It is difficult not to think of science and literature as two distinct ways of knowing and as almost antagonistic in their relationship to knowledge, to the human, and to value. Our attempt to conflate all discourse as fiction becomes an insult to the professional distinctness of the scientific disciplines. By focusing on language and culture we seem to be declaring war.

I discovered, for example, that my rather innocuous reading of Darwin in *Darwin and the Novelists* had become at Rutgers a focal point for controversy about constructivism well beyond the departments at Rutgers in which anyone might have read the book. Our vice president, a polymathic member of another NAS, the National Academy of Sciences, saw me as a flaming postmodernist (which to this day takes my obsolescing breath away) and asked me to instruct the university community about how I could reconcile postmodernism to Darwin's scientific work. It was then, in fact, that Levitt and several of his colleagues emerged snorting from the protesting majority. When, at one point, I assured a heckling scientist that he had nothing to worry about since I would not be asked to referee scientific grant proposals, one woman in the audience shouted out, "Thank God." So this was where my interest in George Eliot had got me.

The experience was disenchanting, if amusing, because it made me aware as I had never been before that the two cultures were really, perhaps incurably, separated. And it was not a Leavis/Snow separation, in which the virtue of "values" was opposed to the virtue of "knowledge," but a disciplinary, turf/economic separation, in which what was at stake was institutional authority and government funding. The separation had little to do with the epistemology that was the ostensible focus of attention; it was embedded deeply in material conditions. I was forced to understand the degree to which the very assumptions that ground the enterprise of humanist or social scientist investigation of science imply what may be an irreducible professional conflict. While I had come to praise Darwin, not to bury him, I was heard to be burying him and the whole noble enterprise of objective science.

Although I am not convinced that the current separation between science and science studies needs to be quite as wide as it is becoming, I am convinced that we are in a turf battle. In science studies we are claiming to say something important about science; we are standing on somebody else's turf; our decision as nonscientists to study science is a provocation to a discipline whose members have undergone strenuous disciplinary training and who are now also having trouble finding jobs, as much trouble as we are having. Their decision to fire back should not be surprising, unless we are insufficiently aware of the degree to which we are in fact trespassing and they feel themselves, from their positions of institutional and financial power, very vulnerable indeed. As we resent, say, Gross and Levitt's unintelligent critique of science studies, so scientists resent Latour, Schaffer, and Shapin. We are not innocent.

There is a lot at stake in a cultural criticism that might discourage public resources from supporting scientific activity. Feminism is not merely the crazy and Stalinist political positioning that Levitt and Gross caricature; theoretically, if people listen to it, it could affect scientific projects, shift around the flow of money, cut it off in other directions. David Berreby points out in *Lingua Franca* that

> the outrage scientists feel for science studies is fueled by more than intellectual disdain. Government and business are funding less basic research in science. Congress has voted to kill the superconducting supercollider. Doubts have been raised about the space station, even about the human genome project. The turf is shrinking, and so, like tribes forced toward the same oasis by a drought, scientists and sociologists of science are starting to threaten and skirmish.[5]

The announcement of the NAS conference is revealing as it shouts that the attacks on science are "dangerous." Here's why: "They undermine public confidence; they alter directions of research; they affect funding; they subvert the standards of reason and proof." Here is very expensive and moralized turf.

Levitt and Gross take the high moral line in talking about what troubles them, although they regard nonscientists' questioning of the authority of scientific knowledge as medieval. What worries them, they say, is that the larger culture, "which embraces the mass media as well as the more serious processes of education," will lose its capability "to interact fruitfully with the sciences, to draw insight from scientific advances, and, above all, to evaluate science intelligently" (6). In a book that mocks everything that doesn't understand science, it is almost funny that the authors imply here that until these SLS-style critiques came along, the public was interacting fruitfully with the sciences. They are, of course, right that people aren't adequately educated in science, but blaming science studies is both unhistorical and scapegoating. Science studies at least encouraged an interest in science that before then required a Sputnik to vivify. The autocratic and arrogant intellectual snobbery that becomes the model for Gross and Levitt's contemptuous dismissal of meddlers in science is far more likely to be responsible for the public's ignorance than the stars or followers of science studies, who constitute a minute and uninfluential minority when it comes to public power, congressional support, funding, and education. Gross and Levitt mean by fruitful interaction the public's total deference to anything that the scientific communities might argue. The interaction is to be all one way.

Turf is Gross and Levitt's real concern: "In order to think critically about science, one must understand it at a reasonably deep level. This task, if honestly approached, requires much time and labor. In fact it is best started when one is young" (5). And why do you suppose so many humanists are involved in these medieval exercises of attack on the authority of scientific knowledge? Because, one might have guessed, humanists are taking their revenge for having been the lowest disciplines on the totem pole of epistemological authority: "Literary criticism, finally, has been looked upon as a species of highly elaborated connoisseurship, interesting and valuable, perhaps, but subjective beyond hope of redemption, and thus out of the running in the epistemological sweepstakes" (12).

Unfortunately, the NAS/Gross-Levitt position, seeming so extreme in its deliberately contemptuous formulations, is pretty much the norm (except for questions of civility) in the communities of science. The assumptions that mark most of our work and that I have quoted from Jonathan Smith's new book are simply beyond the pale. We need to face the fact that these assumptions have barely any life beyond societies like ours, and we should be spending a lot of time and energy on how to move them beyond these conference walls.

Note, for example, what has been happening in the response by science to the new national legislation known as Goals 2000. As many of you are aware, the Department of Education is developing guidelines on national standards for what children should know about the various disciplines, and the NAS was given the responsibility to develop those standards for the sciences. These are issues of enormous importance for the future of education in America. The federal government is responding to the public debate about what has gone wrong with contemporary education, and federally imposed standards for all schools are only five or six years away. To what degree do twenty-five years of science studies have anything to say about these issues?

Let me quote a recent article from *Science* to give you an idea:

> Some scientists were up in arms over the description of the philosophy of science. Instead of saying that researchers make discoveries, the document described science as a "social activity" of "constructing knowledge," and emphasized the "tentative nature of scientific knowledge." Physicist James Trefil of George Mason University in Fairfax, Virginia, says the early version conveyed "the really bizarre postmodern notion that somehow science is just a matter of social convention, rather than analysis of data. Harvard University physicist Eric Mazur, a pioneer in undergraduate teaching, was so dismayed by this section that he resigned from the project immediately after reading it. "Science is much more discovery-based than they seem to think," he says.
>
> But Trefil and other scientists give [Richard] Klausner [the director of the project] high marks for his response. Klausner insists the academy never intended to weaken the rigorous underpinnings of the profession, and he says all hints of the offending philosophy will be excised from the final draft.[6]

We may believe that scientists have postmodernism all wrong, but we cannot ignore the horror most of them feel at the idea of constructivism, which they see as threatening the rigorous underpinnings of science. When I ask who cares about what we do in science studies, I mean not only who is concerned with the dangers of our work, but who can see its full implications and recognize its potential value. Who, for example, understands that constructivism leaves the rigorous underpinnings of science right where they were and is part of a different project entirely? Who understands that the implicit positivism of the educational program that seems to be endorsed by the Academy is not only intellectually suspect but fundamentally damages the relations of science to the culture as a whole?

Contemporary apologists for science are on a crusade for reality. Objects fall through space at a certain rate not subject to deconstructive analysis. It really happens, empirically. Gravity is not constructed. It's out there. It's real. So what do we do with science and reality? As an indication of our difficulties and responsibilities, I want to look briefly at two key texts in our recent conceptions of the study of science and literature and then raise one question about reality.

One of the most important figures for us in the study of science and literature is Katherine Hayles, who does not escape the direct wrath of Gross and Levitt. But when we turn to *Chaos Bound*, we find a book that is pretty neutral politically (a ruse, Gross and Levitt suggest; the politics are really there driving everything anyway). Hayles makes the mistake of beginning her book by arguing that "different disciplines are drawn to similar problems because the concerns underlying them are highly charged within a prevailing cultural context."[7] Such an argument implicitly denies that science is intrinsically, not extrinsically, driven. There is no connection between Gödel's theorem and modernist art; there is no connection between chaos theory and deconstruction. All science moves according to its own internal logic, and the development of particular theories within it are not related to cultural patterns.

I don't want to dwell here on Gross and Levitt's absurd attack on Hayles. She has herself responded effectively, and there is no longer any need to dwell on their well-known reductions and simplifications. My point, simply, is that they serve as a good example of how impossible it has been for those outside our discourse to accept or even understand the assumptions with which we work. They see Hayles as denigrating science because she talks about some scientific work as being akin to postmodern literature. Her analyses, therefore, "in effect derogate the reliability and accuracy of standard science, and snidely disparage those scientists—that is to say, the vast majority of *all* scientists—who have been oblivious to this ostensible revolution in thought" (92). This, I believe, would be a majority position among those outside the humanities and some of the social sciences in the academy: they have not bought into our talk about "langue" and "parole," or into our sense of how culture works beyond the intentions of the individuals who compose it. Here is nonsense or Marxism, or worse.

For our part, however, we need, if we believe in the sorts of culturewide analyses Hayles conducts, to make the assumptions about culture that underlie them available to a world that sees them as counterintuitive if not mad. This is probably harder than most of the work we do. It's crucial that we make it clear that understanding the way scientific and cultural attitudes interpenetrate need do no harm to science and might well lead to that mythic lost ideal condition in which science and the public interact creatively.

Second, let me turn briefly here to Andrew Ross's work, which has developed some reputation among many of us. *Strange Weather* is not strictly a book on science and literature. It belongs rather to the developing world of cultural studies, takes as its dominant metaphor and subject the science of weather prediction and the cable weather channel, and is written overtly in the interest of what Ross calls progressive politics. Unlike Hayles, Ross has a political agenda, and it's not hard to notice that he is not well loved by those outside cultural studies in which he works. Ross makes it clear that he doesn't know much about science, and yet he spends most of the book attempting to undercut, for political reasons, the special authority of science against protest movements like New Age activists.

What is most important about Ross's book is that it is centrally concerned with the question of how a lay public might most effectively relate to science, as institution and practice. This, I take it, is the critical issue. He doesn't, after all, accept the ideas of the protest movements he studies but is interested in the political significance of their antiscientific positioning, and he criticizes these groups because they argue their own authority by adopting some of the very strategies of the sciences they critique. But the very audience we are trying to reach doesn't seem to have the time or the training to understand what books like Ross's are trying to do. What matters for them is that the book seems to be antiscience and that it is driven by a political program.

Valuable and important as it is, *Strange Weather* may, rhetorically, be striking the wrong note. It's the kind of problem we need to be thinking about as we seek to extend our range beyond those initiates who begin with our assumptions and our training. Ross is deliberately playful and theatrical throughout the book—rather too much so, I think. The opening statement seems impossible not to read as a provocation: "This book is dedicated to all the science teachers I never had.

It could only have been written without them."[8] That's cute, of course. And it places Ross rather aggressively, it would seem, in the anti-science camp. The problem is, books written like this depend on readers who like to play with language, but they are often, and I believe most importantly, read by people who don't. Since one of the points of Ross's book is that in order to act meaningfully in the politics of contemporary culture, we need to become *more* literate in science and technology, this opening aggression suggests a more humble and science-favoring reading than it seems at first to allow. For Ross, there is too much at stake to ignore science, and science is too important to leave to scientists. I am not suggesting that Ross's provocations are misguided. For anyone who can read, they open up the complications of the subject. But for those who can't read, those whom we need to engage, those with whom it would be good if we could "fruitfully interact," Ross's wit is an alien language. It is no more intelligible to Gross and Levitt than the technical language of Levitt's topological research is to me. Here, for example, is the problem I'm most concerned with: the arrogance of specialist discourse when we are talking to each other about things that matter.

I am particularly struck by the irony that beyond the ignorances, aggressions, and brilliancies of disciplinary difference, Ross in fact claims as his object something quite similar to what Gross and Levitt claim to seek. Where they had asked for a lay public that might "interact fruitfully with the sciences, . . . draw insight from scientific advances, and, above all, . . . evaluate science intelligently," Ross seeks workable strategies that address "the desire for personal responsibility and control that will allow nonexperts to make sense of the role of science in their everyday dealing with the social and physical world" (29). Here he isn't being playful; this is what it's all about. How can we create a reasonable and strategically effective critique of science that might have some plausibility in the world beyond these walls?

What Ross resists and what needs resisting are the political and economic implications of the scientific culture of expertise. Scientists do know how some things work, so it seems absurd not to accept their judgments, to open ourselves, as a lay public, to scientific authority. But does that entail the pious submission to absolute scientific authority on the part of the lay public whose very lives will be determined by the choices scientists and their material supporters make? If fruitful lay interaction depends on full scientific literacy, I fear that abject, pious submission is all that lies before us—even, ironically, for most scientists, who have no standing outside their own areas of expertise.

Precisely that model provokes Ross to his provocations and me to these reflections. Gross and Levitt show that science claiming universal rationality and the right method and a community of truth seekers gets loaded with a lot of political freight dangerous to anyone who might question it. Of course, we would be fools to behave as though there is no knowledge of the natural world to be had and that science has no better shot at it than any other professionals, or nonprofessionals.

How then to reconcile our respect for science with our resistances to it, to recognize the need for knowledge and to sustain our sense that science is in culture, that it is never any more unpolluted by the society out of which it emerges than any other cultural product, that the power of its knowledge production requires of us more, not less, attention? Being playful while talking to a like-minded group alienates those who need to be convinced. Pretending to ignorance may seem like encouraging it, and the one thing people who engage in serious discussion of difficult scientific/cultural issues cannot afford is ignorance.

Let me focus this problem with a simple exemplary narrative, whose moral is banal and whose particular point is probably well-known to many of you. At Rutgers, we own a stretch of genuinely virgin forest—Mettler's Woods, "the last remaining uncut upland forest in central New Jersey."[9] The forest, as what was known as a "climax" forest, is invaluable for research and teaching. The old oaks in it are, on average, 235 years old. As a result, "no manipulation or human disturbance" has been allowed there. The biologists who supervise the forest believe that it was crucial to keep it at its present equilibrium, and thus even spraying for gypsy moths has been disallowed. Supervision has been careful and thoughtful. The aim has been preservation. Certainly, the refusal to do anything to the forest but let it be without human interference would, prima facie, meet the approval of most conservation groups, most people concerned with ecological issues.

And yet, as several young Rutgers scientists have noted, the forest is falling into decay. The old oaks are failing to reproduce themselves and seem at last to be "senescing." Somehow, the best of intentions and the best available ecological paradigms are not working, and it has since been discovered that some ecologically unsound camping and careless campfire building would have done wonders for the forest. "It is becoming clear," the scientists now say, "that oak species in many parts of the eastern United States require some sort of ground fire for successful regeneration" (69).

Such a story is familiar—not much different, except in scale, from the lesson of the fires in Yellowstone Park. And the moral is that as we move out into large, substantive, cultural/scientific issues such as ecology where most of us have strong feelings and commitments, we simply have to know what we are talking about. Even preliminary social and moral decisions like whether to preserve the woods or to preserve endangered species require some "scientific" information. There are costs to everything; how do we determine what the cost will be? If we decide that preserving the woods is a good thing, we should first know why and second know how to preserve them. That requires serious research, what is now called scientific research. Ecologically, it turns out, it was a *scientific* mistake to treat humans as though they were not part of nature; to save the woods, we need humans to burn them. And mistakes like that, if we extrapolate to, say, the rain forests of South America, or even the timberlands of the Northwest, have profound moral and political implications. We are in a joint enterprise. Somebody is going to have to do the studying and somebody is going to have to listen—on both sides.

In this context, the strategy of Ross's book may seem at times unfortunate. I think that he is by and large substantively right, and adopting the strategy of theater and provocation may in fact be the way to get us talking about our mutual interests. But there is a lot of educating on both sides to be done, and when we say in our various literary, ecological, social, and cultural ventures that science is just another discourse, we need to make it clear that we are not saying that as just another discourse it doesn't have a particular and crucially important role to play in anything we might choose to do.

Our interdisciplinary ambitions have inadequately touched the people whose disciplines we want to talk about; our rhetoric has been far too comfortable in affirming the dominant assumptions of our current work and on the whole too lax in affirming the necessity of the kind of knowledge science produces best. And we have been insufficiently alert to the way our assumptions and the very practices of our field embattle us and alienate the scientific disciplines. Thinking always of science, that nonmonolithic though institutionally powerful concept, as the enemy reduces our capacity to make our own discriminations, but more important, reduces our capacity to speak to those whose activities we most want to engage.

In any truly public battle, those arguing for constructivism in general will lose to those arguing for reality in general. What is necessary is first an at least rhetorical concession to the power of the argument for reality, and second, a demonstration of the way particular uses of the constructivist position are humanly helpful and consistent with a rigorous science. It is crucial to make it clear that even constructivists believe that it is necessary to know what they are talking about, that the preservation of Mettler's Woods depends on sound knowledge rigorously ascertained.

Some of the difference, the failure to achieve appropriate contact, is irreducible, but some of it is not, and if we are committed to a broader and richer conception of the interactions between science and literature, of the place of science in the broader culture, we had better make ourselves more alert and begin developing strategies that honor the necessity for developing the constructivist and discourse-oriented critiques we do so well, insisting on the most rigorous possible acquisition of necessary knowledge, doing whatever we can to make it clear that nonscientists have a place in the conversation, but finding above all a way to make scientists part of that conversation.

Notes

This chapter was originally written for the keynote address of the Science and Literature meeting in November 1994. In the course of revising it, I came to realize that the audience I was addressing at that meeting is an important part of the argument and that any revision that turned that audience into a more generalized reader would lose much of its point. As a result, I have kept the lecture's rhetoric and mean by "we" the members of the Science and Literature Society, predominantly from departments of literature, but including historians and scientists as well.

1   Norman Levitt and Paul Gross, *Higher Superstition: The Academic Left and Its Quarrels with Science* (Baltimore, Md.: Johns Hopkins University Press, 1994), 6.

2   Paul A. Komesaroff, *Objectivity, Science, and Society* (London: Routledge, 1986), vii.

3   Jonathan Smith, *Fact and Feeling: Baconian Science and the Nineteenth-Century Literary Imagination* (Madison: University of Wisconsin Press, 1994), 5.

4   Gerald Holton, *Science and Anti-Science* (Cambridge, Mass.: Harvard University Press, 1993), 153.

5   David Berreby, "That Damned Elusive Bruno Latour," *Lingua Franca* 4, no. 6 (1994): 24.

6   *Science*, 16 September 1994, 1649.

7   Katherine Hayles, *Chaos Bound* (Ithaca, N.Y.: Cornell University Press, 1990), xi.

8   Andrew Ross, *Strange Weather: Culture, Science, and Technology in the Age of Limits* (London: Verso, 1991), i.

9   Steward T. A. Pickett, V. Thomas Parker, and Peggy L. Fiedler, "The New Paradigm in Ecology: Implications for Conservation Biology above the Species Level," in *Conservation Biology: The Theory and Practice of Nature Conservation and Management*, ed. Peggy L. Fiedler and Subodh K. Jain (London: Chapman and Hall, 1992), 68.

Unity, Dyads, Triads, Quads, and Complexity:

Cultural Choreographies of Science

Sharon Traweek

O NE of the goals of this issue, I was told, was to "coordinate views about the current state of the contests between social rationality and scientific rationality." That seems to me completely laudable, if a bit difficult, due in particular to that pesky verb *coordinate*. The word brings to my mind Descartes's orthogonal measuring lines and the social niceties in the late 1950s of getting the colors of one's clothes aligned or even getting the lines of our high school drill team perfectly straight. It seems to emphasize the disciplined side of life. I was reading a lot of papers just now and trying to discipline my thoughts into classifying student work under five of the first six letters of the alphabet and then inscribing my judgment into little boxes with the correct marking implement. As you can probably tell, I am keener on the other goal of the issue: "Drawing attention to the cultural prejudices inscribed in the very epistemology of scientific inquiry." However, I would be much happier if we could drop the word *prejudices* and replace it with something less prejudicial.

I prefer to draw attention to the cultural choreographies embodied in scientific inquiries. In *Choreographing History*, edited by Susan Foster, several of us wrote about how social, intellectual, political, scientific, economic, art, and cultural histories are enacted and performed, produced and consumed by human bodies moving through specific places.[1] Of course, these embodied actors perform their moves in ways that they and others around them understand. Their movements might be rigid or fluid, formulaic or inventive, but they are enacted in the context of cultural codes that make them decipherable to most everyone around them, just as most of the readers of *Social Text* could probably navigate the pedestrian traffic at midday on the sidewalks of big cities. We know our way around gatherings of the sort of people who read *Social Text*. We know the gestures, the tones of voices, the styles, the rhythms, the jokes, the texts, the details that make a difference. I think we should know more of the moves made by scientists, engineers, and physicians as they get around their worlds.

Ex Cathedra Voices

Many would argue that Thomas Kuhn's *Structure of Scientific Revolutions*, published more than thirty years ago (in 1962), launched the new empirical researches into the practices of scientists. (Certainly there were several others who were building similar arguments against the older notions about scientific research.)[2] Nonetheless, Kuhn's work did not, in fact, disrupt the familiar litanies about science. Until the late 1970s, most historical, sociological, and philosophical investigations about science, technology, and medicine continued to assume and celebrate but did not investigate the notion that scientists had invented a perfect way of knowing, quite free of all human constraints.

I want to recapitulate, perhaps too tersely, a few of those older notions we usually first encountered in the pages of high school and undergraduate textbooks, or perhaps in museums and on television, that often survive in our minds.

- Until Galileo invented experimental research, almost all important discussions about the phenomenal world were conducted as theoretical debates.
- Galileo's ideas were rejected by the Vatican because they challenged Catholic religious beliefs of the time.
- Francis Bacon developed the idea of a laboratory and codified the procedure for research now called the scientific method.
- The printing press made possible the accurate reproduction and circulation of experimental data.
- Isaac Newton invented the idea and the means of using mathematics to analyze experimental data.
- Scientific method is based upon skepticism.
- Scientific method identifies and controls all variables in an experiment.
- Scientific analysis is mathematical analysis.
- Scientific knowledge is amassed progressively and cumulatively.
- Scientific theories and data are rejected when subsequent efforts at replication fail.
- New scientific theories are accepted because they explain more experimental data more economically than their predecessors.
- Scientific thinking and methods are incompatible with religious thought and feeling.

—Scientific reasoning proceeds by deduction and induction; hypotheses are deduced from existing experimental data and experimental data are tested against hypotheses inductively.

—Scientific research is made objective by eliminating all biases and emotions of the researchers.

—Scientific research is neutral with respect to social, political, economic, ethical, and emotional concerns.

—Scientific research has an internal intellectual logic; there is an external social, political, economic, and cultural context for science that can only affect which scientific ideas are funded or applied.

—Improvements in the quality of human life and the duration of human life during the past two hundred years are due primarily to the application of scientific discoveries.

—Technology is applied science.

—Basic research and applied research are easily differentiated.

—There is a significant rate of "social return" on scientific research.

In their most rudimentary telling, our traditional notions about science are usually recited in the cultural forms developed by the medieval European Catholic Church: a list of saints' (geniuses') lives, their miracles (discoveries), and holy sites (laboratories). These reverential stories can be found easily on television (especially the Discovery Channel and public broadcasting stations), invoked as documentaries on science and technology, complete with authoritative male voice-over, offering instructive and amusing examples of Derrida's notion of "absent presences."[3]

New Voices

All of these conventional ideas about science have been very powerfully challenged during the last thirty years of scholarship by anthropologists, economists, historians, sociologists, philosophers, and others.[4] To continue in these beliefs is to signal that one is unfamiliar with the massive body of scholarship that has undermined them. The main way in which these ideas are still taken seriously in today's research is to inquire as to the conditions for the circulation of such beliefs in certain specific cultural political economies. Oddly, the people I find most attached to this set of beliefs are faculty in the humanities and students in undergraduate science and engineering courses. That would only be a curiosity if there weren't such serious consequences: those old ideas make it very hard to think carefully about sciences and technologies and their practitioners and consumers.

My own exposure to the notion that these ideas were inappropriate descriptions of scientific work came from scientists engaged in what later became Nobel Prize–winning work. Excellent scientific, engineering, and medical work is being conducted that can in no way be described with the old litanies. When asked why the old ideas are still so widely circulated, these scientists reply that what they really do is too complicated to explain to the public, students, and even lesser scientists. Besides, they muse, what scientist would want to stop doing research long enough to write about stuff like that? (They are all demure when I ask if they might be benefiting from the extramural circulation of those litanies.)

Research on these human activities called science, technology, and medicine has changed from hagiographies (lives of "geniuses") and lists of miracles ("great discoveries" and "inventions") to careful scrutiny of the practices of those who engage in such work. The research is conducted for many reasons: to improve scientific, technical, and medical education and research; to improve policies concerning the funding and application of scientific, technical, and medical research; to organize more effectively scientific, technical, and medical institutions; to understand ways of producing knowledge that is efficacious; to understand the social and cultural organization of the production, distribution, and consumption of knowledge, locally and transnationally, and so on. This research has been conducted in many countries using a very wide range of inquiry modes and analytic strategies. The primary sites for this research include North America, northern Europe, the U.K., Scandinavia, India, Japan, Australia, and Brazil.

Of course, some of the research has been conducted by people who are hostile to science or ignorant about science, just as some anthropological fieldwork or historical archival work or literary interpretation is conducted by researchers who feel hostility to their subjects of inquiry, and some fieldwork and archival work and cultural interpretation is conducted by people who do not learn much about certain crucial practices of the people they study. This work is usually not too interesting to other researchers. In science, technology, and medical studies as a whole, I would hazard as a guess that 50 percent of the researchers have bachelor's, master's, doctorate, and/or medical degrees in science, engineering, or mathematics. Since many of the readers of this journal probably are involved in bestowing these degrees, we know (approximately) what they do and do not mean. In anthropological fieldwork, we often are not already initiated/socialized/experts in the practices we set out to study; in fact, we often think that this would make it difficult to study the shared assumptions and practices of the group under study. That, of course, does not mean that we remain ignorant about the knowledges regarded as crucial in these communities.

There are, by now, some very widely accepted "findings" of thirty years of research on scientific, engineering, and medical practices.

–There are many practices called "science" by their practitioners, not one such practice; there are many methods called "scientific method" by their practitioners, not one such method. That is, each research subfield has its own distinctive research practices. Hence, the proper terms are plural: *sciences* and *scientific methods*.

–The forms used in scientific writing have converged and have not varied significantly over the last couple of centuries. For example, all references to the agency of the scientists involved in the research is minimized. The written presentation of findings have become quite stylized and terse; it would be almost impossible to reproduce an experiment based upon the information provided in scientific articles. I strongly doubt that an article that fully discloses the complete process of conducting an actual experiment or even a "thought experiment" would be published in any field. The purpose of publishing scientific articles is to announce findings and to lay claim to a discovery, and for that purpose a succinct and formulaic literary economy suffices. In some fields the writing of scientific articles is often assigned to the person in the research group with the least status; the power of the claim is not established by a distinctive or original way of writing. In fact, claims are made in a formulaic mode.

–Access to scientific knowledge is highly restricted. That is, there is restricted access to different stages of training; to funding, positions, publication, conferences—the whole infrastructure of knowledge production/consumption; to networks of active researchers; to tacit knowledge—the crucial craft knowledge that is never written into articles but without which one could never really understand or reproduce an experiment; to the groups that define present and future priorities for problems, methods, research equipment; and to the process of establishing and revising reputations of researchers. There are different levels of access: that is, one can get access to certain levels of training, but not others, get no access, get full access and yet accomplish nothing significant, gain access to research sites, but only as a helper, and so on.

–Problem selection is a process highly subject to the available resources.

–Experimental equipment constitutes signals; scientists adjudicate whether those signals correspond to significant information about the phenomenal world.

—The more capital intensive the research process the less likely the research community is to endorse funding research that replicates other experiments. In the most expensive research there is no replication; in such fields data instead are corroborated. Similar data generated by very differently designed experiments with very different forms of data analysis are taken as especially corroborating.

—Adjudicating which experimental data to take as facts and which theories to take as important is a collective process conducted by those who are tacitly empowered with the authority to participate; it does not include all practicing scientists in the particular field.

—Closure of debates about the status of data and theories is not accomplished with definitive findings as to their truth status, but with a consensus that certain data and/or theories are more useful to more of the practitioners who are entitled to participate in the debate.

—The forms of reasoning conducted in research communities as they interpret the signals from their research equipment recapitulate all the known forms of human reasoning.

—Mathematical analysis is a very limited aspect of research. For example, in my fieldwork during more than twenty years among particle physicists engaged in experimental work, the practitioners usually report that they engage in mathematical analysis an average of about three hours per month. By contrast, there is a great deal of time spent in accurate enumeration and measurement.

—For a few centuries scientific arguments have been probabilistic, not causal; some would say that since calculus became widely used among scientists, their mathematical analyses have been approximations.

—Being conducted and constructed by groups of human beings, scientific, technological, and medical practices and ideas are necessarily social and human. Because those practices and ideas are about the phenomenal world, they often, but not always, also require an engagement with that world. What constitutes a satisfactory engagement with the phenomenal world is necessarily open to debate among the practitioners.

—The definition of science is made by those who are empowered to offer resources for work they consider scientific; for example, the work funded by the NSF, SSRC, NIH, or NIMH is science.

There is more, but this list is already far too long. I merely wanted to point out what has already been asserted, debated, and widely adjudicated to be the case about scientific, engineering, and medical practices by researchers in the fields of sci/tech/med studies. Do some people still disagree with some of these findings? Sure. Does that change the fact that most researchers take these statements as a sort of boring baseline of shared knowledge in the field? No.

Did I rehearse these lists in order to open debate on them here or to provide you with bibliographic essays about any of them? No. Nor was my goal in rehearsing all this to urge us to spend a lot of time trying to get our beliefs about the old litanies in alignment. Discussion about scientific, technological, and medical practices unfortunately have relied too often on formulaic, if cherished, general statements about what "science" is or isn't. So why is it that so many people have such turgid notions about science, engineering, and medicine, often spoken with either an ex cathedra voice or a pounding clenched-fist-in-the-face voice? That is the big, interesting question, appropriate for cultural, psychological, historical, political, feminist, economic, anthropological, and social research. At the moment I want my terse list of what a generation of researchers around the world has learned about those kinds of theories about science to help us get beyond the old scripts. I also hope that the second list will open another sort of discussion.

Choreographies: One Step, Two Steps, Three Steps, Four

I am eager to see us discuss our different modes of inquiry, our different modes of producing ideas, our different modes of adjudicating, our different modes of training, our different modes of problem selection, our different modes of writing, our different ways of making sense, and whether this diversity is interesting. I would like for us to teach this. Certainly, some of us sometimes like to think with coordinates, means, norms, lines, boxes, parts, and categories. I know as many people in the humanities and arts who think like that as I do in the sciences, engineering, and medicine. In fact, we all realize that this mode of thought has quite a history in human affairs. The orders of Le Notre's gardens, the periodic table of elements, St. Benedict's guide for governing monasteries, Kyoto's grid plan, and Descartes's arguments have some obvious parallels. They are spatial and temporal orders; they create grand views and privileged sites; they create remote sites and borders; they create insides and outsides. Whatever doesn't fit into the grid contributes to disorder, to mess. The law of the excluded middle prevails.

Of course, there are some variants within this mode of thought.

First, there are dyads, triads, and quadrants. Just itemizing some crucial instances could keep us all busy for a semester. For openers, here are the usual dyads: objective/subjective, reason/emotion, positive/negative, and good/evil. Then the triads chime in: thesis/antithesis/synthesis, father/ son/holy ghost, Lévi-Strauss's triangles, and induction/deduction/abduction. Booming quads are coming to the fore: north/east/south/west, Cartesian coordinates, not to mention the little analytic boxes built by Mary Douglas, A. J. Greimas, and Jürgen Habermas. Then there are the great charts of hierarchies with bifurcations and branches. Linnaeus gave us great diagrams of that line of thinking, as did Darwin, the kinship theorists, and the decision modelers. Of course, the dyads, triads, quads, and decision trees all have their analogs in poetry, music, painting, sculpture, architecture, dance, and prose. I think it can be great fun to run any old idea through the gamut of twosies, threesies, and foursies, with a finale of Busby Berkeley–style ascending and descending of hierarchical steps.

A singular focus on simplicity, stability, uniformity, taxonomy, regularity, and hierarchy can, of course, be limiting. Furthermore, every way of making sense has its cognate forms of obsession. Certainly, there is an aesthetics of purification that can linger over the ways of the mind and body I described in the last paragraph. Swirling around with Occam's razor, slicing away what cannot be categorized, leaves more than order behind. At this point some of my students always say: "What else is there?" I am always fascinated that they have not been taught any other language for thinking carefully.

Blasphemies

Well, what else is there? To begin, I left out the ones: the singularities and the universals. Just naming them begins to reveal one of our problems. Just how did we get to believing in those peculiar singular generics: science, man, woman, state, justice, evil, god, love, truth, beauty, logic? Why is it, in our time, in our country, in our academies, considered so very blasphemous to add an s to those words? Why is it so horrifying to suggest that we might think more interestingly, and perhaps more carefully, if we stopped, just for a while, using any singular generics. Just saying this in a seminar once led to a philosopher announcing that in the future he would refuse to be in the same room with me.

On another occasion I was asked to be the discussant for a presentation by an eminent anthropologist who is sometimes called a postmodernist. I sometimes like his work very much and took care in preparing my comments. As I read and re-read his paper I thought something unusual was going on in his argument. Using some notation left in my head from mathematical logic courses I began to map his ideas; then I tried using some symbols from philosophical logic classes. After that I explored some techniques of rhetorical analysis I had learned in graduate school. All three strategies gave me some insights but did not locate my vague sense of unease. Then I used a device I had hit on in graduate school. I slowly went through the piece putting a little line through every singular generic and I found the problem. The generics were not randomly distributed; they clustered and they disappeared. They piled up whenever he tried to address certain topics about women. Those topics made his argument disorderly, and singular generics were waved over the mess like a protective fetish. Naively, I suggested that he think about not using any singular generics; I thought it would lead him to strengthen an otherwise very interesting paper. He was not amused.

So what is so sacred about the singular generics and what is so outrageous about wanting to defer them, even if just for a moment? I got a clue in Japan. In Japanese there are no definite and indefinite articles, no *a* and *the* to differentiate *a cat* from *the cat*, and without them one cannot differentiate the singular generic *cat* either. There is no way to distinguish *state* from *a state* or from *the state*; *mass* is the same as *a mass* and *the mass*. As I read the drafts of scientific papers written in English by Japanese physicists I needed to explain to them why their uses of *a*, *the*, and neither *a*/nor *the* seemed inappropriate to me. I could not remember the grammatical explanations for the terms, so I was left trying to explain the differences for physics, but I knew that the Japanese had done perfectly interesting physics for a century without recourse to the singular generic, the indefinite article, and the definite article. Obviously, it was not necessary for science.

All that led me to think about what I could remember about the same distinctions in Latin, Italian, and French, languages that in different decades of my life I have once had some skill in reading. I believed that the grammatical rules about these distinctions are not the same in those languages, but I was not certain. And yet when I returned to the United States, I found people just as annoyed with me for making my plea that we skip the singular generics, that we "just say no" to beauty, truth, logic, science, man, woman, state, justice, evil, love, violence, sex, music, art, poetry. . . . Maybe William Blake was right that god(s) and all the rest are in the particulars.

Monotheism and Morphing

What links cultural, gender, and social studies of science, technology, and medicine of the last thirty years? Certainly, there has been no singular theory, method, or way of defining our questions. What can the empiricists, nominalists, postmodernists, feminist epistemologists, actor-network-theorists, post-Althusser/post-Gramscian Marxists, systems analysts, chaos theorists, discourse analysts, ethnomethodologists, postcolonialists, constructivists, and so forth, among us possibly have in common? Collectively (even though most of us would decry being thrown together for even a moment), that whole generation of research can definitely be said to have dislodged the notion of singularity about science and technology, not to even mention their difference. That is the sin we have committed together.

The other singularities have gone too. European arts no longer set the world standard; the "human condition" is no longer defined in Europe or North America alone. Beauty, truth, and logic have multiplied and dispersed. Some have just begun to notice that the same has happened with science. What is the name of that obsession for singularity and unity, for an order that does not divide, for a world of symbiotic union, for a world that begins and ends with an indissoluble ego? What is the name of the rage against a world of particular plurals? Is it like the rage that some felt against a heliocentric universe or the rage that others felt against a Darwinian world? Why should there be only one way to think well, only one way to have fun with our minds? Why is mental monogamy required? Are we still fighting about monotheism, Manichaean fallacies, and Albigensian heresies?

Does thinking without singularities mean we cannot think carefully about ourselves, other human beings, and our phenomenal world? Not only are we doing it, we already know how to be playful and graceful as we think, dance, and sing about ourselves, the other humans, and our world. What is the name of some of these other ways of making sense? I always look to the students in artificial intelligence, graphics, and music to answer that one: I know that they can already talk powerfully about complexity, composition, instabilities, variations, transformations, irregularities, patterns, morphing, and diversity in performance, research design, equipment design, images, software design, scores, and data analysis. They feel these approaches are just as aesthetically and intellectually compelling as some of those physics students do about equilibrium. The law of the excluded middle isn't always interesting and it doesn't always hold, especially in the best compositions. There are new ways to think within and about our sciences and technologies. Let's dance.

Notes

1   During the last fifteen years there has been a great deal of research on the mutual production of bodies and cultures. See, for example, the bibliographies in Susan Foster, ed., *Choreographing History* (Bloomington: Indiana University Press, 1995), including that for my "Bodies of Evidence: Law and Order, Sexy Machines, and the Erotics of Fieldwork among Physicists." The following is a tiny sampling of these cultural studies of embodied actors: Jean Comaroff, *Body of Power, Spirit of Resistance: The Culture and History of a South African People* (Chicago: University of Chicago Press, 1985); Robbie Davis-Floyd, *Birth as an American Rite of Passage* (Berkeley: University of California Press, 1992); Julia Epstein and Kristina Straub, eds., *Body Guards: The Cultural Politics of Gender Ambiguity* (New York: Routledge, 1991); Brigitte Jordan, "Technology and the Social Distribution of Knowledge," in *Anthropology and Primary Health Care*, ed. J. Coreil and D. Mull (Boulder, Colo.: Westview, 1990); Emily Martin, *The Woman in the Body: A Cultural Analysis of Reproduction* (Boston: Beacon, 1987); Sherry Ortner and Harriet Whitehead, eds., *Sexual Meanings: The Cultural Construction of Gender and Sexuality* (Cambridge: Cambridge University Press, 1981); *Representations*, special issue on "The Cultural Display of the Body" (no. 17, winter 1987); Peter C. Reynolds, *Stealing Fire: The Atomic Bomb as Symbolic Body* (Palo Alto, Calif.: Iconic Anthropology, 1991); Elaine Scarry, *The Body in Pain: The Making and Unmaking of the World* (New York: Oxford University Press, 1985); Michael Taussig, *Mimesis and Alterity: A Particular History of the Senses* (New York: Routledge, 1992); *Women and Performance: A Journal of Feminist Theory*, special issue on "Feminist Ethnography and Performance" (vol. 5, no. 1, 1990).

2   For the best-known positions see Thomas S. Kuhn, *The Structure of Scientific Revolutions* (Chicago: Chicago University Press, 1962), and *The Essential Tension* (Chicago: Chicago University Press, 1977); Karl Popper, *The Logic of Scientific Discovery*, 3d rev. ed. (London: Hutchinson, 1968). For summaries and statements of the other positions see Imre Lakatos and Alan Musgrave, eds., *Criticism and the Growth of Knowledge* (Cambridge: Cambridge University Press, 1970); and Ernan McMullin, ed., *Social Dimensions of Science* (Notre Dame, Ind.: University of Notre Dame Press, 1992). Some of the other participants in the debates of the time include, in addition to the work covered in Lakatos and Musgrave and in McMullin, Paul Feyerabend, *Against Method* (New York: New Left Books, 1975); Norbert R. Hanson, *Patterns of Discovery* (Cambridge: Cambridge University Press, 1958); Gerald Holton, *The Scientific Imagination* (Cambridge: Cambridge University Press, 1978); Robert Merton, *Social Theory and Social Structure* (New York: Free Press, 1957); Norman Storer, ed., *The Sociology of Science: Theoretical and Empirical Investigations* (Chicago: Chicago University Press, 1973); and Derek J. de Solla Price, *Big Science, Little Science* (New York: Columbia University Press, 1963).

   For an introduction to earlier debates about the production of scientific knowledge, see Gaston Bachelard, *La formation de l'esprit scientifique: Contri-*

bution à une psychanalyse de la connaissance objective (Paris: Librarie Philosophique J. Vrin, 1980); Pierre Duhem, *La chimie; est-elle une science francaise?* (Paris: Hermann, 1916) and *German Science: Some Reflections on German Science: German Science and German Virtues*, trans. John Lyon (La Salle, Ill.: Open Court, 1991); Georges Canguilhem, *A Vital Rationalist: Selected Writings from Georges Canguilhem*, ed. François Delaporte (Cambridge, Mass.: MIT Press, 1994); and Ludwig Fleck, *Genesis and Development of a Scientific Fact* (Chicago: Chicago University Press, 1979). Some would add Emile Durkheim's reflections on "Science as a Vocation," in his *Selected Writings*, ed. and trans. Anthony Giddens (Cambridge: Cambridge University Press, 1972), although in my opinion it only invokes the conventional litanies.

For a compilation of the kind of pragmatic research about science, technology, and medicine during the 1950s, 1960s, and first half of the 1970s, see Ina Spiegel-Rosing and Derek de Solla Price, eds., *Science, Technology, and Society: A Cross-Disciplinary Perspective* (Thousand Oaks, Calif.: Sage, 1977).

For an introduction to the cultural, social, and gender studies of science, technology, and medicine developed since the mid-1970s, see Traweek, "An Introduction to Cultural, Gender, and Social Studies of Science and Technology," *Journal of Culture, Medicine, and Psychiatry* 17 (1993): 3–25, in a special issue on "Biopolitics: The Anthropology of the New Genetics and Immunology, edited by Deborah Heath and Paul Rabinow. See also Stanley Aronowitz, ed., *Technoscience, Power, and Cyberculture: Implications and Strategies* (New York: Routledge, in press); Laura Nader, ed., *Naked Science: Anthropological Inquiry into Boundaries, Power, and Knowledge* (New York: Routledge, in press); Constance Penley and Andrew Ross, eds., *Technoculture* (New York: Routledge, 1991); and Andrew Pickering, ed., *Science as Practice and Culture* (Chicago: University of Chicago Press, 1992).

3   Jacques Derrida, *Of Grammatology* (Baltimore, Md.: Johns Hopkins University Press, 1976).
4   See note 2.

Making Transparencies: Seeing through the Science Wars

Sarah Franklin

It is a logic of transparency that underlies the introduction of the
fetal image into pro-life political activities. —Faye Ginsburg, *Contested Lives*

Introduction

IT is no coincidence the Science Wars have erupted in British and
American academic circles close on the heels of the Culture Wars
preceding them: they are both about cultural values. The politics
of knowledge at stake in both chapters of intellectual change are the
same. In both skirmishes, the foundations of scholarly inquiry have
been subjected to critique on the basis of their purported universality,
formulaic absolutism, and exclusivity.

These challenges were inspired by several converging trends. One
is the rise of poststructuralist, deconstructionist, psychoanalytic, and
postmodern theory. These are the so-called traveling theories identified by Spivak, through which a decentering of many of the givens
that previously structured assumptions about knowledge, its objects,
and its subjects is effected. The new interdisciplinary fields of feminist theory, postcolonial theory, critical race theory, and queer theory
are all areas of contemporary scholarship derivative of a move away
from previous conventions of objectivity, neutrality, and canonical
tradition. One way to define cultural studies in the United States is
simply as the space in which discussions motivated by all the above
changes can take place. It is both a popular and a contested arena. In
general, the U.S. academy tends to observe strict disciplinary boundaries, whereas in Britain disciplinary boundaries are often more dif-
fuse. In both countries, some disciplines operate as more of a "closed
shop" than others. The significant decrease in public funding for
higher education in both the United States and Britain during the past
ten years has had the tendency to exacerbate tensions between the

post- and interdisciplinary movements and their more discipline-bound corollaries.

All of this has raised the suspicions of traditionalists, who see in the rise of multiculturalism, ethnic and women's studies programs, cultural studies programs, and the anticanonical stance of many prominent contemporary intellectuals the decline of the American and British university systems. These traditionalists fear the loss of standards, conventions, tradition, continuity, and moral values that they believe to be essential to democracy, liberal humanism, rational inquiry, or simply the nation itself.

More recently, this same pattern of critique and countercritique has begun to gain momentum in the context of the Science Wars. Long enshrined as a kind of apex of rational knowledge production, so powerful as to remain largely immune to the vicissitudes of social change, science is now up for deconstruction just like all the rest of the Western canonical fare.

Scientists have become very agitated about this development. They, like the rest of the academy, are experiencing a downsizing in terms of public funding for research, and this fuels their anxiety. E. O. Wilson (1994) put it bluntly: "Multiculturalism equals relativism equals no supercollider equals communism." This is the central dogma for critics such as Gross and Levitt (1994), who argue that the academy is suffering from a "leftist infestation" they describe as "perspectivism": the idea that everything is merely a point of view. Indeed, critics such as Donna Haraway (1988) have explicitly advocated that knowledge be understood as always-already situated (somewhere), and that even objectivity itself is a located perspective. Haraway argues that this situatedness has consequences, and that these consequences must be taken account of in the deployment of knowledge. But this is precisely the view that has caused such a furor among scientists who view such an argument as incompatible with the search for unmediated knowledge about the natural and the physical world.

As Haraway and many other feminist science studies scholars have argued, the privilege accorded the value of objectivity is based heavily on a conflation between seeing and knowing. Detached observation, accurate description, and value neutrality in the pursuit of understanding are the key components of the objectivist epistemology at the heart of modern science (Jordanova 1989; Stafford 1991, 1994). Central to this constellation is the will to transparency, the idea that things can be known in and of themselves through a method of observation and description that does not leave a mark upon its objects:

they are to be rendered transparently (Phelan 1993). As Latour (1987) has argued, nothing is ever strictly transparent in the context of scientific investigation, which depends on a representational and interventional apparatus with its own codes of visibility or clarity. Measurement, instrumentation, recording and monitoring devices, even the conceptual structures guiding data acquisition all bear the marks of particularity rather than of transparency (Hacking 1983). But the will to transparency refuses such critical deconstructions. The value of objectivity is what anthropologists might describe as a "symbol that stands for itself": it is as much an article of faith as a rational precept.

The guarantor of this faith is its instrumentalism, its practice, its efficacy. If science isn't "true" because it is "objective," its power is nonetheless manifest in *what it does*. Richard Dawkins (1994, 17) expressed this view when he declared in the pages of the Higher Education supplement of the *London Times*: "Show me a relativist at 30,000 feet and I will show you a hypocrite." The power of science, of scientific objectivity, of the experimental method, of rational empiricism, is that *it can do things*. In turn, these accomplishments authenticate their origins through the very power of their presumed self-evidentness. But what is so self-evident about the fact that planes can fly? This feat could as easily be described as sophisticated tool use instead of as an indicator of epistemological certainty. And as certain as the fact that planes can fly is the fact that their design is constantly evolving, that experiments such as that of the Wright brothers have as much to do with desire as with established scientific principles, that some airlines show prayer films during take-off to invoke the aid of Allah in order to remain safely airborne, and so on.

The will to transparency is founded on the desire to remove things from their context. To know a thing in itself is the equivalent of radically decontextualizing it. The scopic instruments of scientific investigation depend upon the limitation they impose on their field of vision: the microscope, the telescope, the laparoscope, the endoscope. The effect is of looking through a toilet paper tube: it is an effect of being radically blinkered. Disciplinarity is defined by what is excluded, and so is detached, objective observation. The idea is that you see through a tunnel to the thing itself, the object to be known, the entity which is "thrown before" the gaze, as the etymology of "object" suggests. Both tunnel vision and disciplinarity rely upon not seeing "the rest of the picture" so it does not distract from the object in question. Like objectivity, disciplinarity is about screening things out.

Put this way, it is very obvious why both objectivity and disciplinarity have been keyed to power effects: insofar as they rely upon exclusion and the establishment of a privileged, partial perspective, they operate like spotlights highlighting the main event. It is the hypocrisy of asserting that so partial a perspective *is not a perspective* that attracts the critical eye of other observers. The history of knowledge practices—especially authoritative ones such as science—is a history of perspectives.

There is never only one perspective, and theorists of knowledge as a form of perspectivism point this out quite empirically. Marilyn Strathern (1992a, 3), for example, whose insightful anthropological reading of Euro-American knowledge practices as cultural effects is certainly among the most elegant formulations on this subject to date, argues that "culture consists in established ways of bringing ideas from different domains together." Culturally, understanding is not composed of single perspectives but of their contrast *and their relations to one another.* A technical term for this is *analogy*, and much has been written on the history of the use of analogy in science to direct its inquiry. Donna Haraway's early work, for example, on the history of modern embryology, tracks the competing analogies at stake in the study of embryonic morphogenesis. Mechanism, vitalism, and organicism depended upon the use of specific analogies, for which her book is entitled: *Crystals, Fabrics, and Fields* (1976). The work of these analogies is not merely metaphoric, like so much window dressing. Analogues are central to both the information and the life sciences: they are tools of reasoning. Analogy itself is defined as a form of logic, and in science it is a powerful means of directing inquiry.

My contribution to this special issue of *Social Text* is also guided by analogy. The analogy is taken from Gross and Levitt's *Higher Superstition* (1994), which is the main text critiqued in this account. In defense of the scientific way of knowing, Gross and Levitt invoke an analogy to life itself: it is the unmediated character of the relation between science and its objects, they argue, which is its "lifeblood"(17). I argue here that this is no rhetorical flourish but rather a core analogy to their defense of science. Through a brief historical anecdote on the relation between genius and generativity, I explore this conflation of knowledge with reproductive substance and argue that a particular reproductive politics informs their work. In turn, I make the claim that their pro-life defense of science is not unrelated to other Right-to-Life campaigns in current American politics, particularly in its reliance on a notion of transparency.

Vital Principles

For Gross and Levitt, the privilege of scientific perspective derives from its unmediated relationship to reality. They have no doubt that "reality" is "out there" as an "ocean of truth" awaiting scientific discovery. Would that it *were* socially constructed, Gross and Levitt propose, for if reality could be bought, influenced, pressured, or bullied into toeing the line, science would be a wholly different undertaking! In their view, it is precisely the fact that scientific knowledge *cannot* be imposed on reality that created the inherent, progressive virtue of the scientific enterprise. And reality, in their view, is far from passive. Using a string of animated analogies, they argue that reality is "the overseer at one's shoulder, ready to rap one's knuckles or spring the trap into which one has been led by overconfidence"; it is "the unrelenting angel with whom scientists have to struggle." It is reality that "creates the pain as well as much of the delight of research" and holds "even the boldest imagination" hostage to its authority (234). Embedded in this description of the reality principle as they see it existing for scientists is the image, as Phelan (1993, 3) describes it, of the "Real-real" as a powerful force, or even as a domineering figure. It is the reality principle that keeps scientists on their toes. It keeps them honest, modest, and true. It brings them pain and it brings them pleasure. Reality is a disciplinarian: "an overseer" who is "unrelenting." Reality's authority is absolute, even divine; it is "an angel" in their midst.

What is most important about this complex relationship, according to Gross and Levitt, is its unmediated character. Science may be "influenced" by culture and politics, but it is ultimately a "reality-driven enterprise" (234). This is the bottom line—the point past which the vicissitudes of particular location of identity do not hold. Reality is Reality is reality, and, as the anthropologists say, "it's turtles all the way down." In Gross and Levitt's view, the character of this relationship is not one of give-and-take, but rather of right-or-wrong: when scientists get it right, they are able to match the nature of reality with the precepts of rationality, and this is the distinctive power of the scientific enterprise. It is in turn the nature of this relationship through which scientific knowledge is depicted by them as a source of virtue: there is no playing fast and loose with the Real-real. "God does not play dice with the cosmos," they insist. And, "since individually and collectively we are not God . . . we must inevitably regard the universe as a kind of crap game" (262).

Such statements explain the intolerance of Gross and Levitt, among others, for the claim that science is a cultural or political endeavor. The stern taskmaster that is their figuration of the Real-real ultimately holds the cards, and the best scientists can do is to read them with greater or lesser precision. Science is not a game scientists can ever "win": it is only one they can get better at playing.

The point at which science studies scholars and traditionalists like Gross and Levitt part company is in the matter of what difference it makes to construct knowledge as relational. No amount of rational argumentation, historical documentation, or cultural interpretation is capable of dislodging their view *because it is ultimately one that positions knowers as less powerful than the reality they describe.* At the same time, they argue, scientists have a privileged access to this reality because of the knowledge practices they employ—and, as they see it, it is this relation which has a life of its own.

This is not the view of either knowledge or of its objects espoused by many cultural theorists, for whom knowledge is relational in a much broader sense, invoking different relations of power and authority. For cultural critics too, knowledge may be measured by what it can do, by its power to inspire, to convince, to bring about change, or to defend a point of view. That knowledge relations are fixed in an authoritarian, hierarchical relation of knower to known, a relation which itself exists in some hermetic dimension of unlocated, disembodied, and acultural time and space, can only be accurately described as a very particular way of understanding what it means to know. Above all else, the view that knowledge is only authenticated by a distant, knuckle-rapping, cipherlike agency, with powers commensurate to those of a divinity, can only be described as itself a perspective. According to Gross and Levitt, this construction cannot be interrogated as a metaphysical, cultural, or historical question, which would, by definition, be unscientific and thus insubstantial. Effacing their own perspectivism, science defenders describe their own knowledge practices as the only route of legitimate access to the empirical reality they describe, thereby reentering the loop in which a presumed definitional function of empiricism affirms its own tenets. That there are many forms of empiricism is itself an empirical fact, but not one that makes sense from Gross and Levitt's or Dawkins's perspective. What is most left out of their account is precisely what defines it, namely, a specific perspective on the relation between knowledge and its object. "Perspective," as Phelan (1993, 24) points out, "is essentially a theory of relationships."

Contested Conceptions

Untrammeled by a holy ghost analogous to Gross and Levitt's unrelenting angel of authority, cultural studies scholars move in a different "ocean," more accurately described as fluidly metaphoric than as filled with "truth." "Truth" for most cultural theorists is a contextual entity: it comes in many varieties. As Adrienne Rich (1987, 27) writes of truth, "There is nothing simple or easy about this idea. There is no 'the truth,' 'a truth'—truth is not one thing or even a system. It is an increasing complexity." A different way to consider the arguments presented by the *indignati* of the Science Wars is to invoke a comparison.

The comparison that comes to mind is of a different field of contestation—of reproduction, and of reproductive politics. As Zoë Sofia (1984, 48) so wisely put it, "Every technology is a reproductive technology," and this must also be said for disciplinarity. Gross and Levitt know this all too well. The reproductive implications of the peril they perceive are quickly and literally rendered. Early in the book they warn that "the manifestation of a certain intellectual debility afflicting the contemporary university" threatens "the future of our descendents and, indeed *of our species*" (1994, 7; emphasis added). *Species* is a reproductive term, for species are defined, in large part, by their reproductive borders. The threat to our species posed by the "attacks on science that grow out of a doctrinaire political position" are, significantly "misconceived." The attacks are the result of ignorance—an "ignorance even more profound" than the "self-satisfied ignorance" of the classicists and historians "excoriated" by C. P. Snow at midcentury. This ignorance threatens not merely the scholarly credentials of future generations but our species itself.

Reproductive imagery is liberally sprinkled through Gross and Levitt's text. The aforementioned "ocean of truth" available to science is described as its "lifeblood," its "vital principle." In turn, this uncontainable vitality brought forth "the birth of Western science" and likewise "the birth of its prestige as a uniquely reliable and accurate way of describing the phenomenal world" (17). These and other images convey an organismic impression of knowledge practices that is consistent with a long-standing association between epistemic and organic conceptions.

Historically, such "conceptual" associations have often been remarked on in relation to the notion of scientific "genius," long associated with the "seminal" arguments of the "founding fathers" of the "hard" sciences. The history of exclusion of women from science on the grounds of the "polluting" influence of their reproductive capacity is also well documented (Schiebinger 1989, 1993). Academic life remains terminologically heir to such traditions, with its bachelor's and master's degrees, its seminars and testimonials, and its other androcentric remnants. The convention of copyright, protecting the property of the author's individual inspirations, is modeled on the proprietary rights of other paternal emissions, and authorship is itself closely indexed to paternity. Christine Battersby, in her illuminating study, *Gender and Genius* (1994), describes this as the Virility School of Creativity, and its reference is clearly procreative.

This reference has an ancient pedigree, documented by classicists and historians across two millennia. Richard Onians offers an intriguing portrait of the ancient Greeks' ideas about creation and conception:

> There is varied evidence that the head was holy with potency by which to swear and make appeal and was thought to contain the life. . . . Homer and his audience knew that during life and for some time after death the brain is a fluid mass. . . . It had nothing to do with ordinary consciousness (perception, thought and feeling being the business of the chest and its organs), but instead was the vehicle of life itself, of that which continues and does not die. But life does not persist in the individual; it issues forth. This is the greatest miracle, the holiest mystery. . . . It was natural and logical to think that the "life" issuing from a man must come from the "life" in him, from his head therefore, and, helping that location, to see in the seed, which carries the new life and which must have seemed like the very stuff of life, a portion of cerebro-spinal substance in which was the life of the parent. It will indeed appear that this interpretation of the cerebro-spinal substance as seed is vital to the whole thought. (1951, 109–10; Greek terms omitted)

Referring to changes in the latter half of the fifth century B.C., Onians describes the importance of popular beliefs about the "mindful generative marrow," which was applied to the brain and its fluid as well as to that of the other bony cavities. "For Democritus the 'life' was bound and rooted in the marrow," especially the "divine marrow" of the head. "The seed is enclosed in the skull and spine and explicitly identified with marrow, or, as it is once called, 'generative marrow,' and flows thence in the propagation of new life. It breathes through the generative organ. This appears to be original popular belief" (119; Greek terms omitted).

For the Romans, as H. J. Rose (1923, 135) argues, "the genius is the life, or reproductive power, almost the luck of the family. . . . It is one, and one only, for each *gens*." This leads Onians to conclude that the Greek conceptions of life and mind directly influenced those of the Romans.

> It has been generally recognised that [the Roman view of the head] was used as loosely equivalent to "life". . . . I suggest what does not appear to have been suspected (is) that it was thought to contain the seed, the very stuff of life, and the life-soul associated with it. . . . The use of our *caput* (our 'capital') for money which produces interest can now be seen to arise from this thought; 'interest,' what the 'head' produces, was 'offspring,' *fenus* which the Romans rightly thought to have the same origin. (105–6, 124)

Speculative though such classical accounts may be, the broad parameters of connection between genius, generation, gender, and the *gens* are clearly established, as their etymological proximity in English already makes evident. The links to modern English usage in relation to blood are also evident in antiquity, as they are in Old English.

These linkages have a number of implications not only for the gendering of knowledge as creativity, linked to the notion of the male seed, but also to definitions of kinship and property. In turn, the importance of the concepts of genealogy and consanguinity—so essential not only to Darwin as a means of unifying nature, or life itself, as a system of descent, but to Morgan, Frazer, Taylor, and Maine as a means of theorizing social organization along evolutionary lines—becomes apparent. The definition of species itself, of *Homo sapiens sapiens*, of "modern man," shares this derivation. "If *sapere* meant originally 'to have sap, native juice,' and was applied to the chest, that accords perfectly with what we shall find, that the Greeks and Romans related consciousness and intelligence to the native juice in the chest, the blood" (Onians 1951, 62–63).

This "juice" has been significant to more than genealogy. As the earlier reference to capital suggests, it has also had important implications for definitions of property. Writing of the origins of copyright in the early eighteenth century, Mark Rose describes the protestations of Daniel Defoe against the "plagiarists and pirates" who traded in the (then) unprotected literary works of authors such as himself. Building on the Latin meaning of *plagiary* as kidnapping, Defoe equated such literary appropriations with child stealing:

A Book is the Author's Property, 'tis the Child of his Inventions, the Brat of his Brain; if he sells his Property, it then becomes the Right of the Purchaser; if not, 'tis as much his own, as his Wife and Children are his own—But behold in this Christian Nation, these children of our Heads are seiz'd, captured, spirited away, and carry'd into Captivity, and there is none to redeem them. (cited in Rose 1995, 1)

As Rose notes, this analogy rests on the likeness between "own" in the kinship sense (my own flesh and blood) and "own" in the sense of property, as in "ownership." What is also evident in these debates occasioning the "birth" of literary property is the importance of the analogy to paternity, and its equation with propriety, in the older sense of rights to property. It is the relation of father to child, modeled on the Judeo-Christian (or Greco-Roman) concept of paternity as *begetting*, which is, according to Rose, "the most common figure in the early modern period" to represent the relation between authors and their texts.

The same pedigree applies for science. From the Aristotelian model of "the seed and the soil" has developed a consistent tradition of privileging the generative power of the male seed, thus uniting procreative acts with the power to create in general (see Delaney 1986, 1991; Laqueur 1990). As Rose points out, William Harvey expressed this principle again in his famous refutation of the Aristotelian model of conception, which had held sway more or less intact for two millennia. In "Disputations Touching the Generation of Animals," based on his experiments with the King's Royal Deer in England in the mid–seventeeth century, Harvey drew an analogy between artistic and biological conception: "The generation of things in Nature and the generation of things in Art take place in the same way. . . . Both are first moved by some conceived form which is immaterial and is produced by conception" (cited in Rose 1995, 5).

As these historical examples make clear, it is not "merely metaphoric" to speak, as Gross and Levitt do, about the "lifeblood" of the scientific endeavor, or about its "vitality." Invoking reproductive substance to do service for the inviolability of knowledge practices and depicting this potency as the lifeblood of the scientific enterprise, the protection of which is essential to species survival, is not to make an idle comparison. It is rather to invoke a well-established Western idiom of creation, of which the examples given here provide no more than the most cursory of indications. Gross and Levitt's descriptions of science studies arguments as misconceived are consistent with this same tradition, as are their references to the "degeneracy" that will ensue if such criticisms take greater hold. It is only a small step from such claims to the "future of the species"; as the very notion of the species—the continuity of the germplasm, the consanguinous universe that is the post-Darwinian model of Nature or life itself—is based on the same elementary analogies. From this point of view, the threat to knowledge-as-we-know-it, to its potency and propriety, is rightly perceived as a reproductive threat, and overdeterminedly so.

Transparent Values

The relevance of a comparison to other contemporary American reproductive struggles becomes apparent through the explicit representation of the Science Wars as a vital matter of reproductive rights and wrongs. In her now-classic ethnographic monograph on the abortion struggle in the United States, Faye Ginsburg (1989) presciently noted the importance of the fetal image as a symbol in the war against abortion. The aggressive deployment of this image by antiabortion campaigners since the early 1970s is based on the premise that if women had accurate (scientific) knowledge of fetal development, they would not choose abortion. As Ginsburg notes, "A popular quip summarizes this position: 'If there were a window on a pregnant woman's stomach, there would be no more abortions.' It is a logic of transparency that underlies the introduction of the fetal image into pro-life political activities" (104). Specifically, Ginsburg notes that "the Right-to-Life belief that conversions will take place after seeing the 'truth' about abortion relies on the way knowledge of the fetus is constructed" (105). In other words, the "truth" about the fetus which will have the desired effect of "converting" new members to the Right-to-Life position has to be constructed in a particular way so as to appear logically as well as visually "transparent." Ginsburg describes the formal dimensions of pro-life imagery constituting their central moral and political message:

> Right-to-Life visual material offers two representations of the fetus that are continually "active together." The principal subject is the magnified image of the fetus—for example, floating intact inside the womb, or focussed on tiny, perfectly formed feet held between the thumb and forefinger of the adult. These pictures are usually in warm, amber tones, suffused with soft light, rendered more mysterious by their separation from the mother's body. Jux-

taposed to these photographs are gruesome, harshly lit, clinical
shots of mutilated and bloody fetal remains "killed by abortions"—what pro-life activists refer to as "the war pictures." (150)

As Ros Petchesky (1987, 286) argues, the fetal form has acquired a
"symbolic import that condenses within it a series of losses—from
sexual innocence to compliant women to American imperial might. It
is not the image of a baby at all, but of a tiny man, a homunculus." The
message of the Right-to-Life fetal imagery is clear, and its aim is to
convert. "Once a potential convert witnesses a certain 'truth' and
comes 'under a conviction,' there is only one path to follow" (Ginsburg 1989, citing Harding 1987, 105).

Like the unmediated truth sought by scientists (according to the
model of knowledge espoused by Gross and Levitt), the relation of the
American public to Right-to-Life fetal imagery is premised on transparency, on the removal of interfering context (the woman's body),
and on the direct apprehension of "truth" by modest, righteous witnessing. Like Gross and Levitt's relentless "angel," this truth is a dis-
ciplining force: it is simultaneously awesome, mysterious, and "clear."
The point is that the relationship between the knower and the "reality" to be known produced in both contexts, both of which hover
between secular and religious experience, is identical. Barbara
Duden, in her cogent discussion of the abortion debate in Germany,
uses the term *visualization-on-command* to describe this means of
apprehension. Structured by the rigid dictates of specific representational practices, the eye is disciplined to "see," as it were, "what we
are shown" (1993, 17).

Conclusion: The Politics of Perspectivism

What is at stake in the politics of making transparencies is the politics of perspective. It is not that science assumes a perspective
which is seen to be iniquitous or false by science studies scholars. It
is instead the denial that such a perspective is a perspective which is
seen to carry the most significant political and moral consequences,
for it is this denial that attempts to render invisible and inaccessible
to scrutiny or questioning exactly how that perspective works, what
it includes and excludes, and how that inclusion or exclusion is itself
a cultural effect. It is not inappropriate for scientists to rely upon disciplined conventions of apprehension or analysis. It is not odd or out
of place that they should assume a certain position in relation to the
object of this gaze, embedded as it is in at least two centuries of ded-

icated refinement of such observation techniques. It is, as Gross and Levitt claim, in fact very impressive what such a relationship can do, what it can produce, and what it can enable its practitioners to deduce about the world. I, for one, am in no need of convincing on the point that there is "something special" about scientific ways of knowing. This point is particularly well argued by John Moore in his book about the history of the life sciences, entitled *Science as a Way of Knowing* (1993). But this title says it all: science is *a* way of knowing. It is not necessarily *the* way of knowing by which everything under the sun should be judged for all time because it "works" for scientists. It does some kinds of "work" as a form of understanding; it most certainly does not do other kinds. And, most important, the fact that this perspective depends on an analogy of transparency does not mean that it is itself so self-evidently "clear" as to need no further explanation.

Gross and Levitt accuse their critics of a panoply of evils: from envy, arrogance, and "sheer puffery" to weak-mindedness, debility, and ignorance—and much else in between. To correct the shortcomings of their critics, whose work is most often dismissed by them as merely "silly," and which, by their own admission, they often do not understand, they propose two basic principles. One is that they are not prepared to accept any discussion from nonscientists about the scientific enterprise until those critics know science "in itself," that is, as a scientist would. This principle, as stated in their chapter on feminist critiques of science, reads as follows: "*We would have to be shown that there are palpable defects due to the inadequacies of male perspectives, in here-to-fore solid-looking science and that the flawed theories can be repaired or replaced by feminist insights*" (112; original emphasis). The principle here expressed is that the *only* criterion for evaluating science studies is its ability to improve existing science. It thus rigidly excludes any attention to the conventions that structure scientific criteria, or the relationships constituted by scientific perspectivism. It insists on working *within* criteria as the *only* means of assessing their validity.

The second principle elaborates upon the first, in the form of their advice to social scientists:

> What we wish to emphasize . . . is that the underlying strategy that guides the intellectual enterprises of Smith, Diderot, Locke, Gibbon, Herder, Hume, Jefferson, and (what was until recently) a pantheon of others remains as an ongoing tradition that is unlikely

to disappear within the imaginable future. This is simply, in its most naked [*sic*] form, the strategy of taking the social order, per se, as the object of one's critical investigations, seeing it as describable, in large measure, on the basis of discoverable first principles. It is to be implemented by combining careful and exhaustive attention to solid empirical fact with the construction of a more or less rigorous deductive model.

At their very best, such theories yield chains of propositions which themselves may be regarded as confirmed insights into the social organism, or as tentative hypotheses to be tested in the hard world of experience, as a trial of soundness of the fundamental postulates of theory. (18–19)

Scientific methodology, as these two principles suggest, is not merely for the "hard" sciences. Social scientists too should reconstruct their object of inquiry as an "organism" and their method as the hypothetico-deductive model derivative of the experimental method. Insofar as either social scientists or science studies scholars do not conform to these dictates of the reality principle hypostasized by science, their contributions are not only illegitimate and degenerate, but unscholarly and venomous.

To police and maintain these principles within the academy, Gross and Levitt exhort their scientific colleagues to abandon the gentle tolerance and openness they have shown in the past, and to wake up, smell the coffee, and get involved in tenure decisions. Don't be afraid to speak up, they encourage their would-be converts. Take a stand against the fatuous, self-inflated pomposity of the antiscience zealots, with their glamorous self-presentation, their "cult status" shenanigans, and their dubious pretensions to have anything to say about scientific subjects they clearly do not understand.

The righteousness of this invective, its firm conviction as to the one, true path before us, and its depiction of our species, our nation, and our technologically advanced way of life as threatened by the totalitarian aspirations of the "absolute relativists" clearly comprise, as Gross and Levitt are fond of saying, a species of reactionary demonization. Like Gross and Levitt's origin story for the science troubles, which roots the current generation of "leftist academics" in a state of petulant denial over the failure of social movements of the 1960s and the decline of socialism internationally, these criticisms are as knee-jerk traditionalist as they are fruitlessly ad hominem.

With their "war pictures" out in front of abortion clinics, the Right-to-Lifers continue to wage a battle against the legal right to abortion, which remains a legal option for American women. With the conviction of their singular perspective, and their through-a-toilet-paper-tube cameos of still-life fetal forms, they assert transparency as truth from the standpoint of a radically decontextualized claim on the realities of reproduction. As Phelan (1993, 30) notes, "By displaying the fetus as the single image within the triangulation of reproduction, Operation Rescue attempts to ignore the dilemma of the pregnant woman entirely and to leave unmarked the freedom of (invisible) paternity." But like other forms of truth, the realities of reproductive decision making are not so clear. There is not a singular perspective on pregnancy, any more than there is a singular perspective on anything else. Making the womb "transparent" does not make the problem of unwanted pregnancy disappear.

Gross and Levitt's *Higher Superstition* is a "war picture" for their fellow academicians. It too is a form of Operation Rescue based on the trope of imperiled progeny. Through a careful juxtaposition of imagery aimed at making their message perfectly clear, they exhort their compatriots to rally to the cause of a singular path. Like other conversion stories, theirs is based on a lengthy testimonial of witnessing the threat within. Like Randall Terry and the Operation Rescue campaigners, Gross and Levitt espouse a paternalistic Right-to-Life discourse concerning the vital essence of the scientific ethos, and the importance of its salvation on behalf of our children's futures. They are confident that the will to transparency is the only source of knowledge "hard" enough to see our universities, our nation, and our species safely across the millennial divide, and beyond. Their certainty is especially impressive given that paternity has never been the celebrated context for exactitude or conviction that certain other knowledge practices are said to offer in its stead.

References

Battersby, Christine. 1994. *Gender and genius: Towards a feminist aesthetics*. London: Women's.

Dawkins, Richard. 1914. *Times Higher Education Supplement*.

Delaney, Carol. 1986. The meaning of paternity and the virgin birth debate. *Man* 21: 494–513.

———. 1991. *The seed and the soil: Gender and cosmology in Turkish village society*. Berkeley: University of California Press.

Duden, Barbara. 1993. *Disembodying women: Perspectives on pregnancy and the unborn*, translated by Lee Hoinacki. Cambridge, Mass.: Harvard University Press.

Gilbert, Sandra M., and Susan Gubar. 1979. *The madwoman in the attic: The woman writer and the nineteenth-century literary imagination.* New Haven, Conn.: Yale University Press.

Ginsburg, Faye. 1989. *Contested lives: The abortion debate in an American community.* Berkeley: University of California Press.

Gross, Paul, and Norman Levitt. 1994. *Higher superstition: The academic left and its quarrels with science.* Baltimore, Md.: Johns Hopkins University Press.

Hacking, Ian. 1983. *Representing and intervening: Introductory topics in the philosophy of natural science.* Cambridge: Cambridge University Press.

Haraway, Donna. 1976. *Crystals, fabrics, and fields.* New Haven, Conn.: Yale University Press.

———. 1988. Situated knowledges: The science question in feminism as the site of discourse on the privilege of partial perspective. *Feminist Studies* 14, no. 3: 575–99.

Harding, Susan. 1987. Convicted by the Holy Spirit: The rhetoric of fundamental Baptist conversion. *American Ethnologist* 14: 167–81.

Jordanova, Ludmilla. 1989. *Sexual visions: Images of gender in science and medicine between the eighteenth and twentieth centuries.* Madison: University of Wisconsin Press.

Laqueur, Thomas. 1990. *Making sex: Body and gender from the Greeks to Freud.* Cambridge, Mass.: Harvard University Press.

Latour, Bruno. 1987. *Science in action: How to follow scientists through society.* Cambridge, Mass.: Harvard University Press.

Moore, John. 1993. *Science as a way of knowing: The foundations of modern biology.* Cambridge, Mass.: Harvard University Press.

Onians, Richard. 1951. *The origins of European thought: About the body, the mind, the world, time, and fate.* Cambridge: Cambridge University Press.

Petchesky, Ros. 1987. Fetal images: The power of visual culture in the politics of reproduction. *Feminist Studies* 13: 263–92.

Phelan, Peggy. 1993. *Unmarked: The politics of performance.* New York: Routledge.

Rich, Adrienne. 1987. *Women and honor: Notes on lying.* London: Onlywomen.

Rose, H. J. 1923. On the original significance of the genius. *Classical Quarterly* 17: 133–43.

Rose, Mark. Forthcoming. From paternity to property: The remetaphorization of writing. In *Cultural agency/Cultural authority*, edited by Martha Woodmansee and Peter Jaszi. Durham, N.C.: Duke University Press.

Schiebinger, Londa. 1989. *The mind has no sex: Woman in the origins of modern science.* Cambridge, Mass.: Harvard University Press.

———. 1993. *Nature's body: Gender in the making of modern science.* Boston: Beacon.

Sofia, Zoë. 1984. Exterminating fetuses: Abortion, disarmament, and the sexo-semiotics of extra-terrestrialism. *Diacritics* 14: 47–59.

Stafford, Barbara. 1991. *Body criticism: Imaging the unseen in Enlightenment art and medicine.* Cambridge, Mass.: MIT Press.

———. 1994. *Artful science: Enlightenment entertainment and the eclipse of visual education.* Cambridge, Mass.: MIT Press.

Strathern, Marilyn. 1992a. *After nature: English kinship in the late-twentieth century*. Cambridge: Cambridge University Press.

Wilson, E. O. 1994. Personal reflections on a life in science. Public lecture presented at the History of Science Society Annual Meeting, Forum for the History of Science in America, Distinguished Scientist Series, New Orleans, Louisiana, 14 October.

Gender and Genitals: Constructs

of Sex and Gender

Ruth Hubbard

AT this time, both science scholars and practicing scientists are trying to sort out our concepts of nature and science and the relationships between the two.[1] On all three topics a wide range of opinions exists, and all of them are passionately defended. Most science scholars and some scientists acknowledge that our concepts of "nature" of necessity reflect cultural beliefs about the way human beings fit into and interact with nature. We stress that cultures can look on nature as inimical, friendly, neutral and uncaring, parent (mother), resource, deity, and so on. Indeed, contradictory combinations of such beliefs often coexist within the same culture and even the same person. In addition, we take it for granted that, in the effort to understand nature, the ideological commitments of science as well as its conceptual and practical tools reflect such underlying beliefs. These beliefs determine what is interesting and important enough to investigate and what is trivial and not worth attending to. Both science's holy grails and its garbage pails have their origins in the culture at large.

No matter how difficult scientists find it to reconcile the view that science is culturally grounded with more traditional, positivist notions about the progress of science, what enrages them even more is the so-called strong program in the sociology of science. It is true that, in its extreme form, this social analysis makes "nature" disappear amid scientists' jockeying for positions and power: trying to understand nature becomes the excuse that makes it possible for scientific centers of power to be established, to flourish, or to die out. Like other human activities, the argument goes, science is a struggle to accumulate recognition and status, and in science this goal is best achieved by turning "facts" about nature into scientific papers which in turn can be brokered into university appointments, membership in professional societies, and nationally and internationally recognized honors and prizes. In this extreme view, and especially in the caricature of it which positivist scientists have lately presented, the extent to which science does, or does not, reflect the "truth" about nature is irrelevant to the enterprise at hand.[2]

As a feminist as well as a scientist, I am keenly aware of our embodiment and of the fact that we are part of the natural world. My body exists and nature exists. But that does not mean that I believe the methods of scientific observation give us unmediated access to The Truth about how our bodies or other parts of nature function. Obviously, the way our bodies function affects our work, including what science we do and how we do it. Our sex organs, the color of our skin, our height, weight, muscle and skeletal structures, and what gets summed up as degrees and varieties of "disabilities" profoundly affect our social and economic positions and our short- and long-term commitments. And to the extent that these variables reflect, as well as affect, our biological and psychological "natures," our nature as well as the nature within which we function affects our lives at all sorts of levels and in all sorts of ways. But, the way in which that "nature" gets encoded in "science" depends on who observes and interprets it.

Though nature is material and real, our descriptions and understanding of it are necessarily mediated by the culture of science. Enabling a wider range of people not only to understand science but also to participate in developing its agenda and to practice it would improve our understanding of nature. People who live and work outside the institutions in which science is traditionally taught and practiced might bring in new questions, answers, and ideas about nature that the culture of science has heretofore obscured. In addition, scientists have traditionally been a rather homogeneous group—by gender, race, and class as well as by education and aspiration. The reason they usually feel entitled to consider this fact irrelevant to their efforts to understand nature, of course, lies in their exaggerated confidence in the powers of scientific "objectivity." And, indeed, the methodologies of experimental science minimize bias arising from individual idiosyncracies that might make it difficult to repeat specific observations. But the methods of scientific investigation are not proof against systematic biases arising from commitments shared by significant segments of the profession and the culture.

Built-in biases are usually most blatant (though unfortunately not most obvious) in relation to questions that involve the interplay of biology and society and hence to most questions related to human biology and medicine. They are especially prevalent, but also especially well concealed, when it comes to our understandings of sex and gender, since in Western societies sex and sex differences are linchpins in our conceptions of ourselves and our culture.

In this essay I would like to present a case study of the effects of built-in bias on scientific outcomes, and thus on our views of "nature," by discussing some recent insights into the way the social and biological sciences have constructed sex and gender. In so doing, I accept the usual distinction whereby sex—whether we are male or female, women or men—is defined in terms of chromosomes (XX or XY), gonads (ovaries or testes), and genitals (the presence of a vagina or a penis, or, more usually, merely the presence or absence of a penis). Gender, specified as masculine or feminine, is taken to denote the psychosocial attributes and behaviors people develop as a result of what society expects of them depending on whether they were born female or male. However, as Kessler and McKenna, and Barbara Fried have pointed out, the concepts of sex and gender are often overlapping and blurred, not only in ordinary speech but also in the scientific literature.[3] Thus, note that Money and Ehrhardt's classic *Man and Woman, Boy and Girl* that popularized the distinction between sex and gender, confuses the two concepts in the subtitle—*Differentiation and Dimorphism of Gender Identity from Conception to Maturity*—since, surely, conception is too early to speak of "gender identity."[4]

Not all languages have two different words, sex and gender, but the fact that English does may have encouraged American scientists to try to disentangle the biological aspects from the psychosocial manifestations of sex difference. But, as with all attempts to sort "nature" from "nurture," the confusion is more than linguistic. The point is that many of the manifestations we decide to designate as natural are shaped, or at least affected, by cultural factors, while many of the manifestations we choose to attribute to nurture are affected by biology—genes, hormones, and such. In general, what we attribute to nature is no more immune from change than what we attribute to socialization.[5] In fact, in our technological and medicalized era, supposed biological factors are often easier to manipulate than the forces thought to reflect cultural institutions and traditions or deeply held beliefs. With these caveats, I shall, in what follows, accept the usual blurry distinction between sex and gender.

Sex is usually assigned when an infant is born by looking to see whether it has a penis. If it does, it's a boy; if it doesn't, it's a girl. Gender develops over time and the generally accepted lore in the social science and medical literature is that, for psychic health and to develop a coherent gender identity, children should know that they are a girl or a boy by the time their language abilities are at the appropriate stage—by about age two or two and a half.

Embedded and unquestioned in this developmental formulation of sex and gender, which reflects reproductive biology but not other biological possibilities, is the binary paradigm that, biologically speaking, there are only two categories of people—women and men—to which all people belong; socialization and experience emphasize the characteristics appropriate to each gender. Let us now look at this situation in greater detail.

When it comes to sex, the Western assumption that there are only two sexes probably derives from our culture's close linkage between sex and procreation. This linking, if not a direct result of the teachings of Western religions, is surely reinforced by them. Yet, this binary concept does not reflect biological reality. The biologist Anne Fausto-Sterling estimates that approximately 1 or 2 percent of people are born with mixed or ambiguous sex characteristics, though, for obvious reasons, it is difficult to be sure of the numbers. Such ambiguities can involve frank hermaphroditism—an infant born either with one ovary and one testis or with so-called ovotestes, organs that contain a mix of both kinds of tissues. They can also involve inconsistencies between chromosomal and gonadal or genital sex.

For example, the tissues of some children born with XY chromosomes, who as embryos develop testes, do not differentiate in the usual way in response to the hormones their testes produce. Though "male" according to their chromosomes and gonads, these children develop a vagina, though it is foreshortened in some. In medical parlance, they are said to have "androgen insensitivity," and since they are born looking like girls, they are usually assigned and reared as females. Depending on the kind of medical care they encounter, no one may notice that they have testes or anything else unusual until puberty, when they do not begin to menstruate at the expected time. They may, however, develop breasts, since the liver secretes sufficient amounts of the necessary hormones.

In an analogous variation, some XX ("female") embryos have what is called adrenogenital syndrome, which means that their adrenals secrete excessive amounts of so-called male hormones or androgens.[6] Though as embryos they develop ovaries, their uterus, vagina, and labia may or may not develop as usual, and their clitoris may be enlarged to the point that it looks like a penis. At birth, such children may be categorized as boys or considered ambiguous as regards their sex. The existence of various intermediate forms has led Fausto-Sterling to refer to "the five sexes," though there are likely to be more.[7]

Other types of intermediate forms exist. For example, in several villages in the Dominican Republic a certain number of children who are chromosomally XY and who develop embryonic testes manifest a genetic variation in which the transformation of their testosterone into dihydrotestosterone (DHT) is impeded. Since DHT is the form of testosterone that ordinarily serves to masculinize the external genitalia in XY embryos, these children are born looking like girls and are therefore socialized like girls. However, at puberty their testosterone shows its effects: their testes descend into what have hitherto been thought to be labia, their voice deepens, and their clitoris is transformed into a penis. The U.S. biomedical scientists who first described this situation reported that, though these children have been raised as females, most of them accept their transformation into males and are accepted as males by their society. They change not only their sex but also their gender identity and become biological and social males.[8]

Since this finding was first published, there has been a good deal of debate about this situation. The original team of U.S. scientists seems to have been entirely unaware of their own enculturation in the binary paradigms of sex and gender and apparently did not ask the people among whom this phenomenon occurs any questions about what they thought about sex differences, the immutability of sex, or the relationship between sex and gender.

The fact is that the villagers have special terms for these individuals. They call them *guevedoche* (balls at twelve) or *machihembra* (male female). These words suggest that they do not regard such persons as either female or male, but as a third category, a third sex. The attempt to describe the Dominican Republic system in terms of our own binary sex/gender systems has been criticized by the anthropologist Gilbert Herdt.[9] He notes that the biomedical researchers' lack of self-awareness may have unfortunately distorted the Dominican villagers' viewpoint sufficiently to make it impossible to reconstruct either the way they previously conceptualized this situation or their mode of codification, in terms of either sex or gender (if this distinction is at all valid in their setting).

Herdt has described hermaphroditism of the same biological origin occurring among several peoples in New Guinea who clearly make room for a third sex in addition to male and female. However, Herdt points out that the Sambia—the group he has observed most closely

—make every effort to detect this condition, which shows that it is sometimes possible to identify the male genitals. If they do, they call these children *kwolu-aatmwol* or *turnim-man* (turns into a man) from birth. Though such infants may superficially look female and be coded as a *kwolu-aatmwol* or third sex person, they are reared as male from the start. An occasional especially talented *kwolu-aatmwol* is honored as a shaman or war leader, but most are looked upon as "a sad and mysterious quirk of nature."[10] However, as Herdt emphasizes, there are ways of incorporating differences—beyond that of male or female—into identities that are obscured by our own medicalized system.

Other examples of the acceptance of more than two sexes have long been described among Native Americans, especially the Navajos and Zunis, where a person can be *nadle* or *berdache* (as it was called by the French colonizers), individuals who have a special status and function as neither male nor female. It is not entirely clear whether *berdache* are biological hermaphrodites or transvestites and cross-dressers and, thus, whether their's is a question of sex or gender identity. The point is that, either way, they are accepted as a third sex. This is true also of the *hijras* in India, who are considered neither men nor women in their sex or gender identity and are able to function as a third group.

In our own culture, in the old days, people who were intermediate in their anatomy or physiological functions tried to keep this fact in the closet. If their indeterminate status became known, they lived more or less miserable lives because intermediate forms are not accepted in the West. In the last few decades, in conformity with the binary paradigm, medical interventions have been developed to try to "correct" the genitals of infants who manifest any form of sex ambiguity.

I do not want to pass judgment about whether and to what extent such medical "solutions" benefit the individuals in question. Given the intense social pressure that sex be binary, so that people must be male *or* female, only very unusual parents would choose not to "repair" their child's genital or other sex ambiguities if physicians assure them it can be done. To date, there is little information about how that decision affects such people as adults.[11]

A rule that appears to operate in such medical sex reconstructions—or rather constructions—is to concentrate on the appearance of the external genitalia and to make them look as unequivocally male or female as possible. Since chromosomal and gonadal sex are thus pushed into the background and it is more difficult to construct a credible looking penis than vagina (which is fashioned as a blunt pouch or tube), the majority of children born with ambiguous genitals are turned into girls. Some effort is made to accommodate parents' wishes for a boy, but given the choice of a "real girl" or an "ambiguous boy," most parents will opt for the former.

Another rule is for the physicians to emphasize that, from the start, the infant has been of the sex they have decided to assign it to. The ambiguity is made to appear as a minor mistake of nature that modern medical methods can readily right. Therefore, the physicians try to determine as quickly as possible which sex assignment is technically most feasible and to stick with that decision. If they must revise their assessment, every effort is made to say that the baby all along was the sex to which it is being definitively assigned and that the physicians initially made a mistake. The goal is to make the parents confident in their child's intrinsic male- or femaleness as soon as possible so that they can act on this conviction in the way they raise her or him from earliest infancy and do not "jeopardize" the child's gender identity.[12]

In this way, as Suzanne Kessler points out, "the belief that gender consists of two exclusive types is maintained and perpetuated by the medical community in the face of incontrovertible physical evidence that this is not mandated by biology."[13] In other words, our gender dichotomy does not flow "naturally" from the biological dichotomy of the two sexes. The absolute dichotomy of the sexes into males and females, women and men, is itself socially constructed. The fact that we insist that sex be binary and permanent feeds into the notion that, for people to be "normal," their gender must also be binary and must match their genital sex. Where ambiguities exist, whatever their nature, the external genitalia are taken to be what counts for gender socialization and development. Kessler and McKenna summarize the situation this way:

> Scientists construct dimorphism where there is continuity. Hormones, behavior, physical characteristics, developmental processes, chromosomes, psychological qualities have all been fitted into [sex or] gender dichotomous categories. Scientific knowledge does not inform the answer to "what makes a person either a man or a woman?" Rather it justifies (and appears to give grounds for) the already existing knowledge that a person is either a woman or a man and that there is no problem in differentiating between the two. Biological, psychological, and social differences do not lead to our seeing two genders. Our seeing two genders leads to

the "discovery" of biological, psychological, and social differences.[14]

If, as we have seen, sex differences are not all that clear-cut, the situation is even more confused when it comes to gender. We admit in
our everyday language that both males and females can be more or
less feminine or masculine. And we know from experience that most
of us play with gender, or "play gender." The degree of our masculinity and femininity is not fixed for life but changes over time and in
different social situations. As we construct our persona and revise it
at different times, we allow ourselves more or less leeway in the way
we express gender. Our culture not only accepts, but admires and
enjoys, the ambiguities embodied in a Marlene Dietrich or Greta
Garbo as well as the deliberate gender bending of Grace Jones, David
Bowie, k. d. lang, the Rolling Stones, or Madonna, to name but a few
examples. Movies and the theater have celebrated cross-dressing, and
many people, without ever identifying as "transvestites," enjoy cross-
dressing and do it with verve, even if only at parties and "for fun."
When it appeared in the 1960s, unisex used to appall, but now it is an
accepted part of our culture and, as with cross-dressing, provides
accepted outlets for our desire, or need, to allow our imagination to
roam in the realm of sex and gender.[15]

Lately, however, a more radical change has occurred, as some transgender theorists and activists have begun to insist that the binary
model is hopelessly flawed and needs to be abandoned. They argue
not only for an increased fluidity but also want to have gender
unhooked from genitals and speak of a rainbow of gender. There is no
good reason, they say, for the accident of being born with a penis or a
vagina to prevent one from fully experiencing what it is like to live
like either women or men.[16]

Not surprisingly, transgender activists and theorists want to demed-
icalize decisions about gender and hence to abolish psychiatric categories such as *gender identity disorders* or *gender dysphoria*. On the
other hand, many of them would like easier access to hormones and
surgery so as to make it less difficult for people to transform their
anatomies in ways that blur their sex/gender or change it.

Some, but not all, present-day transgender theorists are what used
to be called transsexuals, though they prefer the term transgendered
or transperson. However, there is a substantial difference between
modern transpersons and classical transsexuals, who by and large
repudiated the genitals with which they were born and spoke of

themselves as men "imprisoned in the body of a woman" or vice versa. Such language was probably forced on them by a medical profession that insisted that they truly feel that they are of the other sex in order to be accepted as candidates for a sex change operation.

Until recently, except for a few public transsexuals such as Jan Morris or Renée Richards, most transsexuals hid the fact that they were living a different sex from the one into which they were born and invented personal histories to go with their transformed bodies ("When I was a little girl, my mother used to . . ." or "In high school, my girlfriends and I . . ."). But as transpersons have come out of the closet, they have acknowledged their life stories and are exploring the personal, political, and theoretical implications of their transformations. As a result, both the theory and the situation are shifting.

Accounts by or about some of the newer transgenderists place less emphasis on actual surgical transformations of the genitals and concentrate more on other satisfactions associated with becoming a transperson. Martine Rothblatt and Kate Bornstein say they never rejected the (male) genitals with which they were born and are not especially focused on the genital aspects of their transformation, whether or not they undergo the surgical procedures. Rothblatt writes: "I learned how one's genitals are not the same as one's sex. And I experienced sex as a vast continuum of personality possibilities, a frontier still scarcely explored after thousands of years of human development."[17] She looks forward to the use of computer technology for cybersex, where people can "try on genders and . . . pave the way . . . [to] being liberated from single birth-determined sex."[18]

Janice Raymond's erstwhile claim that male to female transsexuals merely reinforce gender stereotypes and represent the furthest reach in men's appropriation of women's bodies no longer fits the bill, if it ever did.[19] Sandy Stone, a 1970s transsexual whom Janice Raymond chose to attack by name, writes in 1991, "Besides the obvious complicity of [earlier autobiographical accounts by male to female transsexuals] in a Western white male definition of performative gender, the authors also reinforce a binary, oppositional mode of gender identification. They go from being unambiguous men, albeit unhappy men, to unambiguous women. There is no territory between."[20] Stone's article is an attempt to move beyond.

As a result of the greater openness, the demographics have begun to look different. The fact that most of the earlier public transsexuals had been born male gave the appearance that fewer individuals who had been born female wanted to change their sex. Now, about the same number of women and men approach medical providers about a sex change.[21] Among the female to male transpersons, for whom the techniques of genital reconstruction are fairly inadequate, genitals are assigned even lower priority. In her *New Yorker* profile of female to male transpersons, Amy Bloom quotes some of them as suggesting that the surgeons are keener on the surgery than the clients are. They joke about preferring to save their money for travel, a condominium, and other ways to enjoy life. Neither do they insist on a rigid gender identity. According to one of the transpersons who spoke with Bloom, "The gender issue isn't at the center of my life. Male, female—I don't even understand that anymore. And I find . . . it doesn't matter much."[22]

How different this is from "Agnes," one of the earliest transsexuals, whom Harold Garfinkel interviewed for several years, beginning in 1958, during her sex change. Agnes was disgusted by her penis and her existence revolved around getting rid of the hated object and acquiring a surgically constructed and heterosexually serviceable vagina.[23] Yet, even in this story, it is not clear to what extent Agnes's repudiation of the male genitals she was born with was conditioned by the physicians' insistence that she feel herself to be unequivocally female before they would perform the sex change surgery.

To the extent that transgenderism is becoming just another way in which people construct a gender identity and as gender transformations become more acceptable and easier to achieve, the changes need no longer involve the agonies experienced by people who had to overcome society's and their own sense that they were "disgusting freaks." At the same time, surgical transformations, though still important, are becoming more optional and less central to the transgender experience. As people come out of the closet, it is becoming easier for transpersons to think about what they really need or want, and sometimes that is a public persona (or range of personae) rather than a more private, genital transformation.

The question for social and natural scientists to ponder is how to reconcile these newer ways of looking on sex and gender with the barrage of sex-differences research that claims to "prove" that there are clear-cut differences between women's and men's learning styles, mathematical abilities, brain structures and functions, and so on.

To understand both the motivation and the results of this research, we have to bear in mind that, as I suggested earlier, most Western scientists come to sex-differences research imbued with the binary male/female model. With this model as their theoretical starting point, the scientists begin their investigations by identifying the significant attributes that distinguish the two groups. When they find (as they must) that women and men overlap so widely as to be virtually indistinguishable on a specific criterion, they go on to look for other criteria and to concentrate on whatever differences they unearth. Small wonder they highlight characteristics that fit in with their difference paradigm while ignoring the overlaps that contradict it. And so, the dichotomization into two and only two sexes or genders gets superimposed on a heterogeneous mix of bodies, feelings, and minds.

As far as medical "sex change" interventions are concerned—just as pediatricians, confronted with a "sex ambiguous" newborn, frame the situation in terms of the question of whether they will be more successful in producing a girl or a boy—psychiatrists and surgeons look at their adult clients through the binary spectacles of whether the psyche should be conformed with the genitals or whether the genitals should be conformed with the psyche. There is no middle ground. Faced with genital or gender ambiguities, the professionals see only males or females. By contrast, and largely under the influence of feminist theorizing about sex and gender, transgenderists have begun to see the distortions introduced by the insistence on such a polarity and to color in the rainbow between male and female.

The time is ripe for physicians and scientists also to remove their binary spectacles and, rather than explore what it means to be "male" or "female," look into what it means to be neither or both, which is what most of us are. All of us, female or male, are very much like and very different from each other. Major scientific distortions have resulted from ignoring similarities and overlaps in an effort to group differences by sex or gender. A paradigm that stresses fluidity will generate quite different questions and hence come up with different descriptions and analyses than those derived from the binary view. Social and natural scientists need to move on and explore the implications of the emerging paradigm of a continuum or rainbow of sex and gender.

Notes

1   William Cronon, ed., *Uncommon Ground: Toward Reinventing Nature* (New York: W. W. Norton, 1995); Michael E. Soulé and Gary Lease, eds., *Reinventing Nature? Responses to Postmodern Deconstruction* (Washington D.C.: Island Press, 1995); Lynda Birke and Ruth Hubbard, eds., *Reinventing Biology: Respect for Life and the Creation of Knowledge* (Indianapolis: Indiana University Press, 1995).

2   See, for example, Paul Gross and Norman Levitt, *Higher Superstition: The Academic Left and Its Quarrels with Science* (Baltimore, Md.: Johns Hopkins University Press, 1994).

3   Suzanne J. Kessler and Wendy McKenna, *Gender: An Ethnomethodological Approach* (Chicago: University of Chicago Press, 1978); Barbara Fried, "Boys Will Be Boys Will Be Boys: The Language of Sex and Gender," in *Biological Woman—The Convenient Myth*, ed. Ruth Hubbard, Marry Sue Henifin, and Barbara Fried (Cambridge, Mass.: Schenkman Publishing, 1982).

4   John Money and Anke A. Ehrhardt, *Man and Woman, Boy and Girl: Differentiation and Dimorphism of Gender Identity from Conception to Maturity* (Baltimore, Md.: Johns Hopkins University Press, 1972).

5   For a more detailed discussion, see especially the introduction and chapter nine in Ruth Hubbard, *The Politics of Women's Biology* (New Brunswick, N.J.: Rutgers University Press, 1990).

6   For a discussion of the history of the concept of sex-specific, "male" and "female," hormones from its origin around the turn of the century to its demise by the 1940s, see Nelly Oudshoorn, "Endocrinologists and the Conceptualization of Sex, 1920–1940," *Journal of the History of Biology* 23 (1990): 163–86.

7   Anne Fausto-Sterling, "The Five Sexes: Why Male and Female Are Not Enough," *The Sciences* (March–April 1993): 20–24.

8   Julliane Imperato-McGinley, et al, "Androgens and the Evolution of Male Gender Identity among Male Pseudohermaphrodites with 5-alpha Reductase Deficiency," *New England Journal of Medicine* 300 (1979): 1235–36.

9   Gilbert Herdt, "Mistaken Sex: Culture, Biology and the Third Sex in New Guinea," in *Third Sex, Third Gender: Beyond Sexual Dimorphism in Culture and History*, ed. Gilbert Herdt (New York: Zone, 1994), 419–45.

10   Ibid., 436.

11   For personal statements, see *Hermaphrodites with Attitude* (spring 1995), published by the Intersex Society of North America, P. O. Box 31791, San Francisco, Calif. 94131.

12   Suzanne J. Kessler, "The Medical Construction of Gender: Case Management of Intersexed Infants," *Signs* 16 (1990): 3–26.

13   Ibid., 25.

14   Kessler and McKenna, 163.

15   For an interesting discussion of this subject, see Marjorie Garber's *Vested Interests: Cross-Dressing and Cultural Anxiety* (New York: Routledge, 1992).

16   Kate Bornstein, *Gender Outlaw: On Men, Women, and the Rest of Us* (New York: Routledge, 1994); Martine Rothblatt, *The Apartheid of Sex: A Manifesto on the Freedom of Gender* (New York: Crown, 1995).

17   Rothblatt, 164.

18   Ibid., 153.

19   Janice Raymond, *The Transexual Empire: The Making of the She-Male* (Boston: Beacon, 1979).

20   Sandy Stone, "The *Empire* Strikes Back: A Posttranssexual Manifesto," in *Body Guards: Cultural Politics of Gender Ambiguity*, ed. Julia Epstein and Christina Straub (New York: Routledge, 1991), 280–304 esp. 286.

21   Amy Bloom, "The Body Lies," *New Yorker*, 18 July 1994, 38–49.

22   Ibid., 40.

23   Harold Garfinkel, *Studies in Ethnomethodology* (Englewood Cliffs, N.J.: Prentice-Hall, 1967), chap. 5.

Ten Propositions on Science

and Antiscience

Richard Levins

SINCE radicals began to look to science as a force for emancipation, Marxists both as social critics and as participating scientists have grappled with its contradictory nature. Because there is such a rich diversity of Marxist thought about science, I cannot claim that what follows is "the" Marxist position. I only offer in schematic form some propositions about science that have guided the work of at least this Marxist scientist.

(1) All knowledge comes from experience and reflection on that experience in the light of previous knowledge. Science is not uniquely different from other modes of learning in this regard.

What is special about our science is that it is a particular moment in the division of labor in which resources, people, and institutions are set aside in a specific way to organize experience for the purpose of discovery. In this tradition a self-conscious effort has been made to identify sources and kinds of errors and to correct for capricious biases. It has often been successful. We have learned to be alert to the possible roles of confounding factors and to the need for controlled comparison; we have learned that correlation does not mean causation and that the expectations of the experimenter can affect the experiment; we have also learned how to wash laboratory glassware to avoid contaminants and how to extract trends and distinctions from morasses of numbers. Our self-consciousness reduces certain kinds of errors but in no way eliminates them, nor does it protect the scientific enterprise as a whole from the shared biases of its practitioners.

On the other hand, so-called traditional knowledge is not static or unthinking. Africans (probably mostly women) brought as slaves to the Americas quickly developed an Afro-American herbal medicine. It was put together partly from remembered knowledge of plants found both in Africa and in America, partly from borrowed Native

American plant lore, and partly from experimenting on the basis of African rules about what medicinal plants should be like. The teaching of traditional medicine always involves experimenting, even when it is presented as the transmission of preexisting knowledge. Finally, the criteria for prescribing various herbal therapies in non-European/North American medicine are probably better grounded than those that guide decisions about cesarean sections, pacemaker implants, or radical mastectomies in U.S. scientific medical practice.

Even what is described as intuitive (as against intellectual) knowledge comes from experience: our nervous/endocrine system is a marvelous integrator of our rich, complex histories into a holistic grasp that is unaware of its origins or constituents. Scientific and intuitive knowledge are not fundamentally different epistemologically; they differ instead in the social processes of their production and are not mutually exclusive. In fact, one of my goals in teaching mathematics to public health scientists is to educate the intuition, so that the arcane becomes obvious and even trivial, and complexity loses its power to intimidate.

(2) All modes of discovery approach the new by treating it as if it were like the old. Since it often is like the old, science is possible. But the new is sometimes quite different from the old; when simple reflection on experience is not enough, we need a more self-conscious strategy for discovery. Then creative science becomes necessary. In the long run we are bound to encounter novelty stranger than we can imagine, and previous well-grounded ideas will turn out to be wrong, limited, or irrelevant. This holds true in all cases, in both modern and traditional, class-ridden, and nonclass societies. Therefore, both modern European/North American science and the knowledges of other cultures are not only fallible but are guaranteed to err eventually.

To call something "scientific" does not mean that it is true. Within my lifetime, scientific claims such as the inertness of the "noble gases," the ways in which we divide up living things into major groupings, views as to the antiquity of our species, models of the nervous system as a telephone exchange, expectations as to the long-term outcomes of differential equations, and notions of ecological stability have all been overturned by new discoveries or perspectives. And major technical efforts based on science have been shown to lead to disastrous outcomes: pesticides increase pests; hospitals are foci of infection; antibiotics give rise to new pathogens; flood control increases flood damage; and economic development increases poverty. Nor can we assume that error belongs to the past and that now we've got it right—a kind of "end of history" doctrine for science. Error is intrinsic to actually existing science. The present has no unique epistemological status — we just happen to be living in it.

Therefore, we have to consider the notion of the "half-life" of a theory as a regular descriptor of the scientific process and even be able to ask (but not necessarily answer), "Under what circumstances might the second law of thermodynamics be overthrown?"

(3) All modes of knowing presuppose a point of view. This is as true of other species as of our own. Each viewpoint defines what is relevant in the storm of sensory inputs, what to ask about the relevant objects, and how to find answers.

Viewpoint is conditioned by the sensory modalities of the species. For instance, primates and birds depend overwhelmingly on vision. With visual information objects have sharply differentiated boundaries. But that is not the case when odors are the major type of information, as for ants. An anoline lizard sees moving objects as being the right size to eat or as representing danger. A female mosquito perceives an academic conclave as gradients of carbon dioxide, moisture, and ammonia that promise blood meals, while a sea anemone trusts that glutathione in the water is enough reason to thrust out its tentacles in expectation of a meal. The fact that we live on the surface of the earth makes it seem natural to focus our astronomy on planets, stars, and other objects while ignoring the spaces between them. The timescale of our lives makes plants seem unmoving until time-lapse photography makes their changes apparent. We interact most comfortably with objects on the same temporal and size scales as our own and have to invent special methods for dealing with the very small or very large, the very fast or very slow.

(4) A point of view is absolutely essential for surviving and making any sense of a world bursting with potential sensory inputs. Much of learning is devoted to defining the relevant and determining what can be ignored. Therefore, the appropriate response to the discovery of the universality of viewpoints in science is not the vain attempt to eliminate viewpoint but the responsible acknowledgment of our own viewpoints and the use of that knowledge to look critically at our own and each others' opinions.

(5) Science has a dual nature. On the one hand, it really does enlighten us about our interactions with the rest of the world, producing understanding and guiding our actions. We really have learned a great deal about the circulation of the blood, the geography of species, the folding of proteins, and the folding of the continents. We can read the

fossil records of a billion years ago, reconstruct the animals and climates of the past and the chemical compositions of the galaxies, trace the molecular pathways of neurotransmitters and the odor trails of ants. And we can invent tools that will be useful long after the theories that spawned them have become quaint footnotes in the history of knowledge.

On the other hand, as a product of human activity, science reflects the conditions of its production and the viewpoints of its producers or owners. The agenda of science, the recruitment and training of some and the exclusion of others from being scientists, the strategies of research, the physical instruments of investigation, the intellectual framework in which problems are formulated and results interpreted, the criteria for a successful solution to a problem, and the conditions of application of scientific results are all very much a product of the history of the sciences and associated technologies and of the societies that form and own them. The pattern of knowledge and ignorance in science is not dictated by nature but is structured by interest and belief. We easily impose our own social experience onto the social lives of baboons, our understanding of orderliness in business, implying a hierarchy of controllers and controlled, onto the regulation of ecosystems and nervous systems. Theories, supported by megalibraries of data, often are systematically and dogmatically obfuscating.

Most analyses of science fail to take into account this dual nature. They focus on only one or the other aspect of science. They may emphasize the objectivity of scientific knowledge as representing generic human progress in our understanding. Then they dismiss the obvious social determination and the all-too-familiar antihuman uses of science as "misuses," as "bad" science, while keeping their model of science as the disinterested search for truth intact.

Or else they use the growing awareness of the social determination of science to reject its claims to any validity. They imagine that theories are unrelated to their objects of study and are merely invented whole cloth to serve the venal goals of individual careers or class, gender, and national domination.

In stressing the culture-boundedness of science, these analyses ignore the common features of Babylonian, Mayan, Chinese, and British astronomies and their calendars. Each comes from a different cultural context but looks at (more or less) the same sky. They recognize years of the same length, notice the same moon and planets, and calculate the same astronomical events by very different means.

Social determinists also ignore the parallel uses of medicinal plants in Brazil and Vietnam, the namings of plants and animals that roughly correspond to what we label as distinct species. All peoples seek healing plants and tend to discover similar uses for similar herbs.

Other traditions than our own also have their social contexts. Babylonian priests or Chinese administrators were not bourgeois liberals, but for all that they were not wiser or freer from viewpoint. Nor does the phrase "the ancients say" tell us anything about the validity of what they say. Ancients like moderns belong to genders, sometimes to classes, always to cultures, and they express those positions in their viewpoints. Those ancients whose thought has been preserved in writing were also not a random sample of ancients.

But to be socially determined and conditional on viewpoint does not mean arbitrary. While all theories are eventually wrong, some are not even temporarily right. The social determination of science does not imply a defense or toleration of the patently false doctrines of racial or gender superiority or even the categories of race themselves, whether in the conventional academic forms or the "Adamic man" and the "mud people" of the Christian Identity Movement. Racism is a more real object than race and determines the racial categories.

Therefore, the task of the analyst of science is to trace the interactions and interpenetrations of intellectual labor and the objects of that labor under different conditions of labor and under different social arrangements. The art of research is the sensitivity to decide when a useful and necessary simplification has become an obfuscating oversimplification.

(6) Modern European/North American science is a product of the capitalist revolution. It shares with modern capitalism the liberal progressivist ideology that informs its practice and that it helped to mold. Like bourgeois liberalism in general it is both liberated and dehumanized. It proclaimed universal ideals that it did not quite mean, violated them in practice, and sometimes revealed those ideals to be oppressive even in theory.

Therefore, there are several kinds of criticisms of science. A conservative criticism inherits the precapitalist critique. It is troubled by the challenge that scientific knowledge poses to traditional religious beliefs and social rules and rulers, does not approve of the independent judgment of ideas and values, does not demand evidence where authority has already pronounced, and therefore is disturbed mostly by the radical side of science. Creationists quite accurately identify the ideological content of science, which they label secular humanism, against the liberal formula that science is the neutral opposite of ideology. But no matter how much they search the scientific journals for evidence of conflicts among evolutionists and weak spots in modern evolutionary theory, their challenge is not to make science more "scientific," more democratic, less bound by oppressive ideology, and more open. Rather they propose to return to faith, to the more obvious kinds of authority, and to anti-intellectual certainties. Their gut-level anti-intellectualism is often expressed in delight at the stupidities of scientists as against the wisdom of the "simple man," a delight that at first seems appealingly democratic. But this is not the assertion that everyone is capable of rigorous and disciplined thinking. Instead, it denies the importance of serious complex thinking altogether in favor of the spontaneous smarts of uneducated certainties. They accept the dichotomy of knowledge versus values and opt for their particular values whenever there is conflict.

At the same time, conservative critics reject the fragmented and reductionist aspects of modern science on behalf of a holistic, "organic" view of the world. At an aesthetic and emotional level their holism partly resonates with that of radical criticism, but their holism is hierarchical and static, stressing harmony, balance, law and order, the ontological rightness of the way things are, were, or are imagined to have been.

The most consistent liberal critics of science accept the claims of science as valid goals but criticize the practices that violate them. They approve of science as public knowledge and deplore the secrecy imposed by military and commercial ownership of it. They want democratic access to science determined only by capacity, and they deplore the class, gender, and racial barriers to scientific training, employment, and credibility. They agree that ideas should be judged only on their merits and on the evidence, regardless of where the ideas come from, but they see hierarchies of credibility reinforced by a rich vocabulary for dismissing unorthodox ideas and their advocates as "far out," "quackish," "ideological," "not mainstream," "discredited," "anecdotal," or "unproven." They may be horrified by the uses of science in the production of harmful commodities or vicious weapons or the just as vicious justifications of oppression, without however relinquishing the belief that thinking and feeling should be kept separate.

Because of the increasingly obvious blindnesses, narrowness, dogmatism, intolerance, and vested interest in official science, alternative movements have sprung up, especially in health and agriculture. They must be examined with the same tools that we use to look at "official" science: who owns them, where do they come from, what viewpoints do they express, how are they validated, what theoretical biases do they manifest? Embedded as they are in a capitalist context, these alternatives too are a field for exploitation, produce commodities, and often are clothed in shameless commercial hype. They too have class roots that lead some of them to separate individual from social causation (for instance, criticizing the magic bullets of the pharmaceutical industry but peddling their own miraculous "natural" cures, or promoting holistic cancer treatments but ignoring the industrial origins of many cancers). The alternative communities are domains where insightful radical critique mixes with petty and medium-scale entrepreneurship.

Marxist critique attempts to see science in both its liberating and oppressing aspects, its powerful insights and its militant blindnesses, as a commoditized expression of liberal European capitalist masculinist interests and ideologies organized to cope with real natural and social phenomena. Its ideology is both a product of European liberalism and a self-generated contribution to that ideology, not a mere passive reflection of it.

Particular radical critiques of agriculture, medicine, genetics, economic development, and other areas of applied science point out both the external and internal aspects that limit science's ability to reach its stated goals. The external refers to its social position as a knowledge industry, owned and directed for purposes of profit and power as guided by shared beliefs, carried out mostly by men. The modes of recruitment into and exclusion from science, the various subdivisions into disciplines, the hidden boundary conditions restraining its inquiry become intelligible when we examine its social context. We can approach the dominant modalities of chemical therapy in medicine and farming as expressions of the commoditization of knowledge by the chemical industry. But the reliance on molecular magic bullets is also congenial to the reductionist philosophy that has dominated European/North American science since its formation in the seventeenth century, and that in turn is supported by the atomistic experience of bourgeois social life. (As we trace the connections, we see that "internal" and "external" are in fact not rigidly alternative explanations, another example of the general principle that there are no nontrivial, complete, and disjunct subdivisions of reality. Yet science is still plagued by the false dichotomies of organism/environment, nature/nurture, deterministic/random, social/individual, psychological/physiological, hard/soft science, dependent/independent variables, and so on.)

The internal refers to the reductionist, fragmented, decontextual-ized, mechanistic (as against holistic or dialectical) ideologies and lib-eral-conservative politics of science. Marxist and other radical critics have always called for broadening the scope of investigations, placing them in historical context, recognizing the interconnectedness of phe-nomena, and the priority of processes over things, while conservative ideology usually advocates elegant precision about narrowly circum-scribed objects and accepting boundary conditions without even acknowledging them.

(7) A radical critique of science extends also to the inner workings of the research process. In approaching a new problem, my Marxism encourages me to ask two basic questions: why are things the way they are instead of a little bit different, and why are things the way they are instead of very different? Here "things" has a double meaning, refer-ring both to the objects of study and to the state of the science studying them.

The Newtonian answer to the first question is that things are the way they are because nothing much is happening to them.

But our answer is that things are the way they are because of the actions of opposing processes. This first question is that of the self-regulation of systems, of homeostasis. In the face of constantly dis-placing influences, how do things remain recognizably what they are? Once posed, it enters the domain of systems theory in the narrow sense, the mathematical modeling of complex systems. That disci-pline starts with a set of variables and their connections and applies equations to ask, is the system stable? How quickly does it restore itself after perturbation? How much does it respond to permanent changes in its surroundings? How much change can it tolerate? It asks, when external events impinge on the system, how do they per-colate through the whole network, being amplified along some path-ways and diminished along others? We work with notions such as positive and negative feedback loops, pathways, connectivity, sinks, delays, reflecting and absorbing barriers. In its own terms, this analy-sis is "objective." But the variables themselves are social products. For instance, the apparently unproblematic notion of population density has at least four different definitions that lead to different formulas for measurement and different results when the measurements are com-pared across countries or classes. We could simply divide the total number of people by the total area (or resource):

$$D = \Sigma\text{people}/\Sigma\text{area}.$$

We could ask, what is the average density at which people live? Then we would use

$$D = \Sigma(\text{people/area})(\text{people in that area})/\Sigma\text{people};$$

the unevenness of access to resources or land is then included. Or we could do the same but from the perspective of the resource. The total resource per person is

$$D = \Sigma\text{area}/\Sigma\text{people},$$

the average intensity of exploitation of a resource is given by

$$D = \Sigma(\text{area/people})(\text{area})/\Sigma\text{area}.$$

Thus even what seems to be an objectively given measure is laden with viewpoint, and this is either taken into account or hidden. Nancy Kreiger (1994) has used the metaphor of fractal self-similarity to stress that the inseparability of the social and biological occurs at all levels, from the most macro to the fine details of the micro in epidemiology.

The second question is the question of evolution, history, and development. Its basic answer is, things are the way they are because they got that way, not because they have to be that way, or always were that way, or because it's the only way to be. From this perspective we reexamine the first question and ask, what variables belong in the system anyway, and how did they get there? What do we really want to find out about the system? What do you mean "we"? Who says? Do new connections appear and old ones decline? Do variables merge or subdivide? Do the equations themselves change? Should we use equations or other means of description? And since we know that the models we use are not photographically accurate pictures of reality, how would departures from the assumptions affect the outcomes? When does this matter?

What were the givens in the first formulation now become the questions. It is here that the powerful insights of Marxists dialectic, when combined with substantive knowledge of the objects of interest and the manipulative skills of the craft, have been most productive. Here the familiar propositions of the unity and interpenetration of opposites, universal connection, development through contradiction, integrative levels, and so on, so dry in the listings of the formal manuals, burst with rich implications and scintillate with creative potential.

Finally, these same methods are used reflexively to examine the historical constraints that have acted on Marxism itself as a consequence of its own historical circumstances and the composition of Marxist movements. But these methods should not be used in a mechanistic, essentialist way, rejecting notions because they are European and therefore foreign in Latin America, or male and therefore irrelevant to women, or of nineteenth-century origin and therefore inapplicable to the twenty-first. After all, every idea is foreign in most places where it is held, and in all places in the world most of the current ideas are of foreign origin. Rather, the historical context can be used to evaluate the ideas critically, to discover the insights and limitations and the needed transformations. The insights of feminism and the ecology movement, particularly those branches that have already overlapped with Marxism, are especially helpful in gaining the distance needed for this examination. Themes which had been relegated to the periphery of most Marxist vision can now be restored to their rightful places in historical materialism, and societies studied more richly as social/ecological modes of production and reproduction.

(8) Although different theories use different terms, look at different objects, and have different goals, they are not mutually unintelligible. Linnaeus saw species as fixed at the time of creation, with each particular example being a corrupted version of the archetypal design. Evolutionary biologists see species as populations that are intrinsically heterogeneous and subject to forces of change. The description of the typical is then seen as an abstraction from the array of real animals or plants. Nevertheless, I still use Linnaean Latin names for genus and species, many of which Linnaeus himself would recognize, and I could talk with Linnaeus about plants, argue about their anatomy or geographic distributions. He would be delighted to learn that our technologies have given us new ways of distinguishing among similar plants. We would disagree about the significance of variation within a species, and I don't know how he would react to the shocking idea that similarity often implies a common origin. But we could talk.

This is even true across larger cultural divides. All peoples name plants and animals. Most peoples assign different names to plants that correspond to different Linnaean species, and divide up the botanical world much as we do. They also tend to distinguish more finely among organisms that have to be dealt with differently. And like our own theories, theirs also "work." They guide actions that often enough lead to acceptable results. Whether you are a modern taxonomist who recognizes that half the snakes in Darien are poisonous or a Choco who will tell you that all snakes are poisonous but only kill you half the time, the practical conclusion is similar: when walking in the forest, beware of snakes.

Furthermore, the tools of investigation show a greater continuity than the theories. Galileo would be impressed by our more sophisticated telescopes but would not be completely lost in a modern observatory. While a Marxist economist might not be interested in the input-output equilibrium models of the neoclassical school or the techniques of cost-benefits analysis so dear to the corporate mind, these would be perfectly comprehensible to her. The claim that different outlooks are incommensurate, speak different languages, and find no points of contact is a gross distortion of the understanding of social viewpoint. Theoretical barriers do not mean the existential aloneness imagined by distant observers.

(9) The diversity of nature and society does not preclude scientific understanding. Every place is clearly different and every ecosystem has its unique features. Therefore, ecology does not look for universal rules such as "plant diversity is determined by herbivores" or attempt to predict the flora of a region by knowing its rainfall. What it can do is look for the patterns of difference, the processes that produce the uniqueness. Thus, the number of species on an island depends on the processes of colonization and speciation increasing numbers and the processes of extinction reducing numbers. We can go further and relate colonization to distance from a source of migrants, extinction to habitat diversity and area and community structure, try to explain why the migrants are of a particular type, and so on. The outcomes will be very different on tiny islands where populations do not last long enough to give new species or are so close to the source of migrants as to swamp any local differentiation, from islands that are very remote, with high habitat diversity.

Thus the use of site specificity to reject broad generalizations is misplaced. What we look for is the identification of the opposing processes that drive the dynamics of a kind of system (e.g., rain forest, or island, or capitalist economy) rather than propose a unique and universal outcome.

(10) Radical defenders of science cannot defend science as it is. Instead, we have to come forward as critics both of liberal science and of its reactionary enemies. The present right-wing attack on science is part of a more general assault on liberalism, now that the demise of a worldwide socialist challenge makes liberalism unnecessary and intensified competition during a period of long-term stagnation makes liberalism seem too costly. Although its opposition to liberalism is

opposition to the liberating aspects of that doctrine, the reactionary attack on liberalism often emphasizes the oppressive or ineffectual sides of liberalism.

We have to call for opening science up to those who have been excluded, democratizing what is a very authoritarian structure modeled on the corporation, and insist on the goal of a science aimed at the creation of a just society compatible with a rich and diverse nature. We should not hide behind but rather undermine the cult of expertise in favor of approaches that combine professional and non-professional participation. The optimal condition for science is with one foot in the university and one in the communities in struggle, so that we have both the richness and complexity of theory coming from the particular and the comparative view and generalizations that only some distance from the particular can provide. It also allows us to see the combination of cooperative and conflicting relations we have with our colleagues and ways in which political commitment challenges the shared common sense of professional communities.

We should not pretend or aspire to a bland neutrality but proclaim as our working hypothesis: all theories are wrong which promote, justify, or tolerate injustice.

We should not cover up or only lament in private the triviality of so much published research but denounce that triviality as coming from the commoditization of careers in scholarship and from the agendas of domination that rule out of order many of the really interesting questions.

We should challenge the competitive individualism of science in favor of a cooperative effort to solve the real problems.

We should reject the reductionist magic bullet strategy that serves commoditized science in favor of respect for the complexity, connectedness, dynamism, historicity, and contradictoriness of the world.

We should repudiate the aesthetics of technocratic control in favor of rejoicing in the spontaneity of the world, delighting in the incapacity of indexes to capture life, savoring the unexpected and anomalous, and seeking our success not in dominating what is really indominable but in farsighted, humane, and gentle responses to inevitable suprise.

The best defense of science under reactionary attack is to insist on a science for the people.

Reference

Nancy Kreiger. 1994. Epidemiology and the web of causation: Has anyone seen the spider? *Social Science of Medicine* 30, no. 7: 887–903.

Dispatches from the Science Wars

Joel Kovel

LET us begin with a fact or, at any rate, a finding. Other matters could be adduced to support the line of reasoning I have in mind, but it is better to keep focused for now on the following: it has been found, by "science," that for about thirty to fifty years, sperm counts have been declining, in both numbers and motility, among men in industrialized countries. Recent studies from Paris indicate that the decrease amounts to about 2 percent per year during the last two decades. A 175-page report from the Danish Environmental Protection Agency presents the evidence, along with certain interpretations, to be discussed below. Other reports from Scotland and Belgium point in the same direction. These in turn support a 1992 finding by Elizabeth Carlsen, based on a historical analysis of sixty-two separate sperm count studies. The findings are correlated with others: a marked rise in testicular cancer among young men as well as congenital anomalies of the male reproductive organs; a rise in associated problems among women, especially breast cancer; and similar deterioration among wildlife, including panthers, alligators, birds, bats, turtles, and fish.[1]

There are a number of possible responses to this information. The most obvious would be to inquire as to the causes of these phenomena, their implications, and potential remedies. This would be shadowed by an elementary extrapolation: at the rate of a 2-percent decline a year—and there are reasons to believe that the rate will accelerate —the reproductive capacities of higher animals, at least in certain areas and perhaps across the globe, will at some point sink below a threshold of sustainability. In the meanwhile, an increasing number of beings are going to suffer in one way or another, and an increasing number of genetically damaged organisms are going to be launched into the ecosphere. Thus, if the processes to which these studies are calling attention continue, drastic conclusions for the future of complex organisms on earth are to be drawn. For it would appear that a kind of systematic poisoning is inexorably destroying the genetic legacy of a billion years of evolution.

But let us not be too hasty. The preceding paragraph used conditional and subjunctive modes for more than conventional reasons. That the aforementioned extrapolation takes place is itself based on a number of assumptions, namely:

–That the finding matters. This rests upon a foundational recognition of the obdurate materiality of our body, as that part of nature we directly inhabit. Though any number of texts—of life history, of identity, of gender—can be inscribed on it, the body has to be there in reasonably good shape for said texts to be inscribed. It is logically possible to imagine someone not agreeing with this notion, and therefore having no interest in the above finding, but such individuals must be quite rare. Others—also, I should think, a tiny minority—might positively welcome the news of declining sperm counts as an unforeseen solution to the population explosion. However, these neo-Malthusians should realize that the brunt of the effect, at least for quite a while, will be felt where the effects of industrialization are greatest and not among the allegedly overbreeding masses of the South. The great majority, then, should regard this news as a grim sign that the natural substratum of existence is disintegrating. This fearful prospect commits us to paying serious attention to how the body is made and how it interacts with the world.

–To do so entails another recognition, of the collectively derived knowledge we call science, which we entrust with telling us about this body and the world it inhabits. For the above extrapolation to make sense, we must have coherent ideas of genetics, evolutionary biology, chemistry, physiology, toxicology, epidemiology, and of much else. We also need to have some confidence that "science," that is, some intelligent combination of these various disciplines, can tell us something about the nature of the poisoning process that is destroying our germ plasm. Finally, we need to believe that learning these things will point to a way to reverse or mitigate them. All these assumptions are broadly within the purview of modernity, a central assumption of which is the capacity of disinterested reason to bring about human benefit. In other words, to rely upon the above findings and their interpretation requires, if not full endorsement, at least some recognition of the authority of modernity.

I suspect there are more people who would take some issue with the second set of propositions than the first. There are, after all, many reasons to be skeptical about the achievements of science. Severe critics of modernity abound, whether from premodern, essentially religious grounds, or from the various stances of postmodernity. And, although hardly anyone rejects the accomplishments of science tout court, a great many have chipped away at this aspect or that of its edifice. After all, without science we would not have had Hiroshima, Bhopal, and Chernobyl, or the ecodestructive effects of the so-called Green Revolution.[2] Science as the embodiment of human arrogance, the modernist assertion of lawfulness and grand narratives, the search for identity, the drive to control and dominate nature, and the project of masculinist pride—all these realizations have fed into the rejection of science, or at least the adoption of a sharply critical attitude toward it.

Therefore we have a problem: one must trust in science and hence in modernity if the finding of a drop in sperm count is to be pursued. And yet the impulse of much contemporary inquiry is to distrust science. Does one settle this by saying that here, in this and related instances (such as a life-threatening illness), we bracket the contemporary critique of science for the time being and place ourselves in the hands of "experts," trusting for purely instrumental reasons what we would ordinarily criticize? A feeble, timorous response, this; one would think that a critique so easily set aside in favor of a threat to life is not to be taken that seriously.

Perhaps we can get a better grasp on the subject if we learn some more of what science says about the issue in question. Here we immediately encounter an interesting complication. For it turns out that scientists are sharply divided into two camps on the drop in sperm count. One group, let us call them Camp A, either does not even believe that the sperm counts are dropping in the industrialized nations or, if they do accept the finding, they give it an entirely different interpretation from their opponents, whom we will call, not surprisingly, Camp B.

Here is some of what Camp B says is behind the decline: that many common industrial chemicals mimic sex hormones, especially estrogen; that once ingested, even in tiny amounts, these substances interfere with our endocrine balance; that the various sexual/genetic/carcinomatous abnormalities such as declining sperm counts result from such an interference with the body's exquisitely balanced internal hormonal environment; that this effect is especially severe in utero; finally, that these substances, many of them derivatives of organochlorines that have been introduced into the biosphere for the first time in the past half century owing to the rapid growth of the chemical indus-try,[3] are ubiquitous. They are components of pesticides, detergents, cosmetics, and paints, as well as of packaging materials, including plastic containers and food wraps, i.e., substances not meant for ingestion but often impossible to avoid: the lipstick that goes beyond the lips, the Saran Wrap that rubs a few molecules on the fudge brownie, the tiny whiff of paint. In other words, the poisoning, if that is what to call it, is not esoteric, but takes place at the most immediate and widespread sites of everyday life and in the carrying out of the most elementary bodily functions. Further, the reason this problem is likely to grow much worse according to the hypothesis of Camp B is that what we see now is largely the result of toxic exposure during gestations that took place twenty to forty years ago, when the aggre-gate level of chemical exposure was much less than it is today. Accordingly, we may already have launched a far more injured gen-eration than that tested.

Camp A, as stated, does not agree with the entire scenario. For example, it offers studies that show that Elizabeth Carlsen's statistical methods were flawed and that sperm counts have actually been increasing. Another of its number, a man with the Dickensian name of Stephen Safe, concludes a recent study by asserting that "the sugges-tion that industrial estrogenic chemicals contribute to an increased incidence of breast cancer in woman and male reproductive problems is not plausible." Safe grounds this conclusion on the claim that nat-ural estrogen-mimicking compounds in the plants we eat far outweigh the quantities of industrial chemicals that may enter the body by one route or another.

I am not about to enter into the substantive debate concerning the truth claims of the two camps.[4] I am, after all, no "expert" in this field and have only a smattering of acquaintance with it based on partisan secondary literature. But whether I am competent to judge this mat-ter is beside the point, which is to recognize that positive proof is impossible in any case. As Peter Montague, relentless critic of the pol-luters, has written, there can be no Holy Grail of scientific certainty in questions of this sort.[5] Too many variables abound, too many method-ological questions exist ever to be satisfactorily resolved, the time-scale of relevance is beyond clear definition, and so forth. In other words, there are events and processes in nature and society that are fundamentally indeterminate. Positive knowledge of these things must remain chimerical. Even for questions that no reasonable person would dispute (e.g., the causative role played by cigarette smoking in a wide range of illnesses) apodictic, Cartesian certainty is impossible. Indeed, it is one of the chief accomplishments of the critical and postmodern theories of science to abolish the grandiose claims for certainty, except in limiting and often uninteresting cases.

Reflections of this sort usually eventuate in nihilism or, more commonly, in a call for a kind of liberal toleration. Science must cease its arrogant certitude, it is said, and recognize ambiguity, even the existence of a plurality of perspectives. But instances such as the decline in sperm count reveal the inadequacy of both these alternatives. If the issue is neither disinterested pursuit of knowledge nor power over nature but survival—not just of one's own species but of vertebrate life itself—then the rules have to be written differently, for the simple reason that the living body is in danger.

There can be no pluralist accommodation between camps A and B. If the former's position is true, no basic changes need be made; business as usual can continue. If the latter's is true, then society faces its most profound transformation since the industrial revolution.[6] If there is no positive proof, however, then how are we to settle the matter, and how is science to be drawn in? The urgency of the crisis suggests an approach:

–First, we cannot close our eyes to what is happening. Given the scale of the problem and the concerted effort to minimize, normalize, or otherwise distort its terms, this turns out to be quite a piece of work. Denial and apocalypticism, as well as cynical withdrawal or indifference, are all ways of relieving anxiety; here they can have potentially fatal consequences. Survival defines the present conjuncture, and survival returns reason to its most basic function: the adequacy of means to ends. If extrapolations of the present ecological crisis are true—if only the findings about sperm counts are true, not to mention better-known calamities such as global warming, rampant species-loss, destruction of topsoils, and deforestation—then we are facing something for which human society is utterly unprepared. Three possibilities stand forth in this case: various scenarios of collapse and social/biological disintegration; and/or authoritarian state responses to this; or some emancipatory transformation of society into an ecologically rational form.[7] Unless one argues for collective suicide, the overriding end is survival, and the rational means becomes that which promotes survival in a nonfascistic way—that is, that which struggles against the first and/or second of the above possibilities. In the present case, reason tells us to mistrust, though not necessarily to reject, Camp A's denial of the sperm-count crisis.

There is too much smoke for there not to be some fire. Camp B therefore deserves a full hearing and the presumption of some validity.

—In this conjuncture we cannot dispense with science, any more than people at sea can afford to ignore the skills of navigation. But science can never more be taken at face value. The record of its violations must be held before it: the reduction of the universe to brute, mechanically driven matter, the adjunctive role toward domination. In the present conjuncture, certain aspects of so-called normal science must continue: how, after all, are we to contend with the damage that is being done, or devise appropriate technology for an ecological society, if we reject the collective intelligence embedded in the scientific project? But science has to be reworked for an abnormal conjuncture. The myth of its autonomy from society is gone, perhaps forever. What remains must be refashioned according to the revealed crisis of ecology—science now seen in the light of what it has so far largely expelled. What has been reduced away must be restored: respect for the integrity of complex wholes rather than atomized parts; the primacy of dialectical becoming over static, mechanical being; the recognition of our embeddedness in nature and hence nature's immanent consciousness and vitality.[8] Where problems are developed in less than full connectedness to this perspective—as here, the indispensable research into the endocrine effects of organochlorines[9]—this is to be seen as a bracketing necessitated by the limits of human cognition and not as the confirmation of a mechanical metaphysic.

However, the revision of normal science is not sufficient. Another dimension is required: the critique of science, given unprecedented authority in view of the nightmare wrought by science as usual, that series of horror stories to which the sperm-count findings may now be added. This critique has some familiar contours but an altered context. Does it come as a surprise that the study alleging that sperm counts are not in fact declining was a project of the Chemical Manufacturers' Association? Or that Stephen Safe is sponsored by the same forces? Or, while we are on the subject, that the top fifteen public relations firms received more than $90 million in 1993 to promote an ecologically friendly image; or that the overall "greenwashing" tab for U.S. business in 1990 was an estimated $500 million—that is, half a billion dollars per year to engineer consent to the current economic order's treatment of nature?[10]

The likelihood that a study like Safe's would come from any quarter other than capital is near zero. One is, I should think, on more secure predictive grounds in making an ideological critique of the scientific approach to the environmental crisis than in predicting the outcome of any particular technical issue within that crisis, as, for example, sperm counts. This is because nature in this respect is more intricate than society. The former contains many complex and even indeterminable variables, and it is frankly speculative to introduce any single driving entelechy, such as the Second Law of Thermodynamics, the principle of evolution, or some general field theory of physics. The search for some unified Grand Law seems to be a vicissitude of the monotheistic impulse. It is to be sought more in the special circumstances of the human mind than in external nature, which, though lawful, utterly surpasses our intelligence in its innermost workings.

Society has many complications, too, but the presence of a central dynamic, of capital accumulation, and the arrangements in place to secure accumulation, are demonstrable realities—as is the ideological mystification of reality, such being one of the above-mentioned arrangements that secure accumulation. In nature we identify many lawful tendencies but no discrete force that drives the whole and no impulse to lie about itself. In society such a force and impulse are virtually omnipresent. Since nature is known through science and science is a part of society, the study of nature and the unmasking of the social dynamic that enters into this study are two phases of the same endeavor.

From a broader perspective, the current predicament can be seen in terms of a new moment in the intersection of nature and history. Two poles define this moment, now playing itself out before our eyes. Beginning roughly in 1960, the earth's carrying capacity, that is, its ability to buffer the ecodestructive effects of capital accumulation, became exceeded. The 1962 publication of *Silent Spring*, Rachel Carson's warning of the effects of pesticides on the biosphere, can serve as a signpost of this new phase (as can, of course, the finding of declining sperm counts).

The second pole is defined by the global crisis of accumulation which begins about a decade later, and rages on, with cataclysmic effects on labor, urban spaces, the South, and the environment. The degree to which this crisis was itself spurred by the surpassing of the earth's carrying capacity, an eventuality that necessarily raises the cost of business, is not something that can be accurately described here. However, that the two crises interact is indisputable. They do so in myriad ways, including the current rollback of the modest environmental legislation won during the earlier years of the present phase and, to be sure, Camp A's defense of capital's "natural" right to keep turning out organochlorine-based commodities in the face of mounting evidence that these are profoundly ecodestructive.

No one needs reminding that presently the balance of forces has tipped drastically to the right. The environmental rollback is evidence enough of this. But the more capital succeeds, the more it fails. The more pollution, resource degradation, and so on, the more pressure on accumulation. And so the ideological struggle heats up. In this respect, the Steven Safes of Camp A are a platoon of infantry in the Science Wars, whose strategic objective is control over the discourse of nature. And so, on the other side, are the Peter Montagues, and, perhaps, the community of those who produce critique, including the critique of science—the makers of "social texts." I would submit that the ecological crisis in all its ramifications, and especially in its present phase, places the makers of critique in a conjuncture that is not so much radically new (after all, the nightmare did not begin yesterday) as it is newly radical: radical in the revelation of the depth of the crisis, its engagement of that nature which had been taken too much for granted by the critical philosopher preoccupied with language. Radical, too, in its remobilization of the critique of political economy, for it is demonstrably the profit psychosis that consumes nature. Radical in unifying the entire globe as a single entity under assault. And radical in demanding of the strands of critical inquiry that their individual power/knowledge strivings be set aside in a common cause that immanently sublates their partialities.

Thus Marxism, though very much in shambles, is given new life by the relevance of its critique of a capitalism gone berserk. Yet, like the normal sciences, Marxism will have to learn in this new life to listen to, and not dominate, nature. And this lesson is afforded by the opening up of the ecological crisis itself, which pushes an anguished nature, no less than the wretched of the earth, over the edge.

Similarly, what might be oxymoronically called classical postmodernism is now as obsolete as the high modernism it punctured. Given the gathering threat, the postmodern critique of foundationalism clearly has to be rethought. What rang true when framed against the dominative tendencies of modern science's totalizing claims—including those of Marxism—is now glaringly inadequate when the danger to an actual foundation increases before our eyes. For the ecological crisis is no text, though misshapen and false texts play a major role in its working out. It is, rather, a threat to the life that produces texts. Another way of putting this would be to say that the postmodernist critique of science is true, and necessary, but also reductive insofar as it fails to recognize the material dimensions of the ecological crisis. And being reductive, it reveals its own false totalization, in this case, a crypto-idealism. But this, too, can be sublated. Accordingly, expect a more materialist turn in postmodernist critique as it grows under the pressures of the present conjuncture.[11]

The note of optimism that has crept into these last few paragraphs is not, I think, unwarranted. For the ecological crisis reveals as nothing else the disintegration of the Reality Principle upon which modern reason and science are grounded. The stronger the ruling powers appear on the surface, the more corrupt and irrational their legitimations, and the greater the opening for a thought and praxis outside their chains. Given the way the forces today are stacked and the direction things are heading, this may be the slenderest of reeds to cling to. But we need all the reeds we can get.

Notes

1   For example, in the Belgian study the average amount of sperm with strong motility dropped from 53.4 percent in 1977 to 32.8 percent in 1994, a startling decline. Unless otherwise stated, the research data upon which this article is based can be tracked through *Rachel's Environment and Health Weekly* (*REHW*), nos. 438 (20 April 1995) and 448 (29 June 1995). This invaluable service is the work of Peter Montague and is available free of charge through the Internet: erf@rachel.clark.net; or through Environmental Research Foundation, P.O. Box 5036, Annapolis, MD 21403.

2   Vandana Shiva, *The Violence of the Green Revolution* (Penang, Malaysia: Third World Network, 1991). For a wide-ranging study see Shiva, *Staying Alive* (London: Zed, 1989).

3   Tens of thousands of new compounds have been launched into the ecosphere during this period. Not only have all but a tiny fraction gone untested, but virtually no attempt has been made to begin to understand the *interactions* between these compounds, which is the actual context in which the body encounters them. Nor, given the mathematical facts, can this be done within present or future research capacities of society.

4   I cannot, however, resist pointing out that the statistical critique does not deal with contemporary studies from France, Belgium, and Scotland, which presumably employ methodology sensitive to this question. Also, it should be said that Safe neglects the fact that naturally occurring estrogen-mimicking compounds in plants are metabolized in fairly short order, whereas organochlorines and other industrial chemicals that find their way into the body are often not metabolized at all and hence remain active for very long periods, even, in some cases, a lifetime.

5   *REHW,* no. 440 (4 May 1995).

6   I simplify, to draw a major thesis. In fact, a kind of difference-splitting between Camps A and B can take place in which the latter's essential conclusion

would be held as true, but the time span would be altered (perhaps drawn out), or a narrower interpretation of toxic substances made, and so forth. These refinements, however, would not change the basic conclusion.

7   On the first possibility see Donella Meadows, Dennis Meadows, and Jorgen Randers, *Beyond the Limits* (Post Mill, Vt.: Chelsea Green Publishers, 1992). This study is based on systems theory and its social policy implications are questionable. However, it provides a useful set of computer models that show just how dire the situation is. The third possibility, the emancipatory trans-formation of society, runs against the grain of received opinion, which sees a triumphant neoliberal capitalism dealing with all contingencies . . . with the same environmental wisdom it has displayed so far. For a survey, see Martin O'Connor, ed., *Is Capitalism Sustainable?* (New York: Guilford, 1994).

8   For the development of one line of reasoning within this point of view, see Richard Levins and Richard Lewontin, *The Dialectical Biologist* (Cambridge, Mass.: Harvard University Press, 1985).

9   Which, like all biochemically grounded knowledge, requires the assumption of atoms/molecules as individualized actors in a process that metaphorically resembles mechanism.

10   Joel Bleifuss, "Covering the Earth with 'Green PR,'" *PR Watch* 2 (1995): 1–7.

11   I suspect that this shift is what lies behind Derrida's rethinking of Marx. Jacques Derrida, *Specters of Marx* (New York: Routledge, 1994).

The Politics of the Science Wars

Stanley Aronowitz

Introduction

THE main fronts of the Culture Wars—Western civilization versus multiculturalism, high versus low in music, literature, art, and modernism versus postmodernism—are now joined by what may be termed the Science Wars. The defenders of science have framed the debate in terms of reason versus unreason. While the language and vocabularies of science are different from those of the arts, the animus is the same: as for those safeguarding culture and science, the barbarians are at the gates. Those who would demystify science by showing it is subject to the same cultural and social influences as any other discourse, no less than critics who excoriate science for remaining silent when its discoveries are recruited for nefarious purposes, are charged with being prophets of (take your pick) unreason, mysticism, anti-Enlightenment, and nihilism, and with being promulgators of a higher superstition.

Science controversies are by no means as esoteric as one would think. Consider the bizarre result of an FBI investigation into the identity of the notorious Unabomber who, according to the *New York Times*, has, in the last seventeen years, "killed three people and injured 23 others" (Broad 1995). An agent appeared at the New Orleans meetings of the History of Science Association in October 1994 and subpoenaed its membership records because the FBI suspected the "bomber is immersed in the most radical interpretations of the history of science." According to the *Times* report, "professors have begun reconsidering old suspicions, acquaintances and tracts to help solve the crimes." Except for Langdon Winner of Rensselaer Polytechnic Institute, most of the association members and officials the reporter interviewed were donning their detective hats and Sherlock Holmes pipes or were prone to dismiss the bomber as "marginal" in professional science studies. Winner joked he was disappointed the FBI did not consult him on the case. "I feel left out. It's like being left off the guest list for a really good party" (Broad 1995).

Defenders of science such as Paul Gross and Norman Levitt (1994) write polemics that betray philosophical naïveté; others, like the New York Academy of Sciences (NYAS), are hosting conferences and symposia in which the critical theory of science is represented as a virus that must be purged from any affiliation with science and as something the public must be protected against. Members of the faith are circling the wagons against what they perceive to be a serious threat to the church of reason. The practitioners of the relatively new critical studies of science, which spans philosophy, the social sciences, and the humanities, are labeled mystical, anti-Enlightenment radicals who would turn the clock back to the Dark Ages. It does not seem to matter that many social and cultural critics of science are themselves trained in natural scientific disciplines. This knowledge seems to enrage the gatekeepers even more—after all, there is nothing more dangerous to a church than its apostates.

Behind scientificity stands the awesome and the once unassailed edifice of natural science. Together with the similarly God-like house of medicine, it presents itself as both the guarantor of the Enlightenment and the measure of reason. Its methods and results should not be subject to the same withering criticism as Darryl Strawberry, Bill Clinton, a layer cake, or the promises of a politician—except by scientists themselves, and then only within the precept of rigorously established rules. While everybody, including physicists and molecular biologists, is qualified to comment on politics and culture, nobody except qualified experts should comment on the natural sciences.

Of course, the efforts by the scientific community to stifle outside criticism have a long and painful history in the frequent incursions by totalitarian and democratic states to mobilize and otherwise control science by the use of purse strings, blandishments to fame and even fortune, political pressure or, worse, the consignment of recalcitrant scientists to external or internal exile. The sad history of Soviet and Nazi science and the more recent U.S. government decisions to steer public funding to scientific projects that might prove fruitful to helping U.S. corporations meet international economic competition are sufficient reasons for scientists to be wary of outside scrutiny. Like many opponents of government regulation, scientists have assured the public that they are perfectly capable of safeguarding its interests by means of self-policing.

Although scientists are subject to many influences—cultural as well as economic and political—they also constitute a series of communities, no less than industrial workers or the American Bar Association. These communities are formed in the laboratory but also by a hierarchy of research universities and independent research companies, leading nondisciplinary journals such as *Nature* and *Science*, many disciplinary journals, and associations such as the Federation of American Scientists (FAS) and, especially, the American Association for the Advancement of Science (AAAS). These informal and formal networks serve as an arena to circulate knowledge and provide basic information about where research money is going, as a forum for debates within the community, and as a lobby to get more government money and keep outside critics and funders at bay.

On the whole, the system works on its own terms. Government, conservative and liberal alike, is obliged by the specialized nature of scientific knowledge to rely on peer panels to determine the relative merits of funding proposals. Most science writers and journalists are the willing supplicants of a scientific establishment which passes down authoritative news and opinion about the successes of science, successfully manages its failures and, perhaps most important, marginalizes or silences alternative science both at the level of explanation and at the level of discovery. The key players and their institutions are the recognized gatekeepers of what counts as science and, more broadly, what counts as truth (Birrer 1993).

At the bottom of the brilliantly successful history of science since the seventeenth century is the dogma of method. The elements of this dogma are: (*a*) that the book of nature is written in the language of mathematics (for Galileo, who coined the aphorism, it was geometry), and (*b*) that the way to legitimate and reliable knowledge is through the experimental method, the basis of which is our ability to make both observational and falsifiable statements (Popper 1959).

The history of science is written by its winners and their publicists as the story of the smooth, continuous progress of reason. In turn, according to this story, science does best when it is free of interference from the state and also from private interests of any kind, including those of the public (Merton 1973). But there are always detours, and even reversals. Recent studies of the history of science reveal there are invariably zigs and zags. None of these investigations have impugned the claims of science to have made important and valuable discoveries that have enriched our understanding of so-called natural phenomena. It is difficult to deny that science has produced impressive results: rockets do reach the moon; penicillin can treat syphilis, the once life-threatening flu virus can be rendered relatively harmless; and solid state physics has produced an unparalleled information revolution. These technologies, based on theoretical science, have changed the character of everyday life.

My claim is not that science is uninfluential, only that its discoveries themselves and its influence are not unimpeachable. The import of the new social studies of science is to have shown that none of these discoveries amounts to a steady march toward Truth (Mulkey 1990; Barnes 1974). Nor are they free of economic, social, and cultural preconditions or consequences. Although science has its own history and the more sophisticated students of the social relations of science reject the idea that there is a one-to-one correspondence between its social and political preconditions and the content of discoveries, there can be little doubt that contemporary science, no less than its predecessors, is conditioned by these circumstances (Bloor 1976).

For example, during and after World War II, most basic as well as applied science was funded through the Department of Defense (DOD).While DOD did not apply a mechanical criterion of utility in making its awards, many, if not all, funding proposals justified their requests on practical grounds. The choice of investigative objects and their promised results is ineluctably designed to persuade the funder that the payoff is worth the money. In recent years, as competition for public and private money has become even more brutal, most scientists and their organizations have faced the grim prospect that they might be deprived of the opportunity to perform research by tailoring their science even more specifically to practical ends. A recent deal between MIT and a group of private pharmaceutical and bioengineering companies, following a trend in industry-university relations since the 1970s, acknowledges that knowledge is private property and, in return for corporate research grants, assigns a large proportion of patents over to the companies.

Writers such as Dorothy Nelkin, David Dickson, and Martin Kenney, who have documented the close relationship between academically based scientific research and private corporations, are simply subjecting science to the same ruthless criticism that corresponds to the scientific ideal of self-critical inquiry. The critical theory of science does not refute the results of scientific discoveries since, say, the Copernican revolution or since Galileo's development of the telescope. What it does challenge is the notion that science and its discoveries are exempt from ideology critique, deconstruction, or historical investigation that might be trained on any other discourse: literature and art, politics, social scientific theory, and so forth (Dickson 1984; Nelkin 1979; Kenney 1986).

It is not a question of determining the *truth* value of scientific knowledge, if by that notion we designate the correspondence of a proposition to a reality independent of the knower. In the main, the critical theory of science, in conformity to the relationalism of contemporary physical science, allows that, in every stage of its development, the various natural sciences have generated a *regime* of truth consistent with the *frame of reference* within which their theories are generated (Foucault 1980a). Historians such as Shapin and Schaffer, Joseph Agassi, and Georgio De Santillana have produced case studies of leading events in the history of science, demonstrating the salience of what traditional scholars have learned to call "external" factors as precisely the frame of reference within which key elements of "truth claims" are generated (Shapin and Schaffer 1985; Agassi 1982; De Santillana 1955).

In contrast, from the perspective of orthodox positivism, for which "seeing is believing," the historicity of science must be confined to the idea that scientific theories are, at best, successive approximations of a reality existing outside the conditions of investigation. Most theoretical physicists, for example, sincerely believe that however partial our collective knowledge may be of the nature of the microphenomena that constitute the building blocks of physical reality, one day scientists shall find the necessary correlation between wave and particle; the unified field theory of matter and energy will transcend Heisenberg's uncertainty principle.

Biologists tend to regard Darwin and Mendel as worthy predecessors to a history of continuous progress in our collective knowledge of life and its forms. Clearly, for James Watson and Francis Crick, structure or form overrides function, evolutionary process, or the interaction between an organism and its environment as determinations of its essential characteristics. For these biologists, the fundamental account of the nature of life proposed by the new technoscience of molecular biology—that the DNA molecule provides the framework for life and its characteristics—is no longer a hypothesis; it has become *the* fact from which all further experimental and theoretical work should proceed (Watson 1965; Crick 1981).

Ecology, which does not dispute the importance of the DNA molecule, disputes the significance awarded to it by molecular biology. Biologists such as Richard Lewontin and Richard Levins have adopted an ecological perspective and have proposed an alternative paradigm of life (Levins and Lewontin 1985). The interaction of an organism with its own genetic structure is only one of the crucial determinants of its course of development and transformation. Its two environments—its own species and the ecosystem of which it is a part—are intrinsic to both its survival, growth, and transformation. Thus, contrary to classical genetics, both the spatiality and temporality of life forms is essentially indeterminate from the perspective of the genetic code. But since molecular biology is both discursively hegemonic—its account has won broad acceptance in the scientific community—and gets all the research grants from government and pharmaceutical corporations because it is crucially a technoscience, the ecologists are engaged in a Sisyphean struggle for recognition in theoretical terms.

In what follows, I address three distinctly different but closely linked debates: What are the uses of scientific knowledge? To what does scientific knowledge refer? And, perhaps the most complex of all questionings, what are the economic, political, and cultural influences on the content and the results of scientific discovery?

1

The very public counterattack by the scientific conservatives and their publicists is by no means unexpected. What needs explanation is why the scientific establishment, which for years ignored or curtly dismissed critics such as Nietzsche and members of the Frankfurt School as antediluvian cranks, has chosen this moment to recognize that the challenge is worthy of reply. The immediate reason might be that, whereas earlier criticism was consigned to the wilderness of "provocative" but ultimately unrespectable thought and was sequestered within fairly narrow circles, today critical investigations of science and technology have won some academic legitimacy and, perhaps more to the point, have begun to influence some scientists and the public as well.

Undoubtedly, the growing skepticism about the unqualified benefits of science and technology was fueled by the wholesale corporate and government despoilment of the values of scientific disinterestedness and what Robert Merton once termed "communism"—the ethic of knowledge sharing and the concomitant renunciation of private property in knowledge. The claim of science to social neutrality is subject to increasing incredulity since the veritable subsumption, after 1938, of much of U.S. natural science (physics, chemistry, and biology) under the military; since the despoilment of the environment by science-based technologies such as plastics and genetic engineering; in the shameless use of the social sciences in the service of pacification programs in Vietnam; in the dismantling of the welfare state in the name of "public policy;" in controlling workers through industrial psychology; and since the feminist charge that not only have women been occluded when not entirely excluded by science and technology institutions, but that scientific knowledge is itself "gendered" (Harding 1986; Keller 1985).

The genealogy of these critiques may be traced to post-Enlightenment philosophies such as Nietzsche's and Spengler's but also to the various critical theories whose roots are in the deep pessimism of the interwar period engendered by the slaughter of more than twenty million soldiers and civilians during World War I and the subsequent failure of the great powers to make a durable peace on the basis of the democratic and egalitarian precepts of the American and French revolutions to which they were ostensibly committed (Forman 1971). The recognition, especially by postwar intellectuals, that Western scientific and technological culture had wrought contradictory results— weapons of unequaled mass destruction and the material conditions for liberation from work—raised sharply the question whether science was a force for liberation or human annihilation. Of course, the atomic destruction of Hiroshima and Nagasaki did nothing to restore faith in the ineluctable link between science and progress. In the aftermath of the first and last wartime uses of nuclear weapons in 1945, many scientists, who in the wake of Hitler's rise to power participated in their development and urged President Roosevelt to fund the building of atomic weapons, organized an international movement to persuade governments and public opinion that atomic war was unacceptable. But, the genie had been let out of the proverbial bottle. Nuclear weapons have become a major instrument of power; if no government will voluntarily surrender power, no government will lay down its nuclear weapons.

In the 1930s, one of British Marxism's most important institutions, the Social Relations of Science group, included figures prominent in British science, such as the biologist J. B. S. Haldane, the mathematician Lancelot Hogben, and especially the Oxford physicist John Desmond Bernal. Bernal's *Social Function of Science* (1939) and *Science and History* (1957) stand together with Joseph Needham's *Science and Civilization in China* (7 vols., 1954–1985) as perhaps the monuments to the fundamental thesis that scientific knowledge depends, inescapably, upon the social context within which it is produced. Employing a Marxist theory of historical periodicity, they argued that economic, political, and cultural contexts within which knowledge is acquired could not be viewed within the framework of conventional "external" and "internal" distinctions. For them, the development of science and technology crucially depended on the ideology and practical demands of a given *mode of production* and, insofar as these pursuits were "work," were aspects of it.

The New Left, which played an enormously significant role in the early movement against U.S. intervention in the southeast Asia wars, was not merely an activist core committed to such specific goals as peace, participatory democracy, and university reform. It was also an intellectual movement which, in many respects, carried on the tradition of the Social Relations of Science group without collectively being aware of it.

Of course, there are major differences between the Old Left's view of science and technology and that of the generation that came to politics in the 1960s. Bernal, Haldane, and other leaders of the Social Relations of Science were proud legatees of the Enlightenment's celebration of scientific knowledge as a means of human emancipation. For them, scientifically wrought forces of production would, under socialism, be set free from the shackles of capitalism. In short, despite their effort to link scientific development to its historical conditions, they were definitively not critics of knowledge; they were critics of the economic system that thwarted its full development.

In contrast, the New Left was skeptical of science because of its visible *effects*: Rachel Carson's *Silent Spring* became a central text, in part, because it showed, with remarkable literary skill, how scientifically based technologies threatened the survival of life forms of all sorts; as William Gibson has shown, the Vietnam War was preeminently a technological war, a lesson not lost on antiwar activists (Gibson 1986) and, of course, the ever present threat of nuclear conflagration loomed over the lives of the generations born after 1940 for whom the A-bomb was a defining feature of social and psychological worlds.

What Marxism endows to the critical theory of science and technology is its insistence upon a *social* theory of science. After Bernal, Needham, and the Soviet historian of science B. Hessen's account of the "Social and Economic Roots of Newton's Principia" (1931), which inspired Robert Merton's early work in the sociology of science in the 1930s, it is difficult to imagine a purely internal account of scientific discovery (Hessen 1931). The intellectual legacy of the social relations of science approach has encompassed a wide range of knowledge domains: literature, history, the social sciences, and the natural sciences. In the sciences and technology, organizations that exposed the complicity of scientific institutions with the military and the corporations sprung up in the late 1960s under the rubric of social responsibility. To our own time, groups of physicists, biologists, computer scientists, technicians, and physicians have, in their own disciplines, continued to challenge their colleagues to consider the ethical implications of their work and its uses. They have ferociously criticized the doctrine of scientific neutrality as a ruse and a serviceable alibi.

If there is a common theme to these critiques, it is that the dominant ideology of American science is influenced by what C. Wright Mills once termed "the American celebration"—the idea that in this, the best of all possible worlds, science, literature, and other forms of culture flourish because this is truly a pluralistic society in which European-type struggles linked to class, race, and gender have yielded to compromise and consensus, the vehicles through which social progress may be forged.

I want to insist that the convention of treating natural and human sciences according to a different standard be dropped from the perspective of prevailing science studies. Since *scientism* or positivism dominates both social and natural sciences, I want to treat the controversies within each domain as aspects of the same general problematic: How are the objects of knowledge constructed? What is the role of culturally conditioned "worldviews" in their selection? What is the role of social relations in determining what and how objects of knowledge are investigated?

If my thesis is correct that scientificity permeates *all* knowledge domains, including the humanities, the distinctions between the natural and human sciences are not as significant as their similarities. The rigorous separation between them was the product of nineteenth-century physicalism that *reduced* social life to its biological and physical aspects. From the important theoretical posit that humans and their interactions are part of natural history, and therefore that our biological being could not be abstracted from social theory, writers such as Herbert Spencer and Darwin's cousin Francis Galton concluded that social life was merely an efflux of biological behavior, a theme revived in the late twentieth century by B. F. Skinner, Edward O. Wilson, Robin Fox, and Konrad Lorentz, to name only the most notorious among latter-day social Darwinians.

Rejecting this doctrine, German neo-Kantians such as Wilhelm

Dilthey insisted that the natural and human sciences be separated because they obey different laws. While the predictability of physical phenomena on the basis of precise measurement may be achieved on the basis of strict causality, the indeterminacy of social life produced by interaction and consciousness demanded a different although no less rigorous scientific algorithm. Evidence in the historical sciences was not commensurable with that in the natural sciences, since observation and experiment were not possible, except in animal psychology. The method of the natural sciences was, inevitably, historical and hermeneutic. One reads the texts of the social world rather than relying on the laboratory. Moreover, mathematics has little or no role in the human sciences, because the human sciences necessarily abstract quantity from quality and address themselves to extension.

My argument is not only grounded in the pervasive positivism of all academic disciplines but depends on one of the more important interpretations of quantum theory suggested by physicists and philosophers as diverse as Niels Bohr, David Bohm, and Roy Bhaskar. In opposition to interpretations of the Heisenberg uncertainty principle, according to which the problem of measurement of the electron is chiefly epistemological, that is, following Kant's famous doctrine that knowledge of the real is inevitably mediated by the categories of mind, Bohm, for example, offers an *ontological* interpretation of quantum theory. According to Heisenberg, the conditions of measurement prevent a precise determination of position and velocity at the same time. Hence, even if the particle exists objectively independent of the process of knowledge, we cannot derive *meaning* without taking account of the measuring instrument. There is no warrant to predict simultaneously with *certainty* the velocity and the position of a particle, since calculation is inevitably probabilistic.

Bohm's solution to the posit of a split between knowledge and its object, which dominated classical physics and remains among the most influential interpretations of quantum mechanics, is to argue for

> the *undivided wholeness* of the measuring instrument and the observed object. . . . Because of this it is no longer appropriate, in measurements to a quantum level of accuracy, to say that we are simply "measuring" an intrinsic property of the observed system. Rather what actually happens is that the process of interaction reveals a property involving the whole context in an inseparable way. Indeed it may be said that the measuring apparatus and that which is observed *participate irreducibly* in each other, so that

the ordinary classical and common sense idea of measurement is no longer relevant. (Bohm and Hiley 1993, 6)

For Bohm and others who attempt to overcome the dualism of the observer and the observed, the field is not constituted by objects whose antinomy is the subject-observer. While carefully framing his statement to refer to *measurement* rather than an independent subject, it is clear that the knower and the means by which the known become intelligible are intrinsic to objects or processes. For this reason, Bohm argues that probability is not a "defect" of science or its instruments but a property of the universe, a position that corresponds to recent work on complexity in which order and chaos are understood as inextricably linked.

This interpretation of quantum theory has profound implications for the human sciences. In the first place, the attempt to model social sciences on the methods of the old natural sciences is entirely misplaced. Second, the neo-Kantian presuppositions of social constructionism, which, under the sign of anti-essentialism, have refused to acknowledge that biological and physical being are aspects of social being, must similarly be refused. Rather, the continuity as well as the difference between natural and social history, which are the foundations of the biological theory of *integrative levels*, constitute an alternative to that of both physical reductionism and social constructionism. That is, life, in both its evolutionary and structural aspects, is organized within each organism by its physical, biological, and social levels, and the higher levels integrate but also *negate* the so-called lower levels. Here, negation is used in the Hegelian sense: that, in human societies, for example, the social and cultural do not *cancel* physical and biological aspects. After all, through our physical, chemical, and biological organization we are in an unavoidable relationship with the universe which, as we know, is a condition of our existence.

Roy Bhaskar's orientation is, in its essentials, congruent with Bohm's insofar as he agrees on an ontological solution to the problem of knowledge raised by quantum theory. Bhaskar's polemic with positivism and empiricism has been among the most vigorous in contemporary philosophy of science. He has argued against the Kantian constructionist view that science refers to its own conditions of knowledge production rather than to an independent external reality. However, consistent with his philosophical Leninism, he insists on epistemological arguments for the objectivity of the material world.

That is, against the idealism's "irrealism," which Bhaskar ascribes to the Kantian idea that science refers exclusively to the conditions of *knowledge,* he retains objective reality as a(n) (indeterminate) referent independent of the processes of knowing—a distinction which, like Louis Althusser and Gaston Bachelard, remains sympathetic to this Cartesian formulation upon which all epistemological accounts of science ultimately rest (Bhaskar 1975, 1986, 1993).

Bhaskar's "transcendental realism" explains the source of knowledge as the "generative mechanisms" of the objects, but he is not in basic agreement with conceptions such as Bohm's that place processes of investigation within the "real." This leaves him with a self-described realist metaphysics. And, although he invokes historical as well as natural determination to explain the characteristics of scientific knowledge, there is no content to this claim; Bhaskar does not refer to the Marxism of the Social Relations of Science group, to the wealth of historical accounts of the close link between Enlightenment scientific knowledge and power, or to the relation of science and domination suggested by the Critical Theory of the Frankfurt School. In short, Bhaskar is firmly ahistorical, entrenched in the conventions of the Popperian philosophy of science. His own discourse abounds in scientistic formal logic. Even when he writes somewhat sympathetically of Hegel's dialectics, the influence of English positivism and empiricism remains heavily on the page.

2

Critical investigations into the history, philosophy, sociology, and anthropology of science did not, at first, have the powerful effect of parallel efforts in literature and the social sciences for fairly clear reasons: the natural scientific community was far more unified, it was amply funded for what it was doing, and it claimed not merely a nationalist legitimation but the mantle of universal validity. Since the seventeenth century, when Robert Boyle vanquished his opponents by building a scientific community that counterposed the evidence of "seeing" to that of speculation as the criterion of truth, science has seemed to be the common sense of legitimate knowledge. Moreover, especially in Britain, New Left historians just did not see scientists as important social and cultural agents. You don't, for example, find any discussion of science in essays like Perry Anderson's "Consequences of the National Culture." Nor did Raymond Williams, perhaps the most important figure who linked the New and Old Lefts, write about science in relation to culture. However, Williams did study the cultural and industrial impact of television and communications, but from the standpoint of their reception.

Scientists and their organizations have been on the defensive since the antiwar movement exposed the complicity of perhaps the entire science establishment with the Department of Defense (DOD) and since others exposed the degree to which scientists and their institutions had been subsumed by large corporations concerned only with practical applications of scientific theories to the bottom line of profit. Why did scientists accept funding from the military and permit corporations to own the patents to their discoveries? With private as well as state funding, why have they engaged in massive experimentation with and production of artificial organisms in the face of evidence generated by other scientists that genetic engineering—a form of industrial production—in densely populated areas could endanger public safety and health? Why have many scientists remained silent in the wake of the decline of ecosystems essential for life? Why did the majority of scientists fail to follow the post-Hiroshima advice of some of their most eminent leaders—Albert Einstein, Leo Szilard, and Phillip Morrison among others—not to engage in research that would lead to the further development of nuclear weapons? (Lanouette 1992, 259–80).

Indeed, many of these questions have a long history. Under the influence of publicity given to major environmental controversies—most recently, the widely reported cases of many cancer deaths in the western region of the country resulting from radiation contamination during postwar nuclear tests—issues once confined to national security circles are today taken more seriously by broad layers of nonexpert opinion. Although I would not want to deny that an element of panic attends the emergent survivalist ethic that has accompanied frequent reports of imminent or long-term environmental disasters, there is no doubt that practices that went unnoticed and unopposed as late as thirty years ago—such as using incinerators to dispose of garbage; dumping toxic waste, including nuclear materials, into bodies of water and landfill sites; pollution connected with ordinary industrial production techniques; and environmental threats, such as global warming and the holes in the ozone layer—are today under microscrutiny.

The favorite response to these questions by scientists and their academic and journalistic acolytes is that the public has been entirely misled. Most scientists maintain the public position that science is not political but is in the main a disinterested inquiry into the nature of things. Universities and other research institutions claim they remain committed to scientific *discovery* and pay little or no attention to the commercial, military, or other uses to which their work is put. When some research scientists contract patents for their discoveries to corporations in return for funding for pure as well as applied research, the intention remains to advance the single cause of science: knowledge. Scientists portray themselves as vehicles facilitating the advancement of learning and, by application, the progress of humankind. If corporations are willing to provide resources to advance knowledge, scientists have little choice but to accept these resources if they wish to continue their investigations in the face of severe cutbacks of government funding.

No doubt many scientists were, and are, ambivalent about the role of the military in postwar scientific research. Scientists point out that, since the military-industrial complex was (and is) the main source of funds for basic research, many recent breakthroughs in physics and biology that have made life better would have been impossible without this money. For example, they say, even when establishment science agreed to cooperate with the looney tunes Star Wars program, most of the money advanced pure scientific knowledge and had little to do with the actual antiweapons project. In short, scientists are not in power; they cannot control who and what is made of their discoveries. Yet, somehow scientists believe in the integrity of their work, of the scientific community which, despite violations of its autonomy, remains the ultimate arbiter of the worth of a theory or empirical finding. This contradiction plagues science: while it is not in power it controls the conditions of the production of knowledge. We can ascribe this paradox either to naïveté or to bad faith. If the latter, it may help explain the profound animosity of some scientists to their critics.

Of course, some scientists are enthusiastic supporters of U.S. foreign policy; some sincerely believe in the social value of bioengineering and even minimize its dangers just as others have vehemently denied that the industrial uses of scientific research may be injurious to the health of workers and their communities. But the bulk of scientists remain self-styled neutrals, depicting themselves as civil servants of knowledge.

Writers like Dickson (1984) and Kenney (1986), who have profusely documented the extent of the subordination of the scientific community, including university-based research, to instrumental ends dictated by the exigencies of policy or profit, have focused on the issues surrounding the *uses* of science and scientifically based technologies.

In the main, their work belongs to the genre of literature that addresses the question of whether the selection of the scientific object is free of considerations having to do with its applications.

3

While the history of science can be told through "internal" accounts of scientific discovery—this or that experiment or this or that mathematical equation providing the basis for this or that "breakthrough"—putting a finger on economic, political, or cultural *influences* on, let alone *determinations* of, the content of knowledge is a far trickier business. Until recently, many historians and philosophers of science, negligent of the work of Bernal and Needham, denied any significant relationship between the historical and social context of scientific discovery to the key elements of what constitutes science itself. Karl Popper, whose *Logic of Scientific Discovery* (1935) remains among the most influential works of the philosophy of science in this century, and Robert Merton, perhaps the leading figure in the development of the sociology of science, acknowledge the importance of the political and cultural environment for encouraging or repressing free inquiry. Merton argues that democratic regimes provide the best environment for science because they are committed to pluralism and freedom. He allows that the constraints of the scientific community, the state, and economic interests might retard the forward progress of science, at least temporarily. Conversely, economic changes might place new demands on science. For example, Merton's first major study in the social relations of science supported Hessen's contention that seventeenth-century science, of which Newton's magisterial discoveries were the crowning achievement, responded to three crucial historical developments: the demands for a new science of navigation spurred by the global adventures of the British navy and merchant fleets; mining to provide raw materials for the new manufacturing; and the emergence of new mechanical weapons and means of transportation of war. But, it is one thing to argue that war, industry, and commerce, which accompanied the bourgeois revolution in England, provided an underlying cultural framework for the development of science and another to claim that this framework was, itself, crucial for determining its substance. While necessity might be the "mother of invention" insofar as it provides the motive force, is it not science itself, its spirit, algorithms, theoretical debates, and the interchange among members of the scientific community that constitute scientific knowledge?

Popper is prepared to stipulate that scientists are frequently motivated by ideological considerations: they want to improve the human condition by choosing biological research in order to find the basis for treating diseases which shorten life, and their nationalist feelings might inspire them to participate in their country's war effort by offering their talents to weapons development. Certainly, in the face of the rise of fascism, many physicists were recruited to help develop the atomic bomb, often against their own pacifist beliefs. But, he argues in the last analysis, in order to be taken seriously by the scientific community, every scientific proposition must be subject to the criterion of falsifiability, since (he acknowledges) verification suffers many pitfalls.

The "new" social studies, philosophy, and critiques of science have focused as much on scientificity as on politics, as much on processes of scientific work as on the uses of professional science, especially by industry and the state. But social research on science has been extended to a close examination of what goes on in the laboratory, the iconic site of the production of natural scientific knowledge. Bruno Latour argues, for instance, that the laboratory is, in addition to a major site for knowledge production, among the new sites of social power and that culture reflects what happens there as much as the reverse. Latour, Andrew Pickering, Sharon Traweek, and others have accumulated a fair amount of ethnographic documentation about what happens in physics and biology laboratories (Latour 1985; Pickering 1984; Traweek 1988). They have discerned that the abstracted picture of the process of scientific discovery according to which agreement is reached on the basis of rigorous algorithms of proof is, in fact, quite messy.

Following the precept that laboratory life is directly relevant to the results as well as to the process of discovery, Latour and Woolgar (1979) identified three important types of interaction: conversations between scientists, their interpretations of written reports, and the machinery that mediates their perception of what they have actually "seen." All are relevant to what counts as legitimate knowledge or truth, to which the informal influence exercised by leaders of the research project must be added. Indeed, as Thomas Kuhn and others have shown, the judgments that notables in the scientific community pass on results of research are as important as formal procedures for determining validity. The criteria for rejecting scientific theories or discoveries depend as much on relations of power and influence within the scientific community as they do on procedures (Kuhn 1962, 1969).

Rejected theories might offer equally plausible explanations to those that are ultimately accepted. But, lacking the political support

or the context of a large university or of independent scientific organizations, the work of many falls by the wayside. Rejected or marginal
sciences such as parapsychology, the study of clairvoyance, and, in
the wake of the triumph of molecular biology, ecological and evolutionary biology, are just a few examples of the evidence that the scientific "community" as a site of power determines what counts as
legitimate intellectual knowledge, even when the results of the marginalized sciences are obtained by traditional methods.

Working in the late 1950s, Kuhn suggested an alternative theory of
what he called scientific revolutions to contest the commonly held
belief in the idea of scientific progress. Following the arguments of
American philosopher Charles Sanders Peirce, Kuhn argued that
paradigm shifts occur when the leaders in the scientific community
accept a new set of explanations for anomalies that appear in the
course of doing what science always does, puzzle solving. When
enough puzzles, such as the Michaelson–Morley experiment and
Brownian motion, cannot be solved inside the prevailing paradigm, a
new paradigm is proposed. But its adoption by the scientific community depends, ultimately, on whether the major figures accept it. The
new paradigm does not necessarily stand on the shoulders of the old
one but eventually displaces it by its capacity to explain data that
were in conflict with the accepted view, but only on the condition that
leaders in the scientific community accept it. Kuhn remains agnostic
with respect to the truth value of the new paradigm but, tacitly,
accepts a highly relativistic version proposed by Peirce: the truth is
what those who are qualified say it is.

Philosopher Paul Feyerabend (1976) took skepticism one step further. Contrary to the accepted view, he argued that a dominant paradigm of scientific truth may not owe its success to its superiority over
previous theories when measured by traditional Baconian and Popperian criteria but to a variety of factors that are external to experimental or mathematical calculation. Perhaps his most controversial
claim is that the "Copernican revolution" and its crowning achievements, Galileo's "proof" that the earth was in constant motion around
the sun, and Newtonian mechanics were not "superior" to Ptolemaic
science. Using the criteria (a) that knowledge be obtained through
observation and experiment and (b) that theory explain a broad range
of phenomena, he showed that Ptolemaic cosmology was not radically different from that of Copernicus, and that far from being purely
speculative, Aristotelian physics obeyed much of Karl Popper's rule
that scientific theory be refutable by an inductive algorithm.

The sum of these investigations is to bring science and scientificity down to earth, to show that it is no more, but certainly no less, than any other discourse. It is one story among many stories that has given the world considerable benefits including pleasure, but also considerable pain. Science and its methods underlie medical knowledge, which, true to its analytic procedures, has wreaked as much havoc as health on the human body; and it is also the knowledge base of the war machine. Science has worked its precepts deep into our everyday life. Science as culture is as ubiquitous as is science as power.

4

Heisenberg's inclusion of the observer in the field of observation corresponded, as Paul Forman (1971) has argued, to developments in philosophy and culture during the Weimar period of German history. Forman's thesis is that the development of quantum mechanics was crucially linked to the loss of confidence of intellectuals, among them scientists who, during World War I, had given their hearts and their minds to the imperial aims of the German government. The transformation of the zeitgeist was evident in art, politics, and culture. Forman shows that Oswald Spengler's best-selling *Decline of the West* was not only emblematic of the pessimism that afflicted culture but was deeply influential on its development in the 1920s. In particular, the first (and most controversial) chapter of that book argued, on the basis of available anthropological evidence, that the conception of numbers characteristic of Western culture was entirely conventional and, contrary to the common sense belief that our system of numbers was universal, Spengler insisted it be viewed a cultural creation.

The originality of Forman's work is placing the scientific community in the context of these cultural/historical shifts. Rather than view them as a self-contained knowledge community, Forman insists that scientists, especially in turn-of-the-century Germany and Austria, were a major component of their respective intellectual classes. The most difficult aspect of Forman's account is his attempt to find the mediations between culture and scientific knowledge. Amassing a large quantity of historical documentation, he shows that scientists shared in the pessimism that afflicted the rest of the intellectual elite and links this pervasive mood among them to the shift from the old quantum mechanics, which retained large elements of classical physics, to a theory that, in its own reflection, was pushing Einstein's own relativity theory to its limit.

In their study of the controversy between Thomas Hobbes and

Robert Boyle in the seventeenth century, Shapin and Schaffer (1985) argue that the proposition according to which reliable knowledge required as its precondition the evidence of the senses and a procedure whereby such evidence may be tested by means of experiment was severely contested. They show that the modern scientific method is entirely conventional and forged in the context of heated intellectual debates, and that Boyle's triumph over Hobbes and other alchemists and hermeticists was due not only to his ability to constitute a scientific community that supported his thesis that his opponents' claims were suffused with speculation masked as scientific knowledge but also to his construction of a machine, the air-pump, which abetted Boyle's claim to have adduced objective results from experiments. Boyle's triumph was also due to his great capacity to use the tools of popular writing and speech to extend his influence. Despite the difficulties with the experiments conducted by Boyle and Robert Hooke, they were able to prevail because they captured intellectual hegemony, not so much on the basis of their experimental results which proved to be significantly flawed but perhaps more by their social and literary skill. Shapin and Schaffer (1985, 77) argue that

> the role of Boyle's literary technology was to create an experimental community to bound the discourse internally and externally, and to provide forms and conventions of social relations within it. The literary technology of virtual witnessing extended the public space of the laboratory in offering a valid witnessing experience to all readers of the text. The boundaries stipulated by Boyle's linguistic practices acted to keep that community from fragmenting and to protect items of knowledge to which one might expect universal assent from items of knowledge that generated divisiveness. Similarly, his stipulations concerning proper manners in dispute worked to guarantee that social solidarity that produced assent in matters of fact and to rule out of order those imputations that would undermine the moral integrity of the experimental form of life.

"Seeing is believing," the common sense of everyday life, became the common sense of science as well. The burden of the study of the Hobbes-Boyle controversy is to historicize the requirement that scientific propositions contain observational statements. As De Santillana (1955) showed, Galileo's rise to prominence at the turn of the seventeenth century owed its success to his literary and social abilities much more than to what turned out to be fairly dubious experimental work. That is, he was able to disseminate his *ideas* to other practitioners who supported his struggle against the academic establishment, which derided his argument that the earth was not stationary, and he achieved his greatest notoriety only after he resigned his academic position in Padua and placed himself under the patronage of the powerful Florentine Crown.

These examples from the history of modern science may not satisfy those for whom science is the single inquiry whose methodological rigor makes its actual results, as distinct from its motive force, immune from social and cultural influences. As Sir Karl Popper retorted to similar arguments advanced by Theodor Adorno, perhaps the quintessential critic of positivism, the question is not whether culture or history influences science but whether such influences can be filtered out of scientific theories and experimental results by scientific method (Adorno et al. 1971). Forman is acutely aware of this challenge and shows that one may render a plausible account of the uncertainty principle that Bohr insisted inhered in both inquiry and the processes it studies by understanding the historically specific frame of reference of Weimar culture.

Forman's work takes the social relations of science hypothesis to the level of culture or Weltanschauung (worldview). His meta-argument is that profound changes in scientific knowledge are produced not only in the laboratory and by mathematical calculation, or, as Latour and Shapin claim, by social, literary, and material technologies, but also by the zeitgeist. While Popper and the positivists scoffed at such assertions, since the posit of cultural despair is not subject to refutation, the great mathematician David Hilbert, the leading wave mechanics theorist Ernst Schroedinger, and Einstein were all outraged and ultimately renounced the Bohr-Heisenberg thesis of a probabilistic universe, which became one strong interpretation of the uncertainty principle. But, there is little doubt that Bohr and Heisenberg finally won the day. Was it because the theory was unrefutable? Or is it that the "proof" of such a theory ultimately resides in whether influential scientists accept it? What are the multiple determinants of acceptance?

One could speculate in a similar fashion about the triumph of molecular biology. Manuel de Landa (1991) has argued that the history of scientific and technological advances is intimately connected to militarism. No war gives greater credence to this thesis than World War II. The fundamental meta-advance was the close integration of science and technology. From von Neumann's explorations of mathematics, which were linked to the work on radar and atomic weapons, to the emergence of information sciences by Shannon and Weaver emanating from Turing's work and the development of plastics, by the end of the war, technoscience was already an adolescent. But, molecular biology and solid state physics bring technoscience into full maturity because the theory itself is a technology. Like Ernest Lawrence's development of "machine physics" in the work on the atomic bomb, Watson and Crick's discovery of the double helix structure of the DNA molecule was intrinsically linked to the production of new forms.

Perhaps the most sweeping critique is advanced by feminists who claim that science tends to be a masculinist discourse. In the words of Elizabeth Fee, feminist critiques "[expose] how the foundations of our [scientific] knowledge have been built on the assumptions of male domination" (quoted in Bleier 1986). It is not only Evelyn Fox Keller's historical account of the marginalization and exclusion of women from the natural scientific disciplines or the more or less systematic male denial of the value of the work of those who breached the barriers. Philosopher Sandra Harding, biologist Donna Haraway, and others claim that aspects of scientific knowledge are gendered. Science, in Haraway's phrase, is "politics by other means," a proposition that strikes to the heart of the guiding ethic of science: its cultural neutrality and disinterestedness (Haraway 1989).

The question is whether science can evade what every other discourse must face: its dependence on, as well as struggle for autonomy from, culture. Feminists, ecologists, AIDS activists, and those who, from a scientific standpoint, have examined the use of imputed racialized genetics to explain differences in school performance are acutely aware that much of established science remains in a state of deep denial regarding these issues. The debates between established science and its activist critics (including some scientists) are increasingly well-known. The theory of global warming is still hotly contested in scientific circles; the relation of the HIV virus to AIDS and, perhaps more urgently, whether research aimed at finding a "magic bullet" to stop the disease is more effective than broadly based epidemiological investigations remains uncertain; and even though most biologists scoff at the evidence for racialized genetics, many still accept the underlying idea that some genetically derived intelligence exists, independent of the social and cultural conditions that affect school performance.

At the end of the day, the many questions of scientificity and science and its influences cannot be settled by means of a fail-safe method of inquiry. Einstein's relativity theory was subjected to official skepticism twenty years after the publication of his Special Theory article in 1905; and equally passionate partisans of wave and matrix explanations for the behavior of electrons were unable to reach agreement for decades. Similarly, the dogmatists of internalism—Gross and Levitt and so on—who posit scientific method as ruthlessly self-critical and science as ultimately self-contained, are likely to remain unconvinced, even though some of the more prominent writers in science studies have trained in physics (Pickering, Kuhn, Keller, Feyerabend) or biology (Haraway). But the vituperation that accompanies the dogmatists' defense, not only of science but of empiricism and positivism, is rather quaint. Unlike Sir Karl, they are not usually sufficiently philosophically sophisticated to grasp, let alone refute, their opponents' claims.

Underlying many of these disputes is the pervasive fear, by many scientists as well as by the economic and political elites, that the experience of ACT-UP will spread. ACT-UP, an organization of gay and straight activists concerned with the spread of the AIDS epidemic, began by demanding more funds for AIDS research and now intervenes in scientific disputes as to the best treatment for the disease. By force of circumstance as much as by conscious knowledge, it was obliged to question many of the precepts of its own membership, namely, that the doctor knows best. Gradually, members became sophisticated in many areas of the history and philosophy of science. Imagine a polity capable of challenging the uses and truth claims of scientific and technological research. Imagine a new scientific *citizenship* in which democratic forms of decision making were shared between the scientific community and the public. With ghosts of Nazi and Soviet calumnies in their imagination, scientists tend to cringe at the prospect. But they have to face their own conviction that, as far as their work goes, democracy is only appropriate for the few.

References

Adorno, Theodor, et al. 1971. *The positivist dispute in German sociology.* New York: Harper Torchbooks.
Agassi, Joseph. 1982. *Faraday as philosopher.* Cambridge, Mass.: Harvard University Press.
Barnes, Barry. 1974. *Scientific knowledge and sociological theory.* London: Routledge.
Bhaskar, Roy. 1975. *A realist theory of science.* London: Routledge and Kegan Paul.
———. 1986. *Scientific realism and human emancipation.* London: Verso.
———. 1993. *Dialectics.* London: Verso.

Birrer, Frans. 1993. Counteranalysis: Toward social and normative restraints on the production and the use of scientific and technological knowledge. In *Controversial knowledge,* edited by Thomas Brante, Steve Fuller, and William Lynch. Albany: State University of New York Press.

Bloor, David. 1976. *Knowledge and social imagery.* London: Routledge and Kegan Paul.

Bohm, David, and B. J. Hiley. 1993. *The undivided universe.* London: Routledge.

Broad, William. 1995. Esoteric edge of academia roiled by hunt for a bomber. *New York Times,* 5 August, 85.

Crick, Francis. 1981. *Life itself: Its origin and nature.* New York: Simon and Schuster.

De Landa, Manuel. 1991. *War in the age of smart machines.* New York: Zone.

De Santillana, Georgio. 1955. *The crime of Galileo.* Chicago: University of Chicago Press.

Dickson, David. 1984. *The new politics of science.* New York: Pantheon.

Fee, Elizabeth. 1986. Critiques of modern science: The relationship between feminism and other radical epistemologies. In *Feminist approaches to science,* edited by Ruth Bleier. New York: Pergamon.

Feyerabend, Paul. 1976. *Against method.* London: New Left Books.

Forman, Paul. 1971. Weimar culture, causality, and quantum theory. In *Historical duties in the physical sciences,* vol. 19. Philadelphia: University of Pennsylvania Press.

Foucault, Michel. 1980a. *Power/Knowledge.* New York: Pantheon.

———. 1980b. *History of sexuality,* vol. 1. New York: Vintage.

Fox Keller, Evelyn. 1985. *Reflections on gender and science.* New Haven, Conn.: Yale University Press.

Gibson, William. 1986. *The perfect war: Technowar in Vietnam.* Boston: Houghton Mifflin and Co.

Gross, Paul, and Norman Levitt. 1994. *Higher superstition: The academic left and its quarrels with science.* Baltimore, Md.: Johns Hopkins University Press.

Haraway, Donna. 1989. *Primate visions: Gender, race, and nature in the world of modern science.* London: Routledge.

Harding, Sandra. 1986. *The science question in feminism.* Ithaca, N.Y.: Cornell University Press.

Hessen, B. 1931. The social and economic roots of Newton's *Principia.* In *Science at the crossroads.* London: Kniga.

Kenney, Martin. 1986. *Biotechnology: The university-industrial complex.* New Haven, Conn.: Yale University Press.

Kuhn, Thomas. 1962, 1969. *The structure of scientific revolutions.* Chicago: University of Chicago Press.

Lanouette, William. 1992. *Genius in the shadows.* New York: Macmillan.

Latour, Bruno. 1985. *Science in action.* Cambridge, Mass.: Harvard University Press.

Latour, Bruno, and Steve Woolgar. 1979. *Laboratory life.* London: Sage.

Levins, Richard, and Richard Lewontin. 1985. *Dialectical biologist.* Cambridge, Mass.: Harvard University Press.

Merton, Robert. 1973. *Sociology of science.* Chicago: University of Chicago Press.

Mulkey, Michael. 1990. *Sociology of science.* Bloomington: Indiana University Press.

Nelkin, Dorothy. 1979. *Controversy.* London: Sage.

Pickering, Andrew. 1984. *Constructing quarks: A sociological history of particle physics.* Chicago: University of Chicago Press.

Popper, Karl. 1959. *The logic of scientific discovery.* New York: Science Editions. Originally published in 1935 as *Logik der forschung.*

Shapin, Steven, and Stephen Schaffer. 1985. *Leviathan and the air-pump: Hobbes, Boyle, and the experimental life.* Princeton, N.J.: Princeton University Press.

Traweek, Sharon. 1988. *Beamtimes and lifetimes.* Cambridge, Mass.: Harvard University Press.

Watson, James. 1965. *The molecular biology of the gene.* New York: W. A. Benjamin.

Werskey, Gary. 1978. *The visible academy.* New York: Holt Reinhart and Winston.

Consolidating the Canon

N. Katherine Hayles

A sure indication that the cultural and social studies of science have come of age is the attack that Paul Gross and Norman Levitt launch on them in *Higher Superstition: The Academic Left and Its Quarrels with Science*.[1] Whatever else the book does, it demonstrates convincingly that constructivist arguments can no longer be ignored. Gross and Levitt call attention to these emerging fields so they will be exterminated, but those of us who work in them can draw a different lesson from their book. I suggest we use the opportunity to articulate the ideas that have been so extensively documented by different researchers working with various materials that they constitute a canon of the cultural and social studies of science. For our purposes, a canon does not imply that its ideas are beyond dispute, only that they are firmly enough established so they will not easily be displaced.

Thinking about a canon should not be mistaken for drawing the wagons into a circle. The inflammatory rhetoric in which Gross and Levitt engage attempts to set up an us-against-them situation. They imply that if the community involved in the cultural and social studies of science will only purge offending members from their midst, the remainder may perhaps be acceptable to the arbiters that Gross and Levitt set themselves up to be. Surely the world has seen enough of this tactic to know what to make of it. To accept such a proposal (the idea has been discussed with apparent solemnity on certain electronic discussion lists devoted to STS [Science, Technology, and Society]) is to follow the agenda that Gross and Levitt want to set, which in itself should be good reason to be suspicious of it. Instead of acting like naive lambs accepting the wolf's suggestion for our own good, I suggest we think critically about the essential claims at the heart of Gross and Levitt's arguments.

To a surprising extent, the vitriolic nature of their attack has succeeded in camouflaging the basic weakness of their evidence. Some readers might assume that, given the sheer quantity of sarcasm and rejoinder that Gross and Levitt unleash, at least some portions must be true. But as Roger Hart's devastating analysis of the book demonstrates, careful examination of their arguments on a case-by-case basis reveals how shoddy their scholarship is.[2] Hart reviewed works cited by Gross and Levitt and compared the arguments, quotations, and summaries that appear in Gross and Levitt book with passages in their original form. He discovered not merely a few isolated errors but a systematic pattern of misleading and unfair quotation, a failure to read accurately, a failure to grasp an argument's main thrust, opportunistic and biased use of sources, and the use of character assassination and verbal abuse rather than reason to discredit an opponent's work.[3] Because the scope of Gross and Levitt's canvas is so broad, readers are apt to grant the authors' case on the basis of works that the readers do not know, even if they recognize the distortions in discussions of works with which they are familiar. Thus, in circles sympathetic to the cultural and social studies of science, one is still apt to hear such opinions as "Gross and Levitt have done their homework" or "Certainly some of the attacks are valid." Anyone inclined to make these concessions should read Hart's article before granting Gross and Levitt the benefit of any doubt. If there are reasonable cases to be made against constructivist positions, *Higher Superstition* has not made them.

Beside camouflaging shoddy scholarship, the virulent rhetoric of *Higher Superstition* has worked to obscure the positive claim made for scientific inquiry as distinct from the scurrilous negative attacks on individual scholars. Because the attacks are sensationalized, they tend to be what readers remember, rather than the view of scientific inquiry that Gross and Levitt present. If we put aside the ad hominem attacks, misreadings, and distortions (which admittedly constitute the bulk of the book), we can clear the air enough to make their central claim visible. Whereas they are willing to admit in principle (though never, it turns out, in practice) that scientific inquiry may be influenced at the organizational level by sexism, priorities set by funding sources, and so forth, the "core of scientific substance" in science remains entirely unaffected by the cultural institutions in which it is embedded (40). One advantage of articulating a conceptual canon is to demonstrate that, in view of what is known about scientific practice, theorizing, and conceptualization, such a claim simply cannot be sustained.

For the sake of brevity, I will limit my candidates for canonization to three main points. In different ways, each shows why the claim for a pristine "core" of science is unsustainable. First, *it matters what questions one asks and how one asks them.* The relation between how questions are posed and how they are answered is important because it is often in the formulation of questions that cultural factors play a constitutive role. Donna Haraway, in her study of Clarence Ray Carpenter's work with rhesus monkeys on the island of Cayo Santiago, shows that Carpenter's formulation of questions proved decisive for the kind of observations he made and therefore for the conclusions he reached.[4] Asking how groups are formed and maintained, Carpenter conceptualized group formation in terms of male dominance. In this context, it made sense to remove the alpha male from a group and observe what happened to the group's social organization as a result. When the group appeared disorganized and lost territory, it seemed natural to conclude that male dominance was the primary means by which social organization was maintained. It did not occur to Carpenter to remove females and see how their absence affected group organization. The nature of the questions thus influenced the experimental design. In addition, assumptions implicit in the questions also helped to determine what counted as an observation. Carpenter observed primarily dominance behavior rather than nurturing, food-gathering, or grooming.

The wedge that Gross and Levitt seek to drive between science's "core" and the contexts out of which questions arise, research protocols emerge, funding priorities develop, and institutional practices solidify is akin to the distinction drug manufacturers make between a drug's "effects" and its "side effects." In reality, of course, there are only effects. The distinction lies not in nature but in the economic, discursive, and institutional systems that make it profitable to relegate some effects to the periphery while putting others at center stage. A similar strategy informs the distinction, often made in the philosophy of science, between the "context of discovery" and the "context of justification." This distinction continues to be invoked, long after its close cousin, the dichotomy between theory and observation, has been relegated to the Attic of Untenable Ideas. The argument goes as follows: while all kinds of factors, including cultural ones, may be at work in the context of discovery, the context of justification represents a purer realm where the mechanisms of falsification and replication operate to ensure that the knowledge produced is independent of cultural biases. As numerous studies have shown, however, the context of justification is shot through with presuppositions carried over from the context of discovery.[5] The division between the two contexts is as leaky and artificially constructed as the division between a drug's effects and side effects or between the posing of a question and the formulating of an answer.

Ironically, as Steven Shapin and Simon Schaffer document in *Leviathan and the Air-Pump: Hobbes, Boyle, and the Experimental Life*, the distinction between discovery and justification was constituted through social institutions that would later be erased, with the help of that very distinction, from the scene of justification.[6] The documents and practice produced by the nascent Royal Society made an important distinction between "trying," the actual experimental work in the laboratory, and "showing," demonstrating the experiment before an audience at the Royal Society. Only people of the appropriate social class had the power to witness; what the laboratory workers saw did not count toward the production of scientific knowledge. Here, at the founding moments of the distinction between discovery and justification, we can see that the distinction was constituted through class assumptions that made "showing" more productive of consensual scientific knowledge than "trying." Cultural assumptions, far from being excluded from the scene of justification, were instrumental in constituting it as a superior epistemological realm.

The distinction between discovery and justification, like the division between a conceptual "core" and a culturally influenced periphery, make it possible to relegate many of the results of Western science into the limbo of "side effects." The content of the knowledge is judged to be value neutral; values only enter the picture when the knowledge is applied or used in certain ways. But the knowledge would not have been produced in the form it was if certain questions rather than others had not been asked. Question and answer, and content and context mutually produce each other. Prophylactic barriers between them cannot assure safe epistemology. I take this to be Sandra Harding's point when she writes about Western science as an ethnoscience.[7] From the viewpoint of other peoples around the globe, some of the assumptions that mark Western science as an ethnoscience are the distinctions used to separate the content of knowledge from the ways in which that knowledge is discovered, disseminated, and deployed.

A standard objection to the view that Western science is an ethnoscience is the claim that it follows the Scientific Method, whereas ethnosciences do not. This objection leads to the second candidate for canonization. Researchers in the social and cultural studies of science have found that *there is not one but several scientific methods*, each adapted to the requirements of the field in which it is used. In fact, it has proven impossible to articulate a single method that is valid for all scientific fields. As David Hess succinctly summarizes, "There is no single Scientific Method to which all scientists can refer; instead, laboratory procedures are opportunistic and contingent on social factors."[8]

The continuing improvisation that leads to methodological diversity can be seen in the new computer visualization programs used to simulate complex phenomena. When the Danish Hydraulic Institute was asked to evaluate the environmental effects of a proposed eleven-mile bridge across the Great Belt, they devised a simulation program that served as a testing arena for experiments impossible to conduct in real life because of the scale of the operations and the speed and complexity of the interactions.[9] This simulation was not used to create a mathematical model which was then tested against actual experimental results. Rather, the simulation was the experiment. Similar uses of simulation and visualization programs are taking place in a variety of fields. Biochemists use them to study complex stereochemical interactions between macromolecules; and artificial life researchers use them to create the "organisms" that they hope will provide crucial information for formulating a theoretical biology of life.[10] No one disputes that hydraulics, biochemistry, and the study of complex adaptive systems are sciences, but clearly they are using methods that depart in significant ways from standard accounts of the Scientific Method.

What is the significance of having multiple methods rather than a single Scientific Method? Obviously, the multiple-method situation makes it more difficult to define what is and is not a science, for one can no longer assume that a field is a science if it employs the Scientific Method. Equally significant are implications for the claim for science's universality. One sense of that claim is that science is universal in having a methodology that applies to all fields of scientific inquiry. When there is no single method for the production of reliable knowledge, scientists come to seem more like *bricoleurs* than practitioners of a rigorously logical and universally applicable Method. To paraphrase Forrest Gump, "Science is as science does," a position that puts more emphasis on practical results than on methodological purity.

The issue of science's universality (or lack thereof) leads to the third candidate for canonization. As Humberto Maturana and his collaborator Francisco Varela put it, *everything said is said by an observer.*[11] Implicit in this aphorism is the realization that any observer sees from a perspective, and the perspective determines in obvious and subtle ways what is seen. The interfaces through which we construct our worlds include language, culture, and physiological processes that allow us to make a world out of what I have elsewhere called the unmediated flux. To call our interactions with the flux "information," "signals," or "data" always already presumes prior sensory processing, for what counts as data for one species—vision in the infrared region, for example—are undetectable for another. Researchers in linguistics, physiology, anthropology, and cultural studies agree (although for different reasons) that the world as it exists for us is the result of active construction, not simply passive perception. In scientific work, enculturation into a perspective is extensive and precise, ranging from learning how to read instruments properly to developing an intuition about how natural phenomena can plausibly be expected to act.

In what sense, then, is one entitled to say that scientific knowledge is universally valid? Certainly not in the sense that it applies to constructions made from every perspective, for there are perspectives, both human and extrahuman, from which any given form of scientific knowledge would make no sense. Rather, it means that *someone who is properly enculturated and instantiated into a certain perspective* can confirm scientific knowledge claims over a wide range of sites. Stating the case like this makes clear that scientific knowledge is not truly universal, since it depends on an appropriate positionality and embodiment. Rather than universality (an impossible requirement, as Donna Haraway implies when she calls it "the god-trick"), scientific knowledge can, in the best case scenario, be reliable and consistent.[12] An important difference between universality and consistency is that the latter does not depend on absolute truth claims. It occupies the more modest (and humanly possible) position of providing constructs which provide reliable knowledge over the range of perspectives for which the constructs hold good.[13]

In calling this instantiated and situated knowledge reliable, I mean both to reference and modify John Ziman's argument in *Reliable Knowledge: An Exploration of the Grounds for Belief in Science.*[14] Ziman rightly eschews the idea of science's universality in favor of its reliability and consistency, which he links to what he sees as the main advantage of modern Western science over other knowledge practices: its ability to establish consensus among disparate observers. In going beyond Ziman's argument, I want to emphasize that consensus re-

quires enculturation. Not any observer but only appropriately enculturated observers, will be able to agree to the knowledge claims at
issue. Enculturation contributes to the constructive interfaces through
which we make our worlds. These interfaces, precisely because they
are constructive, both enable and limit what can be seen. To paraphrase Sandra Harding, they produce both systematic blindnesses as
well as systematic insights.[15] To illustrate, suppose we ask what light
is. John Milton constructs one version in his passionate address to it
in *Paradise Lost*; a very different construction defines it as that portion of the electromagnetic spectrum visible to humans. What it
means cannot be separated from the position from which one constructs it, whether as a blind male poet, as a comparative anatomist, as
a feminist philosopher of science, or as a visionary physicist. Even
this range, broad as it is, accounts for only a small portion of the
human spectrum, to say nothing of the much broader range of non-
human perceptions that stretch beyond what any of us can take in.

What are the systematic blindnesses that enculturation into modern
Western science produces? In one way or another, they relate to the
doctrine of objectivism that we have inherited as part of the Western
tradition. For my purposes here, I will define objectivism as the
premise that we know the world because we are separated from it.
There are indeed systematic insights to be gained from this premise,
but there are also systematic insights to be gained from accepting the
inverse proposition that we know the world because we are connected to it. Each position incurs liabilities. The systematic blindnesses produced by objectivism include the divisions it introduces
between content and context, knowledge and values, and knowledge
and knower. Behind or underneath these dichotomies, connections do
exist. Consider the wedge that objectivism drives between knowledge
and emotion; emotion is ruled out of bounds, for it springs from connection rather than separation. Does this mean scientists do not use
emotion in their professional work? Obviously not, but it does affect
how emotion can be represented and acknowledged. The struggle resonates throughout Evelyn Fox Keller's biography of Barbara McClintock, with its revealing title, *A Feeling for the Organism*.[16] Since
expressing emotion is a gender-inflected activity in our culture,
McClintock walked a delicate line. She wanted to self-identify as a
scientist (coded male), but she also wanted to affirm the importance of
her empathic connection (coded female) to the organisms she studied.
The complexities of how to do both these things at once is one indication of the toll that objectivism exacts from those who work under

its sign. Nor is the price limited to individual practitioners. If we know the world because we are separated from it, it follows that knowledge practices *about* the world should be insulated from their consequences *in* the world. Here objectivism drives a wedge between science and society, leading to the peculiar belief that funding priorities responsible for setting research agendas can somehow be separated from the knowledge produced by those agendas.

Lest these remarks seem like science bashing, I want to note that the cultural and social studies of science have been no less affected by the tradition of objectivism than science itself. For the past several years, prominent researchers in the cultural and social studies of science have chosen to bracket the question of reference. An assumption that often informs this bracketing is the notion that if knowledge is culturally contingent, it cannot be reliable. I want to argue, on the contrary, that culturally contingent knowledge is the only knowledge available to us as finite, embodied, culturally situated human beings. It is a fallacy, born of objectivist tradition, to think that culturally contingent knowledge is not reliable. If we grant that knowledge is not universal, does the sky fall as a result? Since we are not universal either, what would we do with universal knowledge, even assuming such a thing could be imagined? Knowledge is useful to us because, not in spite of, the fact that it is limited, partial, and perspectival. The important point is to understand, as fully as we can, the limitations and enablings that the perspectives from which we see affect what we see.

Let us suppose that we agree on consolidating a canon. How should it be used? Whereas Gross and Levitt want to foster an us-versus-them mentality, I urge that we find out who our allies are in various disciplines, including the sciences, and work to forge alliances with them. I believe that some constructivist positions, particularly the ones I have argued for here, are fully compatible with the beliefs that a working scientist must adopt to carry on her or his work. Most scientists that I know are fully aware of how constructed their results are, and they do not see this awareness as incompatible with their belief that the results also say something about how the world works.

What factors are necessary to make alliances with (some) scientists possible? For the last ten or fifteen years, academic discourse in the humanities, particularly in literary studies, has been dominated by a rhetoric of resistance. Too often, this rhetoric has been practiced in a vacuum. Session after session at Modern Language Association Conventions are filled with speakers who proclaim their resistance to hegemonic power. The rhetoricians style themselves as freedom fighters staking everything to oppose oppressive forces, but in fact they are risking very little in these forums where the audience nods in agreement. The people who would really disagree with these arguments are not present in the room, will not in all likelihood read whatever publications come out of the sessions, and often are not even aware that such arguments have been made.

This is the kind of situation in which a book like *Higher Superstition* can thrive, for it titillates its audience with horror stories that, though they are little more than caricatures, are nonetheless effective because many readers, especially in the sciences, have no independent experience of the research by which they can judge the distortions and misrepresentations. I believe that we need to do a much better job than we have of reaching out to scientists and engaging them in dialogue about the issues at stake in our research. Instead of posturing resistance, we need to forge alliances. To this end, it would help enormously if we were willing to make arguments that did not place constructivism in opposition to reliability. Many scientists feel they have little stake in defending science's universality, but most believe they have very definite and specific stakes in defending the idea that science can produce reliable knowledge.

One of the grotesque exaggerations in which Levitt and Gross indulge is the fantasy that the cultural and social studies of science are responsible for cuts in funding for basic scientific research. In reality, the cultural and social studies of science are far too minuscule to effect large-scale changes of this kind, even if they wanted to (and many people who work in the field have no such intention). The real causes are not difficult to find. Chief among them is the end of the Cold War and the resulting cuts in defense contracts, especially with universities, the aerospace industry, and weapon manufacturers; global competition to which U.S. corporations have responded by downsizing or eliminating funding for basic research; and a political climate in Washington that has swung sharply to the right, with resulting cuts not only to the NEA and NEH but also to the NSF and to higher education in general.

We might ask ourselves what political purposes it serves for Gross and Levitt to lay at the door of the cultural and social studies of science the results of these much larger economic and global changes. Their (mis)representation serves two mutually reinforcing political ends. It depicts the cultural and social studies of science as a subversive force with dangerous consequences for science. At the same time, it deflects attention away from the real problems confronting the sciences as wide-ranging structural changes take place that significantly affect how Big Science gets funded. In their final chapter, Gross and Levitt reach new heights of absurdity when they imply that if only the obstreperous people in literature departments can be brought to heel, the threat to science will disappear and science can continue its business as usual. Surely this is a dangerous illusion for any scientist to accept, for it invites a misdirection of effort at exactly the time when science can afford it least. Does anyone really believe that if all work in the cultural and social studies of science were to stop tomorrow, that the deep structural problems science faces would diminish to any significant extent? A far more constructive approach is to seize the moment to rethink what the relation between science and society ought to be. Properly framed, the challenge that the cultural and social studies of science pose to objectivism should make science stronger, not weaker, by clarifying its connections to the complexities of instantiated and situated human life.

From the point of view of someone who works in the cultural and social studies of science, the irony in *Higher Superstition's* grotesque exaggerations is the sad fact that to date, the cultural and social studies of science has had little effect, if any, on the way science is done. We get the blame without getting results. If the cultural and social studies of science allows itself to be demonized in the ways that Gross and Levitt intend, we will miss out not only on an opportunity to forge alliances but also on the chance to build bridges between constructivist arguments and practical politics. One of the implications of a constructivist position, as I have argued above, is the possibility it offers to break out of the false dichotomies imposed by objectivism. This means recognizing that there is a relation between how knowledge is produced and how it is used, between the myth of science's universality and the decisions that result in a constantly degrading environment, and between the political and social contexts from which questions emerge and the ways in which they get answered. To engage in *real* resistance to the pernicious effects of objectivism, we cannot afford to speak into a vacuum. We should argue with colleagues within the cultural and social studies of science who claim that their research has no implications for how science should be done.[17] We should talk with our scientific colleagues in ways that are not condescending, not unnecessarily obscure, and not off-putting. We should grant that science does work and seek clearer and more satisfactory explanations of why this is so. We should go beyond bracketing the question of reference and begin to engage the deep episte-mological issues that constructivism raises. Most of all, we should speak to others besides ourselves. God resides at least as much in the connections as in the details.

Notes

1   Paul R. Gross and Norman Levitt, *Higher Superstition: The Academic Left and Its Quarrels with Science* (Baltimore, Md.: Johns Hopkins University Press, 1994).

2   Roger Hart, "The Flight from Truth and Reason: Higher Superstition and the Refutation of Science Studies," unpublished manuscript.

3   Donna Haraway, *Primate Visions: Gender, Race, and Nature in the World of Modern Science* (New York: Routledge, 1989), 84–111.

4   Ibid.

5   Among the influential studies making this point are Bruno Latour's *Science in Action: How to Follow Scientists and Engineers through Society* (Cambridge, Mass.: Harvard University Press, 1987); Evelyn Fox Keller's *Secrets of Life, Secrets of Death: Essays on Language, Gender, and Science* (New York: Routledge, 1992); Bruno Latour and Steve Woolgar's *Laboratory Life: The Social Construction of Scientific Facts* (Princeton, N.J.: Princeton University Press, 1986); and a volumed edited by Andrew Pickering, *Science as Practice and Culture* (Chicago: University of Chicago Press, 1992).

6   Steven Shapin and Simon Schaffer, *Leviathan and the Air-Pump: Hobbes, Boyle, and the Experimental Life* (Princeton, N.J.: Princeton University Press, 1985), 55–65.

7   Sandra Harding, "Is Modern Science an Ethnoscience? The Challenge of Universal Ethnosciences," in *Science and Technology in the South: Sociology of the Sciences Yearbook 1995*, ed. Terry Shinn, Jack Spaapen, and Roland Waast, (Dordrecht, The Netherlands: Kluwer, forthcoming).

8   David J. Hess, *Science and Technology in a Multicultural World: The Cultural Politics of Facts and Artifacts* (New York: Columbia University Press, 1994), 3.

9   Jacob Steen Moller, "Virtual Water: Modeling Rivers and Seas" (Paper presented at the Virtual Nature Conference, Humanities Research Center, Hollufgaard, Denmark, June 1995).

10   Timothy Lenoir discussed computer modeling of macromolecules in "Machines to Think By: Visualization, Theory, and the Second Computer Revolution" (Paper presented at the Society for Literature and Science Conference, Los Angeles, California, November 1995). Discussion of artificial life simulations can be found in N. Katherine Hayles, "Narratives of Artificial Life," in *Future-natural: Nature, Science, Culture*, ed. George Robertson, Melinda Mash, *et al.* (London: Routledge, 1996).

11   Humberto R. Maturana and Francisco J. Varela, *Autopoiesis and Cognition: The Realization of the Living* (Dordrecht, The Netherlands: D. Reidel, 1980), xxii.

12   Donna J. Haraway, "Situated Knowledges: The Science Question in Feminism and the Privilege of Partial Perspectives," in *Simians, Cyborgs, and Women: The Reinvention of Nature* (London: Free Association, 1991), 183–202, esp. 189.

13   The important of the distinction between *true* and *consistent* is developed at more length in N. Katherine Hayles, "Constrained Constructivism: Locating Scientific Inquiry in the Theater of Representation," *New Orleans Review* 18 (spring 1991): 76–85.

14   John Ziman, *Reliable Knowledge: An Exploration of the Grounds for Belief in Science* (Cambridge: Cambridge University Press, 1978).

15   In "Is Modern Science an Ethnoscience?" Sandra Harding writes, "while modern science has produced systematic knowledge, it also produces systematic ignorance."

16   Evelyn Fox Keller, *A Feeling for the Organism: The Life and Work of Barbara McClintock* (San Francisco: W. H. Freeman, 1983).

17   Writing from the point of view of a practicing scientist, Jay Labinger argues that the social and cultural studies of science *should* have implications for how science is done. ("Science as Culture: A View from the Petri Dish," *Social Studies of Science* 25 [1995]), 285–306. In the same issue, see responses to Labinger, 306–41, and his final reply, 341–48.

Detoxifying the "Poison Pen Effect"

Michael Lynch

MORE than forty years ago, Gilbert Ryle used the phrase "poison pen effect" to describe a literary device with which a writer enlists the authority of science in an effort to show that everyday knowledge is illusory.[1] Ryle was objecting to the way a few prominent scientists in his day would astonish their audiences by claiming, for example, that the apparent solidity of familiar objects like tables and chairs is really an illusion created by our sensory receptors and brains. In fact, the scientists would say, science tells us that such objects are composed entirely of tiny atoms surrounded by vast amounts of empty space. Ryle attacked this claim first by questioning its scientific authority: He pointed out that "there is no such animal as 'Science.'" By this he meant, there is no single discrete system of scientific knowledge which can be contrasted with everyday knowledge: "There are scores of sciences. Most of these sciences are such that acquaintanceship with them or, what is even more captivating, hearsay knowledge about them has not the slightest tendency to make us contrast their world with the everyday world."[2] Ryle observed that the so-called world of science which notable scientist-metaphysicians contrasted to an illusory everyday world was "not of science in general but of atomic and sub-atomic physics in particular, enhanced by some slightly incongruous appendages borrowed from one branch of neurophysiology."[3] To make absolutely clear that he was not attacking scientific knowledge—namely the particular hypotheses and findings in specialized branches of physics, chemistry, meteorology, astronomy, palaeontology, entomology, or geology—Ryle added that "I am questioning nothing that any scientist says on weekdays in his working tone of voice. But I certainly am questioning most of what a very few of them say in an edifying tone of voice on Sundays."[4]

Ryle's essay develops an analogy to expose the absurdity of such

Sunday school lessons. He invites us to imagine a confrontation between a student and a college auditor. The auditor keeps a detailed record of all college revenues and expenditures, and he informs the student that these records explain everything that happened during the academic year. The student's college life is accounted for by room charges, dining fees, purchases at the college store, library fines, and so forth. The auditor then tells the student that any educational experiences which cannot be documented by the account book should be dismissed as illusions. Ryle's auditor is analogous to an all-knowing neurophysiologist who correlates what a person sees with the activated pattern of neurons in the projection areas of the brain and then declares that the "conscious experience" in question is no more than the behavior of the neurons.[5] Without challenging the accuracy of the recorded facts, Ryle attacks the metaphysical claims with which they are associated.

The poison pen effect has become especially toxic in recent writings by prominent spokespersons for natural science. In this essay I will discuss two of the more notable examples: Lewis Wolpert's *The Unnatural Nature of Science*, which has stirred controversy in Britain, and Paul Gross and Norman Levitt's *Higher Superstition*,[6] which has been widely debated in North America. Wolpert's book is the lighter of the two, both in tone and content. It carries forward the tradition of science writing that Ryle criticized, and consistent with Wolpert's role as a prominent science popularizer in the British media, the book makes an effort to encourage public awe and faith in science. Gross and Levitt's book abandons "the edifying tone of voice" in favor of fire and brimstone denunciations of academic heresies interspersed with gestures of ridicule and parody. Although Wolpert also criticizes those who do not share his reverence for science, Gross and Levitt mount a far more sustained and explicitly political attack on "antiscience" movements in North American universities. Although Gross and Levitt claim not to be politically conservative, their spirited attack on what they call "the academic left" has appealed to politicians, academics, and journalists in the United States who have carried on a crusade in recent years against "political correctness" in the schools, museums, media, and other public institutions. Like others who have opposed leftist and libertarian elements in the arts and humanities, Gross and Levitt present themselves not as voices of a powerful establishment but as spokesmen for an embattled minority of serious and concerned scholars. The rhetoric of marginality and its inversion (sometimes dubbed "reverse McCarthyism" by critics) is also characteristic of the polarized "culture wars" in the United States. The battle lines are recognizable in Britain, and less so in non-English speaking countries, but I believe that the poison pen effect is inscribed with more severe and explicit political intent in the United States than elsewhere.[7]

For obvious reasons, Wolpert's and Gross and Levitt's books have created dismay in science-studies circles, not only because of what they say but also because of the polarization between "scientists" and "sociologists" that their publication and reception has encouraged.[8] In light of this polarization, I believe that it may be worth developing an updated version of Ryle's antidote to the poison pen effect. In this essay I shall argue that the antidote is simple and can be administered orally. It consists of understanding, and not losing sight of the fact, that Wolpert's and Gross and Levitt's projects are overt exercises in philosophy, history, and sociology, which happen to have been written by natural scientists. Such writings have poisonous effects only when the authors' occupational backgrounds are assumed to provide an authoritative foundation for comprehensive academic and cultural knowledge.

Wolpert's (Un)Naturalistic Philosophy of Science

Wolpert gives a dismissive and perfunctory treatment of the philosophy, history, and sociology of science. As far as he is concerned, practicing scientists seem to manage quite well despite the logical problems raised by Quine, Popper, and Kuhn about theory choice, empirical confirmation, and scientific progress. He also finds it difficult to swallow some of the theses he attributes to social and cultural studies of science: that discoveries are really inventions, that theories invariably reflect vested interests, and that scientists mislead themselves and the general public when they employ the rhetoric of objectivity.[9] In contrast to these theses, Wolpert promotes a "common sense realist" philosophy which, he argues, arises from the practical experience of doing scientific research and is typical of practicing scientists like himself. He ignores that many nonrealist, antirealist, or irrealist philosophers, historians, and sociologists also have backgrounds in the natural sciences, and he does not address how his own research in embryology led him to hold the commonsense realist philosophy he advocates. Wolpert does not obscure the fact that his commonsense realism *is* a philosophy; it is about the general nature of science, and it is remote from his embryological research. In brief, Wolpert rejects established philosophy of science, grounds an alternative philosophy in the commonsense experience of practicing scientists, and says next to nothing about such practical experience. As the title indicates, his book is an exposition on the "unnatural nature" of scientific knowledge, which sets out to demonstrate that scientific thinking is essentially different from ordinary thinking. The book includes examples from the history of science and cognitive psychology, but following Ryle, we should be alert to the fact that what Wolpert is advancing is nothing other than a philosophical thesis of the most general kind. Even if we assume that Wolpert is a knowledgeable and competent scientist, this should not lead us to think that this qualifies him as an authority in the philosophy of science any more than an ability to think qualifies a person to be an authority in the philosophy of mind.[10] Instead, we should feel free to examine his claims about knowledge, everyday reasoning, and the history of science, without supposing that these claims are in any way "scientific." When the veil of scientific authority is pulled away, Wolpert's claims can often seem unsupported, incoherent, and even absurd.

Wolpert elucidates his global contrast between scientific and everyday thinking by reciting a series of problems that people often get wrong unless trained in physics or another relevant field. So, for example, as elementary physics educators can testify, students often give erroneous explanations of why dye spreads in water or why a ball tossed in the air moves upward until it reaches an apogee and starts to fall. He also mentions the findings of psychologists who, in the tradition of Piaget, identify common judgmental errors made by children about object relations. Having laid out a series of such examples, Wolpert then puts on a "Sunday tone of voice" and asserts that scientific reasoning qualitatively differs from everyday thinking: "doing science . . . requires one to remove oneself from one's personal experience and to try to understand phenomena not directly affecting one's day-to-day life, one's personal constructs." He tells us further that "common sense provides no more than some of the raw material required by scientific thinking."[11] For Wolpert, doing science requires a unitary kind of thinking associated with a general "scientific process," a process that he also associates with distinctive personal virtues: "The explicit or formal nature of scientific theories is not only important in its own right but points to a crucial feature of the scientific process: the self-aware nature of the endeavour. This self-aware aspect of doing science, as distinct from other activities, makes science different from common sense almost by definition, since, again almost by definition, common sense is unconscious."[12] "Almost by definition" commonsense reasoners proceed in "the comfortable ignorance of never having considered that things could be otherwise" whereas science elevates its practitioners to "a continual self-aware evaluation of the evidence and subsequent modification of views." Wolpert asserts that "only a minority (about 15 percent)" of us have the capacity to critically reflect upon what we believe to be true.[13] This astonishing claim implies that more than half of the students who pursue higher education in Britain and the United States do not have the capacity to critically reflect on their beliefs. Those of us who attempt to teach such students to "critically reflect" apparently have a hopeless task (I confess that it can seem that way at times), and those who, like Wolpert, are involved in efforts to enhance public understanding of science apparently can convey genuine understanding only to a small segment of the public, whereas a science popularizer can hope to encourage only a benign *misunderstanding* of science among the rest of the populace. All of this is said on behalf of *science* by a scientist, but, to say the least, it is a dubious, disturbing, and largely undocumented story about the relationship between scientific and ordinary knowledge. Wolpert's science lesson is questionable for a number of reasons.

First, Wolpert uses ordinary terminology in a very puzzling way. What can he mean by saying that common sense is "unconscious"? Surely, he cannot mean that when crossing the street I am not conscious of the signal lights, pavement, and traffic (perhaps I would become unconscious if I had the misfortune of being struck by one of the vehicles).[14] And why would I be less "self-aware" when preparing my morning cup of coffee than a lab practitioner is when following a standard molecular biology protocol? I might be unaware of the chemical structure of caffeine and I may not be able to give a physiological explanation of why I get a buzz from the coffee, but would that make me less *self*-aware? Is a mathematician writing furiously at the blackboard any more self-aware than a dancer rehearsing a new routine? Wolpert's strange assertions only make sense if we equate *consciousness* or *self-awareness* with an ability to describe our actions and surroundings in terms of specific physico-chemical relations. Even then, however, it becomes unclear if a mathematician at the blackboard should become self-aware by thinking about patterns of neuronal activity in her brain and physical forces at the interface between the chalk and blackboard in addition to concentrating on the coherence of the equations and anticipating possible rebuttals to the proof being developed. To clarify this point, Wolpert would have to address the philosophical problem of relevance, about which he seems "unaware."

Second, how can Wolpert say that science involves a unitary kind of thinking that differs from an equally undifferentiated common sense? Elsewhere in his book, Wolpert agrees with contemporary philosophers who argue that there is no single method of science. But while he acknowledges a plurality of methods, he also asserts that *one kind of thinking* runs through the various disciplines, presumably holding for field sciences as well as experimental sciences, sciences at different points in their historical development, and theoretical as well as experimental specialities. Somehow or other, this way of thinking does not extensively overlap with various modes of technical practice in medicine or industry. The examples Wolpert gives of popular misconceptions of physical or biological relations demonstrate that students do not grasp certain physical and biological phenomena until they are taught lessons about them, but similar things could be said about an entire range of cultural activities. Persons trying to learn a language or to master elementary arithmetic also make characteristic mistakes, but do such mistakes demonstrate that speaking a language and calculating with numbers require an unnatural knowledge which is entirely remote from common sense? Wolpert also does not explain why he equates science with individual thinking, when so much of contemporary science is performed by large organizations with specialized divisions of labor, complex technologies, and massive record-keeping systems.

Third, how can Wolpert square what he says about commonsense thinking with his avowal of a commonsense realist philosophy? Is he implying that his commonsense philosophy is an unconscious philosophy? If so, should it not be replaced by a philosophy that is more self-aware? I would argue that Wolpert would benefit from becoming more aware of the philosophy, history, and sociology of science, but I would not want to equate this with becoming more conscious or self-aware.

Finally, Wolpert sometimes acknowledges that many of our everyday activities do not require us to take scientific explanations into account. Nevertheless he repeatedly applies the man-as-scientist model to diverse activities. Wolpert seems to realize that an artist can paint a sunset without having studied astronomy, atmospheric physics, and optics. I doubt that Wolpert would deny that a brewer can make a fine ale without knowing the details of biochemistry and a violinist can play Vivaldi without knowing how a physicist would explain the harmonic resonances involved. This is not to deny that study of these scientific subjects would yield valuable lessons for some purposes, but it seems to be backward to suppose that an artist, brewer, or musician who has not had the benefit of such lessons nevertheless dwells unconsciously in their shadow.

The main problem with Wolpert's treatment is that it places non-scientific activities in a scientistic frame. In his account, common-sense reasoners (including children) entertain theories and hypotheses, and their thoughts are organized into mental models. This may be the case when students struggle in the classroom to learn physics, but what about when they play catch, sing songs, and tell jokes? Surely, judgments, choices, and various kinds of thinking are involved in such activities, but why presume an analogy with a generalized scientific procedure? For Wolpert, it seems, all of our everyday activities are primitive versions of science, and the main difference is that when we act as commonsense reasoners we hold our theories unconsciously and our hypotheses remain untested.

Wolpert does more than present arguments. He discusses historical cases (phlogiston, molecular biology, Newtonian mechanics, Mendel and Mendelism, etc.) and even gives a few examples from the history of his own field of embryology. He also cites some examples from cognitive psychology. He does not, however, go into any detail. The examples serve mainly as illustrations for general maxims about the nature of science, and so it seems fair to say that his book is philosophical in style and scope. But at this point things get confusing. Wolpert acknowledges having a philosophy (commonsense realism), but he quickly adds that holding his "position will have made not one iota of difference to the nature of scientific investigation or scientific theories. It is irrelevant."[15] As a commonsense philosopher, he makes no claim to philosophical expertise or sophistication, and he observes that practicing scientists have little use for his own or anyone else's philosophy. Unlike philosophers, he says, scientists are not much interested in defining the general nature of science.[16] Presumably, Wolpert is an unusual scientist who is willing to define the nature of science on behalf of his less-interested colleagues. Unlike the philosophers, historians, and social scientists he criticizes, Wolpert does not rest his claims on technical logic, archival research, or ethnographic investigation. Instead, without presenting much evidence other than his own testimony, he confidently speaks on behalf of what practicing scientists know or assume. So, for example, at one point, he remarks about Quine's underdetermination thesis that "in practice scientists are not concerned with such minute differences [between possible theoretical accounts of the same evidence] except in cases where they will have a real impact on their theories and predictions."[17] He takes this further when discussing Popper and Kuhn, arguing that their philosophies misrepresent what the practitioner knows. Regarding Kuhn's incommensurability thesis he argues, "scientists change theories because the new ones provide a better correspondence with reality; because, like Darwin's theory of evolution, they provide a better explanation of the world." As for Popperian falsificationism, he says, "[i]n real life, scientists often do not conform to this formula for doing science."[18] He is even less complimentary about what sociologists have to say:

> it is misleading to think, as some have claimed, that science is really nothing but rhetoric, persuasion and the pursuit of power. No amount of rhetoric is enough to persuade others of the validity of a new idea, but it can make them take it seriously — that is, follow it up and test it. But persuasion ultimately counts for nothing if the theory does not measure up to the required correspondence with nature. If it does not conform with the evidence, if it is not internally consistent, if it does not provide an adequate explanation, the authority and all the other social factors count for nothing; it will fail. Such a failure is undoubtedly culturally determined, the culture being one that adopts a scientific approach.[19]

Wolpert's general claims about the correspondence between theories and nature, the limits of cultural determination, and the justifications for theory choice in science resemble familiar positions (and sometimes straw positions) in the philosophy of science, but he does not simply assert them as philosophical theses, he offers them *on behalf of practicing scientists*. Somehow, according to him, scientists are able to determine when their theoretical terms truly correspond to natural reality. Although this may be so, Wolpert does not support this assertion with evidence or argument. Instead, he presents it as common knowledge among practicing scientists, even though he also recognizes that this commonsense realism is akin to the much-maligned protophilosophical position of naive realism.

One would not want to begrudge Wolpert his rights to philosophize. Professional philosophers do not, after all, hold a monopoly over questions about reality, truth, and the nature of science, and Wolpert would hardly be the first interloper from other fields to take up these topics. Many sociologists and historians of science, for example, also make general ontological and epistemological claims. The problem is

that Wolpert is quick to dismiss philosophy, history, and social studies of science in favor of a typified scientist's grasp of the general
issues.[20] Worse, he confuses this scientist's philosophy with a scientific validation of its theses. At one point Wolpert cites Wittgenstein,
saying that "What we find out in philosophy is trivial: it does not
teach us new facts, only science does that."[21] Wittgenstein did advise
his readers not to confuse philosophical investigations with scientific
researches,[22] but it was far from his purpose to suggest that philosophy should turn to science for solutions to its problems. In contrast,
Wolpert seems to figure that philosophy's inability to teach us new
facts should be replaced by a scientific determination of "the facts" of
metaphysics. Wolpert lists several general assumptions made by scientists: there is an external world separate from our perception of it;
the world is rational and can be analyzed locally; there are regularities
in nature; and the world can be described by mathematics. He adds
that "these assumptions may not be philosophically acceptable, but
they are experimentally testable and they are consistent with the ability of science to describe and explain a very large number of phenomena."[23] He does not say how such metascientific assumptions can
be tested scientifically. Even if we were to agree that many scientists
hold such assumptions, and that they manage to sustain successful
scientific careers while doing so, it would be an elementary and categorical mistake to say that these assumptions are "experimentally
testable" in the same sense that the effects of a specific hormone on
embryonic development in mice is testable. Although few of us may
doubt that embryologists and molecular biologists are able to "think
scientifically" about the mechanisms of cellular replication, it is
another matter altogether to credit them with a correct scientific theory of the general relationship between empirical evidence and reality. If, as Wolpert says, practicing scientists are indifferent to philosophical questions, one can surely wonder why he would think that
scientists (even the most famous of them) have a more adequate grasp
of "the problem of reality" than do historians, philosophers, sociologists of science, or for that matter, machinists, midwives, sailors, and
secretaries.

It would be tempting to criticize Wolpert for failing to take into
account empirical studies of everyday reasoning, and it would be
equally tempting to berate him for the lack of evidence he adduces for
his pronouncements about what practicing scientists know and
assume. Although I think he would find little support for his assertions about the differences between scientific and everyday thinking

if he were to study the matter more closely, I think there is a deeper problem. This problem is even more relevant to Gross and Levitt's book, which I shall discuss shortly. The problem is the attribution of general epistemic (and moral) virtues to the thinking of scientists. Following Ryle, we should not be too astonished if what brilliant scientific minds say about philosophical, historical, and sociological matters is ignorant and badly researched. We should not be too surprised when Nobel prize winners express sketchy, partisan, and poorly understood versions of the history, philosophy, and sociology of science. Unfortunately, the promotion and self-promotion of individual scientists is often paired with a denigration of the intellectual and moral qualities of persons who are not natural scientists. Like Wolpert's commonsense reasoners, such persons are portrayed as though their minds are continually filled with confusion and dogma. Worse, when such ignorant persons seem reluctant to accept edifying lessons about individual scientists and general science, they are made out to be ideologically driven true-believers who should have no place in a university classroom. The other side of the coin is a reciprocal tendency among some nonscientists and lapsed scientists to indict "science" or "technoscience" for being responsible for all the evils of modernity. I actually agree with Wolpert that sociological critics of science offer "no real threat to science," and I also agree that undiscriminating pronouncements about what science is, or is not, can have "an unfortunate influence on the study of science and its history."[24] But I also think a great deal of the animosity would dissolve if a more resilient lesson were taken from philosophy, history, and sociology of science: the sciences are *not* completely different from other activities, and scientists do not necessarily possess extraordinary metaphysical insight about the nature of reality, mind, intelligence, or "Science." This is not to deny the technical difficulty of many theories, experiments, and proofs, nor does it suggest that scientific innovations are unimportant. Instead, it suggests that it may be time to tone down the almost-religious connotations that are so often ascribed to the practical achievements and implications of science.

Gross and Levitt's Antisocial Sociology of Knowledge

Like Wolpert, Gross and Levitt write as practicing scientists (or, in Levitt's case, as a mathematician) even though, like Wolpert's book, *Higher Superstition* says little about the authors' own research. Gross and Levitt do occasionally cite particular items of scientific and mathematical knowledge, like the Heisenberg uncertainty principle, Einstein's theory of relativity, Cantor's proof, and the Reynolds number, in order to correct mistakes and misunderstandings they attribute to writings by Stanley Aronowitz, Bruno Latour, and others, so it can be argued that their technical expertise does enter into the book. But had they limited their criticisms to correcting specific errors by social scientists and humanists, I doubt their book would have attracted much notice. Its notoriety seems due to its more sweeping claims about the political origins, ideological unity, and antiscientific agenda of the "academic left." In other words, *Higher Superstition's* provocative appeal arises from its social, political, and cultural analysis. In a way analogous to Wolpert's unphilosophical philosophy of science, Gross and Levitt produce an antisociological mockery of the sociology of knowledge. They are explicit about this undertaking, as a key chapter in their book presents itself as a case study of "the cultural construction of cultural constructivism."[25]

Early in their book, Gross and Levitt perform an ideological analysis. They sketch a portrait of the "academic left" (their term for a broad assemblage of radical feminists, relativists, deconstructionists, and deep-green environmentalists), and characterize this group as a coherent ideological subculture.[26] Gross and Levitt acknowledge that one can fairly doubt that "a traditionally Marxist analysis, a hard-core postmodern epistemological critique, and an ecofeminist harangue from the Goddess-worshipping camp" have much in common, but without extensively arguing the matter, they assert that such diverse positions can be considered as "extreme points in a contiguous ideological field."[27] To accomplish their cultural construction of cultural constructivism, Gross and Levitt select a sample of books and articles (and, in some cases unpublished sources) to represent different positions in this ideological field. Then they give a severely uncharitable reading of the texts, focusing on factual errors, tendentious remarks, and outrageous claims. Finally, they ascribe the cultural constructivists' textual errors and rhetorical excesses to general ideological tendencies pervading the "academic left." Perhaps in an attempt to mirror the rhetorical excess they impute to constructivism, Gross and Levitt's own exercise in "cultural construction" makes ample use of sarcasm, mockery, and parody. Under Gross and Levitt's not-so-light literary touch, Latour becomes a "self-appointed heretic and gadfly" whose writings are replete with doodle diagrams, "snide aphorisms," and parodied rules of method which disguise his severe ignorance of mathematics and deep hostility to science.[28] They assert that "Latour fervently minimalizes and trivializes formalization, abstraction, and mathematization," and they dismiss his writing as "a series of flip-pancies, whose intended point is that the deep and surprising pre-dictions about the real world that emerge from exacting logical analy-sis of abstract models are really no more than tautological parlor tricks."[29] One can only conclude that Latour's considerable influence in the science-studies fields is entirely a result of credulous readers being duped by his illusionist tricks. Apparently, readers should also conclude that, faced with the hyperbole and verbal trickery of a mas-ter rhetorician, Gross and Levitt find it necessary to fight fire with fire before dousing the entire conflagration with a wet dose of realism.

Although they are deadly serious at times, Gross and Levitt some-times exude the attitude of a couple of guys having a good time while slumming in the low-rent districts of the academy. Their narrative voice is animated by a presumptive mastery of science which stands in judgment of the ignorant ideas of the social scientists, historians, and philosophers they discuss. In other words, as Gross and Levitt construe the constructivist stance (or, rather, as they mock it up), it is asymmetric.

Asymmetry provides both a warrant for Gross and Levitt's analysis and a basis for their advice on how universities should deal with the antiscience rabble who have crashed the gates of the academy. On the basis of "personal experience," Gross and Levitt assert that the rigor-ous training and superior intellectual powers of scientists allow them to comprehend diverse fields of knowledge: "scientists are deeply cul-tured people, in the best and most honorable sense." They add that members of humanities departments are not quite so versatile, and to demonstrate the point they propose a "fanciful hypothesis" to the effect that if the members of the humanities department of M.I.T. were to walk off the job, the scientists on the faculty would, with some dif-ficulty, be able to patch together a humanities curriculum, but if the scientists walked out, the humanists would have no chance of meet-ing the demand for science education.[30] This thought experiment leads us to believe that scientists know their own practices and are competent to discuss the history, philosophy, and sociology of those practices, whereas humanists and social scientists who speak criti-cally about science are ignorant, not only about specific facts and techniques but also about the general nature of science. This asym-metry sets up Gross and Levitt's claim that members of university sci-ence departments are in a position to mind the affairs of the humani-ties and social science departments, whereas the latter have no basis for interfering with the scientists' business.

As Gross and Levitt are quick to point out, their view is a mirror image of a reciprocal prejudice among nonscientists to the effect that scientists are uncultured technophiles who have no insight into the broader implications (or even the intrinsic meanings) of their own practice. They recognize that their argument transgresses disciplinary boundaries, but they allege that the humanists threw the first stone:

> It will be argued immediately that this is an asymmetric, and therefore inequitable, proposition. If physicists are to judge schol-ars of English, why shouldn't English professors judge physicists? The fallacy here is that the asymmetry originates from the pre-tentions, legitimate or otherwise, of members of the English (or sociology or cultural studies or women's studies or African-Amer-ican studies) department to qualification on scientific questions. If, say, a member of the mathematics department were to engage in the (most unlikely) scholarly project of analyzing the rhetori-cal and stylistic elements of certain mathematical papers, it would be entirely legitimate for literary scholars to pronounce judgment on that work, and for the promotion process to take that judgment into account. To put it bluntly, it is humanists of the academic left who have transgressed the boundaries—as they are eager in most circumstances to proclaim. That's their privilege; but they are not (or should not be) exempt from customs duties![31]

Ryle's account of the poison pen effect should remind us that Gross and Levitt are not writing a biological research report or a mathemat-ical argument, but in this case are developing a "constructivist" analy-sis of rhetorical and ideological tendencies in a large segment of the academy. Those of us who also contribute to social and cultural stud-ies of science should perhaps greet Gross and Levitt as colleagues. Taking them at their word, we may also assume that it is "entirely legitimate" for us to "pronounce judgment" on their venture into our interdisciplinary field. Consequently, the question to ask is how does Gross and Levitt's book stand as a contribution to social and cultural studies of science? Assessments may differ on the quality of their scholarship[32] and the extent to which they have canvassed the rele-vant literature, but in at least one respect they make our assessment easy by implying that, as far as they are concerned, cultural analysis involves an asymmetric orientation grounded in (presumed) cultural superiority. Their "constructivist" account of "cultural constructiv-ism" implies that such analysis (whether performed by the social sci-entists and humanists whose studies they review or by themselves when reviewing them) involves an asymmetric, all-knowing, and hostile stance toward the target groups. What is striking about this implication is that an asymmetrical orientation is, at least in principle, completely out of keeping with the strong program in the sociology of knowledge and the ethnographic orientation of the laboratory studies that have featured so prominently in the development of constructivist science studies.[33] As any student of sociology and anthropology should be aware, sociologists of knowledge and ethnographers go to great lengths *not* to presume cultural superiority, and particularly not to presume that a commitment to "science" (or social science) provides the basis for comprehending the rationality or irrationality of the practices studied, regardless of whether those practices are magical rituals performed by members of a tribe of hunter-gatherers or informal routines developed among a group of industrial biochemists. Far from advocating an asymmetrical stance of cultural superiority, ethnographers generally emphasize the necessity of an extended field experience for gaining insight into important aspects of collective life that are not apparent from a more formal, distanced, or prejudiced vantage point, and sociologists of scientific knowledge distinguish their attempts to explain true as well as false beliefs from a "sociology of error" that presumes the relevance of social influence only in cases of error and ignorance. This, of course, does not imply that sociologists of knowledge and ethnographers always maintain a symmetrical epistemic posture nor does it imply that it is *possible* to maintain such a posture. Whether or not it is possible to understand the "rationality" of an exotic practice like the Azande poison oracle, or to analyze a "belief" without thereby discrediting it as "mere" ideology, and whether it is possible to speak of the social construction of knowledge without implying a diminished epistemic status, all are vexed questions, but they are questions nevertheless, and they are frequently discussed and debated in the science-studies fields.[34]

By equating historical and ethnographic (constructivist) analyses of the production of scientific knowledge with antiscientific attacks on such knowledge, Gross and Levitt (and some other prominent spokespersons for "Science") have confused the issue.[35] Their arguments have force only to the extent that social and cultural studies of science investigate "the very content and nature of scientific knowledge"[36] without undertaking the difficult task of coming to terms with the relevant configurations of knowledge and practice. Some writers in the field attend to these difficulties carefully and responsibly, and some do not, but the main problem, as I see it, is that Gross and Levitt

equate the "cultural constructivist" stance with an attempt to deni-
grate what it investigates. It may be possible to argue that ethnographies of science, and the sociology of scientific knowledge more generally, do not live up to their programs, but Gross and Levitt do not
argue this. If their "constructivist" account is any indication, they do
not even acknowledge the methodological relevance of "understanding" (as opposed to hostile engagement, denunciation, and sarcasm)
in the fields they mock.

Conclusion

I have argued that Gross and Levitt do not simply try to correct mistaken accounts of scientific and mathematical concepts; they perform
a far more ambitious "constructivist" analysis of humanities and
social-science knowledge. Similarly, Wolpert's book does not simply
relate a scientist's reflections about his field of practice, it provides a
general argument about the difference between ordinary and scientific
reasoning. There is nothing intrinsically wrong with these authors
writing about such matters, but it also should not be surprising that
Wolpert's philosophy of science is unsophisticated and incoherent,
and Gross and Levitt's cultural construction of cultural constructivism
follows the lines of a sociology of error which has been relegated to
the prehistory of the contemporary fields they criticize. Such alleged
failures by Wolpert and by Gross and Levitt to produce sophisticated
and up-to-date philosophical and cultural analyses should not count
too heavily against them. After all, sweeping claims about science,
dubious ontologies, crude forms of ideology critique, and denunciations parading as analyses are abundant enough in studies by more
seasoned professionals in the social and cultural studies of science
fields.[37] It would also be misleading to assume that such overblown
claims and denunciations result from inept or unsophisticated efforts
to adhere to a program. When "Science" is in question, the analytic
orientations of ethnographers and sociologists of knowledge exist in
an uneasy tension with disciplinary rivalries in the academy. Such
rivalries are compounded by the fact that, far from being an exotic
belief to be investigated in the traditional anthropological fashion, science is a widely discussed and debated topic as well as a methodological resource in the social sciences, and more generally it is both
a widely valued and feared cultural phenomenon. Consequently,
there may be no easy solution to disciplinary conflicts and transgres-
sions. Ryle's stated intention—to question nothing that any scientist
says on weekdays in a "working tone of voice," while reserving all

criticism for scientists who sermonize *about* "Science"—would be convenient to maintain if we could simply propose that biologists stick to biology, philosophers stick to philosophy, and sociologists stick to sociology. This is untenable because, among other things, sociologists make a living by investigating how others (and these others include scientists) make a living, and philosophers (at least some of them) investigate problems associated with a whole range of practices and inquiries, including those in the sciences. And, of course, scientists who believe that journalists, philosophers, or sociologists mischaracterize their practice have every right to register their objections and correctives. In addition, natural scientists contribute to the production of things that others find interesting, valuable, and sometimes horrifying, and the financial costs of scientific research as well as the benefits and sufferings that accrue from it are public matters. So, from every side, there is no avoiding engagement with public discussions and debates about science. To put accusations about transgressing boundaries in perspective, the problem is that some people presume to write knowingly about practices that they actually know very little about. Gross and Levitt complain that members of the academic left speak authoritatively about science without knowing what they are talking about, and I have, in turn, complained that Gross and Levitt disregard the programmatic aims of ethnography and sociology of knowledge when they construe cultural constructionism as a relentless ideological attack on natural scientists by people who simply dislike science. It is easy to imagine a simple, if unglamorous, solution to the problem of transgression. Stated as a maxim: if you're going to address philosophical questions, pay attention to history of philosophy and do not confuse philosophical questions with testable hypotheses; and, if you're going to study a group sociologically, try to gain an adequate conception of that group's unity, coherence, and history, and also try to understand what members of the group know and find intelligible.[38]

I'm afraid that such plain advice will not quite do the trick. As I have suggested, Wolpert's and Gross and Levitt's books are not about what scientists do. They are attempts to establish how "Science" should be portrayed in general. Especially as Gross and Levitt envision the matter, their effort to portray "Science" is part of a pitched battle with rival antiscientific factions. Advice about the importance of ethnographic understanding is likely to seem meek and irrelevant when there is a war going on. However, it may be worth asking what this war is about. Again to recall Ryle, if there is "no such animal as 'Science,'" then the controversy is literally about nothing. If there is no such thing as "the general nature of science," then, like other religious conflicts, the present one is less about what "Science" is and more about who should control the portrayal of "It." Should the "nature" of science be portrayed as "unnatural" (and therefore accessible only to an elect few) or "natural" (universally accessible, at least in principle); is this "Science" a force of good or evil; is it male, female, or neuter; is it really objective or is it a grand illusion? Cynically understood, the battle only seems to be about a metaphysical "Something," whereas the real basis is a struggle over the tithes that support the monasteries.

Many of my colleagues in social and cultural studies of science wish the whole debate would go away. Like agnostics in the middle of a religious war, they are inclined to see the debate about "Science" as an empty exercise in hostility between factions they would prefer not to join. At the same time, it seems clear that there is no neutral ground and something important seems to be at stake (although nobody seems to know quite what it is). I believe that the battle is over what frequently is called "the public understanding of science." Wolpert, who serves as Chairman of the Committee for the Public Understanding of Science in the U.K., provides arguments that undercut the very idea of a public understanding of science. If, as Wolpert says, scientific thinking is unnatural except for the relatively small proportion of the population that has an aptitude for it, then the vast majority can only trust the word of the experts. Nevertheless, like many popularizers, Wolpert works hard to encourage reverence for science, high regard for scientists, and appreciation for the astounding feats of scientists. This task is no longer as easy as it once was, as there currently are many indications of popular disenchantment and student disinterest in natural science and mathematics, and the old lessons that "science is good" and "science is neutral" have been attacked repeatedly by environmentalist, feminist, and other popular movements. Given the difficulty of maintaining the slogan "science is good," it may be tempting for movement participants to opt for the opposite slogan "science is bad," and it can be no less tempting for scientists like Wolpert, Gross, and Levitt to police the academy against popular movements they hold responsible for promoting a simplistic, negative lesson. The source of the difficulty is the polarized terms of the debate. It would be better for all sides to promote the idea that there is no single lesson, or, if one wants such a lesson, that there is no single entity called "Science." Taken alone, as a proposition, this is not much of a lesson. It does not teach us anything new, and it does not resolve the big questions about what "Science" essentially is, and it does not tell us if it is good or bad. If, however, we view the lesson primarily as a reminder, it can encourage us to pause and perhaps even to laugh at ourselves, when we're tempted to join sides in a war of words over "big questions" that might better be dissolved than fought over.[39]

Notes

A version of this paper entitled "The Unphilosophical Philosophy of Science," was presented at the conference on Science's Social Standing, organized by the Centre for the History of Human Sciences, University of Durham, Durham, U.K. (2–4 December 1974). I would like to thank Steve Fuller and Irving Velody for inviting me to write the paper and Jeff Coulter for his helpful comments.

1   Gilbert Ryle, "The World of Science and the Everyday World," in *Dilemmas: The Tanner Lectures 1953* (Cambridge: Cambridge University Press, 1954), 68–81.

2   Ibid., 69.

3   Ibid., 73. For a contemporary example of such a collision, see Francis Crick, *The Astonishing Hypothesis: The Scientific Search for the Soul* (London: Simon and Schuster, 1994).

4   Ibid., 75. For a striking example of the "edifying voice" of science preaching, note how "science" becomes the agent of its own progressive history in the following remarks by James Lovelock, a lecturer in physical chemistry at Oxford, published in *The Times Higher Education Supplement*, 30 September 1994, 19: "Science proceeds by progressive, never regressive, evolution, gradually accreting information and concepts that jointly enable its web to be extended. True, concepts are discarded as they are found to be unnecessary, inappropriate, or misleading; but their discarding is based on experimental evidence or theoretical argument, not on political, legal, or social expediency. Truth will out despite the consequences and despite the social milieu."

5   Crick, for example, declares (in *The Astonishing Hypothesis*, 3) "The Astonishing Hypothesis is that 'You,' your joys and your sorrows, your memories and your ambitions, your sense of personal identity and your free will, are in fact no more than the behavior of a vast assembly of nerve cells and their associated molecules."

6   Lewis Wolpert, *The Unnatural Nature of Science* (Cambridge, Mass.: Harvard University Press, 1992); Paul Gross and Norman Levitt, *Higher Superstition: The Academic Left and Its Quarrels with Science* (Baltimore: Johns Hopkins University Press, 1994).

7   My impression about the distinctively American context of Gross and Levitt's book was underlined during a discussion session at the summer conference of the International Society for History, Philosophy, and Social Studies of Biology, at Leuven, Belgium (19–23 July 1995). Several members of the audience from non-English speaking countries objected that the focus on Gross and Levitt's book imposed a North American agenda on the conference audience. The implication was that worries about the book's political implications were

not salient in other national cultures where intellectual and academic matters are not so polarized.

8   When used in the context of debates in, for example, *The Times Higher Education Supplement*, the term *sociologists* is a gloss that covers various social, cultural, and literary approaches to the natural sciences.

9   For a more sympathetic and well-informed objection to such treatments of science in social and cultural studies of science, see Jay Labinger, "The View from the Petrie Dish," *Social Studies of Science* 25, no. 2 (1995): 285–306.

10   Wolpert acknowledges this point in the concluding chapter of his book, but the earlier chapters are replete with authoritative pronouncements about the "nature" of scientific reasoning and the limits of common sense.

11   Wolpert, 17.

12   Ibid., 18.

13   Ibid., 19.

14   For an incisive account of the confusions associated with the notion of "consciousness" see Jeff Coulter, "Consciousness: The Cartesian Enigma," in *Wittgenstein's Intentions*, ed. John Canfield and Stuart Shanker (New York: Garland, 1993), 173–94.

15   Wolpert, 106.

16   Ibid., 101.

17   Ibid., 102.

18   Ibid., 104.

19   Ibid., 108.

20   Other scientists who have written or spoken in defense of science in the face of alleged "antiscience" trends include Nobel physicist Steven Weinberg, entomologist E. O. Wilson, and geneticist Richard Dawkins. An organization that calls itself the National Association of Scholars (not to be confused with the National Academy of Sciences) has promoted the backlash against "antiscience." See, for example, Scott Heller, "At Conference, Conservative Scholars Lash out at Attempts to 'delegitimize science'," *The Chronicle of Higher Education*, 23 November 1994, A18–19.

21   Wolpert, 106. The quotation is from Ludwig Wittgenstein, *Lectures: Cambridge, 1930–1932*, ed. Desmond Lee (Oxford: Blackwell, 1980), 26.

22   For example, see Ludwig Wittgenstein, *Philosophical Investigations* (Oxford: Blackwell, 1958), 109.

23   Wolpert, 107.

24   Ibid., 101.

25   Gross and Levitt, chap. 3.

26   Ibid., 10.

27   Gross and Levitt, 10–11.

28   Ibid., 264, n. 26.

29   Ibid., 62.

30   Gross and Levitt, 243.

31   Ibid., 256.

32   For example, in a review of *Higher Superstition*, Michael Ruse praises Gross and Levitt for doing "a fine job, covering an enormous literature with a care that must have taken a huge investment of time" (Michael Ruse, "Struggle for the Soul of Science," *The Sciences* [November/December 1994]: 39–44, esp.

40). Note that in the same review, Ruse criticizes Gross and Levitt for using "academic left" as a "blanket term" and for "deliberately" slanting "the picture both to make humanists look bad and to avoid awkward questions about why serious scientists are part of the attack" (42).

33   Much of the orientation in the sociology of scientific knowledge derives from a selective "strengthening" of Mannheim's sociology of knowledge by extending his program to cover mathematics and the natural sciences. See, Karl Mannheim, *Ideology and Utopia* (New York: Harvest, 1936); and David Bloor, *Knowledge and Social Imagery* (London: Routledge and Kegan Paul, 1976). The best-known ethnographies of science are Karin Knorr, *The Manufacture of Knowledge* (Oxford: Pergamon, 1981); Bruno Latour and Steve Woolgar, *Laboratory Life: The Social Construction of Scientific Facts* (London: Sage, 1979); Michael Lynch, *Art and Artifact in Laboratory Science* (London: Routledge and Kegan Paul, 1985); and Sharon Traweek, *Beamtimes and Lifetimes: The World of High Energy Physics* (Cambridge, Mass.: Harvard University Press, 1988). Although most often associated with anthropological studies of exotic cultures, ethnography has for a long time been used in sociology (principally in the subfields of ethnomethodology and symbolic interactionism) for studying the life-worlds associated with specific groups, occasions, institutions, and organizational activities. Beyond the fact that ethnographies of science are prominent in the sociology and anthropology of science, important historical studies of science also draw upon the methodological and ethical stance of ethnography. See, for example, Steven Shapin and Simon Schaffer, *Leviathan and the Air-Pump: Hobbes, Boyle, and Experimental Life* (Princeton, N.J.: Princeton University Press).

34   See, for example, Bryan Wilson, ed., *Rationality* (Oxford: Blackwell, 1970); Graham Button, ed, *Ethnomethodology and the Human Sciences* (Cambridge: Cambridge University Press, 1991); Karin Knorr-Cetina and Michael Mulkay, eds., *Science Observed: Perspectives on the Social Study of Science* (London: Sage, 1983); and Andrew Pickering, ed., *Science as Practice and Culture* (Chicago: University of Chicago Press, 1992).

35   Gerald Holton presents a more differentiated picture of "antiscience" movements, but he also links prominent work in social and cultural studies of science to a broader effort to "delegitimate" science. In contrast, I would argue that it is important (both for Holton and for those he criticizes) to carefully distinguish efforts to demystify "Science" from efforts to delegitimate the sciences. Gerald Holton, *Science and Anti-Science* (Cambridge, Mass.: Harvard University Press, 1993), 153ff.

36   This injunction is a key tenet of the strong program in the sociology of science. See Bloor, 1.

37   Some readers may wonder why I direct criticism at Wolpert and Gross and Levitt while slyly mentioning (without citation) that the faults I attribute to them can be found in abundance in the fields I defend. One might figure this to be an example of the "soft pedaling" of fundamental differences that, according to Gross and Levitt (11), helps maintain solidarity among the diverse constituencies of the academic left. Maybe so, but there are other rationales to consider. In the context of the present debate, if I were to single out examples of bad constructivist work, I could fairly be accused of scapegoating.

I prefer to consider bad work redeemable rather than subject to purge, and thus I believe that my assessments of particular studies are best made in the form of editorial judgments, advice to colleagues, and critical commentaries, all as part of the give and take of academic dispute. Such assessments and arguments do presuppose a certain solidarity—a commitment to a field (however vaguely articulated) and a respect for colleagues in it—but they do not amount to an uncritical effort to hide differences.

38   This is a sketchy formulation of what Harold Garfinkel has called "the unique adequacy requirement of methods" in ethnomethodology. Harold Garfinkel and D. Lawrence Wieder, "Two Incommensurable, Asymmetrically Alternate Technologies of Social Analysis," in *Text in Context: Contributions to Ethnomethodology*, ed. Robert Seiler and Graham Watson (London: Sage, 1992), 175–206.

39   The idea that philosophy provides reminders, and the agenda of dissolving questions that are conceived at too high a level of abstraction, comes from Wittgenstein, *Philosophical Investigations*, sec. 122–27.

The Flight from Reason: *Higher Superstition*

and the Refutation of Science Studies

Roger Hart

F UNDAMENTAL to Gross and Levitt's central claim that science
is under attack from what they term the academic left[1] is the
series of refutations they present to demonstrate that recent critical studies of science are nothing more than flawed and even fatuous
political diatribes against science. Reviewers and critics of *Higher
Superstition* have neglected analyzing these refutations, choosing
instead to focus on the larger issues to which the book is itself
addressed, apparently in the assumption that the refutations meet the
most basic standards of logic and evidence.[2] Thus Gross is hardly
exaggerating in his recent claim that "so far no scientific howler
exposed in *Higher Superstition* has been shown *not* to be one. Maybe
our portrayal of science studies is 'patently unfair'; but nobody . . .
has shown what exactly is unfair and why."[3]

This essay proposes to evaluate their refutations not for fairness but
simply validity and then to analyze their thesis that the academic left
is a political threat to science.[4] The first section will examine in detail
three of their most important refutations—Shapin and Schaffer, Harding, and Haraway—to assess whether Gross and Levitt accurately present the central arguments from these works and whether their analysis and their evidence (from primary and secondary sources) refute
these arguments. If their criticisms fail to refute these works, the next
section analyzes the types of evidence they present to ascertain to
what extent *Higher Superstition* represents their attempt to demonstrate that the works criticized frequently misstate scientific fact. And
if they present few examples of errors of scientific fact, the third section examines three more of their criticisms—Ross, Hayles, and Derrida—to analyze in detail the approach Gross and Levitt adopt in
seeking to achieve their claim to have refuted the critical studies of
science. The fourth section returns to analyze their various definitions
of the academic left, in particular the incongruity between their defi-

nitions and the conventional one. The last section analyzes how their use of these incongruous definitions is combined with their claim to have refuted the critical studies of science to create the political threat that the academic left was to have posed to science.

Higher Superstition as a Refutation of the Cultural, Gender, and Social Studies of Science

The (unstated) thesis of *Higher Superstition* is that what Gross and Levitt term the academic left—comprised primarily of humanists ignorant of science and resentful of its success—seeks to undermine the content, methodology, and epistemological validity of science through attacks supported by little more than the moral authority that the left claims for itself. The core of *Higher Superstition* is their claim to refute the academic left's critiques of science. In contrast to the central thesis of their book, this claim is stated explicitly, along with their methodology:

> We synopsize and, as we see it, refute some of the representative work in each area. Our space for doing so is limited. Consequently, the synopses are brief. Simultaneously, we extract from the critic's work certain crucial arguments which, in our view, exemplify methodological weaknesses and expose the fallacies of the underlying viewpoint. (13)

On the next page they delineate precisely the scholarship that they claim to refute.

> To exemplify cultural constructivism, we have chosen sociologists and historians of science: Stanley Aronowitz, Bruno Latour, Steven Shapin and Simon Schaffer. For postmodernism, we have settled upon the philosopher Steven Best, the "cultural critic" Andrew Ross, and the literary critic N. Katherine Hayles. The feminist theorists we consider include some of the best known: Sandra Harding, Donna Haraway, Evelyn Fox Keller, Helen Longino. As for the radical environmentalist attack on science, we concentrate on academics like Carolyn Merchant, but also on theorists—Jeremy Rifkin and Dave Foreman. (14)[5]

Comprising chapters 3 to 6, these refutations are more than just the main content of the book; they are the central claims on which the other chapters depend. It is on the basis of these refutations that Gross and Levitt can claim that the left-wing critiques of science "amount, individually and collectively, to very little in strictly intellectual terms" (15), that "in scholarly quality, it [these critiques] ranges from seriously flawed to hopelessly flawed" (41), and that its authors offer "wrongheaded, even fatuous, theories about matters in which their knowledge ranges from shallow to nonexistent" (254). It is on the basis of these refutations that they justify their own critique in chapter 8 of the social, ideological, and psychological roots of the academic left through their counterfactual statement that "if their [the academic left's] ideas were sustained by decent arguments and adequate evidence, this would be an unfair attack ad hominem" (217). And it is on the basis of these refutations that they accuse "a powerful faction in modern academic life of intellectual dereliction" (239), warranting their proposed interventions in the humanities. This section will analyze the validity of their refutations of Shapin and Schaffer, Harding, and Haraway.

Shapin and Schaffer's Leviathan and the Air-Pump. Gross and Levitt's most rigorous attempt to follow their stated methodology of synopsis and refutation is in their criticism of Shapin and Schaffer's historical study of the rise of experimental science in the seventeenth century, *Leviathan and the Air-Pump*—this is the work they take the most seriously.[6]

The central focus of Gross and Levitt's synopsis and refutation is Hobbes's exile from the experimental community and exclusion from the Royal Society. Their synopsis asserts that *Leviathan and the Air-Pump* is an attempt to demonstrate, using the controversy between Hobbes and the Royal Society as an example, that modern science was from its inception elitist and insular. Such exclusivity, they argue, is held by Shapin and Schaffer to have been enforced through scientific credentials that depended on the extrascientific factors of rank, wealth, religious belief, and political loyalties; Hobbes was perceived as a political and religious threat, and his exile was an act of "political prophylaxis" (64). Gross and Levitt conclude their synopsis of Shapin and Schaffer's views stating, "Given this perspective, the scientific community led by Hooke and Boyle, which echoed the aspirations of a moneyed class that sought immunity from the whims of royalist autocracy while casting a suspicious eye on the tumultuous mass of the unpropertied, had no place for the likes of Thomas Hobbes" (64–65).

In their refutation, Gross and Levitt assert that Hobbes's exile from the scientific community and exclusion from the Royal Society were related to his stubborn incompetence in mathematics. Citing Descartes, Spinoza, Halley, and Newton as examples, they assert that "there is more to be said about rigidity and latitudinarianism, intolerance and liberty of opinion, in the seventeenth-century scientific community than that the Royal Society constituted a kind of thought police" (66). They contend that the controversy between Hobbes and Boyle was an episode in a protracted conflict over mathematics between Hobbes and the mathematician Wallis; this, they assert, is a "concrete and substantive reason, *in contrast to an ideological one*, for Hobbes's notoriety in scientific circles" (67, emphasis in Gross and Levitt). Shapin and Schaffer ignore this objective reason for Hobbes's exclusion, Gross and Levitt assert, because of their ideological commitments to relativism and "radically antielitist politics" (68). They further contend that it is "shortsighted to condemn them [the Royal Society] out of hand for addressing the difficulties in language that occasionally smacks of snobbery or political insecurity" (68), and "it is false to read their rejection of Hobbes as a blanket denial of the value of speculative and deductive thought" (68). Ultimately, Gross and Levitt argue, its ideological blinders render *Leviathan and the Air-Pump* no more than an "exercise in tunnel vision" (68).

To what extent, then, have Gross and Levitt summarized and refuted the central arguments of the book? One indication is the book's title: they briefly discuss the air-pump only once and fail to discuss Hobbes's *Leviathan* at all.[7] Another indicator is the book's table of contents, where *experiment* appears in four chapter titles.[8] Still another indication is the very first sentences of the first chapter: "Our subject is experiment. We want to understand the nature and status of experimental practices and their intellectual products" (3).

In the book, Shapin and Schaffer analyze the historical process through which experimental facts became accepted as reliable knowledge in natural philosophy by examining Boyle's air-pump experiments and the mechanical, literary, and social technologies he employed; by interpreting Hobbes's *Leviathan* "as *natural* philosophy and as epistemology" (19); and by analyzing Boyle's defense of his program against criticisms by Hobbes, Linus, and More. Shapin and Schaffer conclude by suggesting that Boyle's new experimental "form of life" (a phrase they borrow from Wittgenstein) offered solutions to both the problems of knowledge in natural philosophy and to the maintenance of social order.[9] The central arguments of the book are outlined in detail at the end of the first chapter (18–21); here Shapin and Schaffer make no mention whatsoever of the question of Hobbes's exclusion from the experimental community or the Royal Society.

Not only do Gross and Levitt miss the book's central arguments, they also miss the central assertions on Hobbes's controversies with Boyle over the role of experimental facts in natural philosophy. Hobbes, Shapin and Schaffer argue, claimed that the air-pump was an unreliable instrument for natural philosophy—it could not produce a vacuum because it leaked—and thus his own plenism provided a better account of the phenomena produced by the air-pump than Boyle's. Hobbes ascribed to Boyle a belief in the metaphysical vacuum, despite Boyle's claimed disinterest, and argued that belief in the vacuum was based on philosophically absurd language and was politically dangerous. Hobbes further claimed that Boyle's experimental knowledge was not public, was produced under the authority of a "master," and could not guarantee assent. In short, Hobbes asserted that Boyle's experimental program could guarantee neither certain knowledge nor civic order, and thus was not philosophy.[10]

Hobbes's relations with the Royal Society—Gross and Levitt's central focus—is addressed by Shapin and Schaffer on pages 131 to 139 "parenthetically," as an "excursion" that becomes significant to them only in its relevance to their central focus on forms of knowledge production.[11] Their argument is presented against Quentin Skinner's explanation that Hobbes was rejected as a "club bore." (Skinner argues that it is anachronistic to view the Royal Society of the period as a modern scientific society in which membership is determined by scientific credentials: the Royal Society was, he asserts, a gentlemen's club which happened to include many prominent scientists and excluded many others; Hobbes was excluded neither because of his religious views nor his incompetence in mathematics or science, but simply because of his dogmatic personality.)[12] Shapin and Schaffer contend that what Skinner interprets as criticism of Hobbes's personality as dogmatic is in fact inextricably linked to programmatic criticism of Hobbes's philosophy as dogmatic. The explanation Shapin and Schaffer offer for Hobbes's exclusion is his rejection of the experimental program—its philosophy, manners, and social relations—as a form of life:

> We now have an answer to the question, "Why was Hobbes excluded from the Royal Society?" It is an answer that does not attempt to distinguish assessments of Hobbes's personality from judgments of his philosophical programme. The connections among personal characteristics, social relationships, and philosophical practices were perceived, as Sprat's polemic shows, to be substantial and vital. The modest and humble Boyle was jux-

taposed to the intolerant and confident Hobbes, just as the modest and humble experimental programme was contrasted to Hobbes's overweening rationalism. Each philosophical programme was predicated upon its distinctive social relationships, and each valued a characteristic philosophical persona. The social order implicated in the rationalistic production of knowledge threatened that involved in the Royal Society's experimentalism. Thus our excursion into Hobbes's relations with the Royal Society is not, in fact, peripheral to our major concern with conflicting knowledge-generating strategies. Hobbes's anti-experimentalism, as expressed in the *Dialogus physicus* and elsewhere, gave grounds for his exclusion.[13]

It may seem remarkable, given Gross and Levitt's lengthy synopsis and refutation, that in a paragraph which begins with the question "Why was Hobbes excluded from the Royal Society?" Shapin and Schaffer provide an explicit and succinct explanation for Hobbes's exclusion—his anti-experimentalism—and that this explanation is never even mentioned by Gross and Levitt. What is more remarkable is that in their refutation, Gross and Levitt quote from this very paragraph.[14]

Logic has proved of little help to Gross and Levitt since they have missed the central arguments, but what about evidence? In their refutation, *they make no claim whatsoever to have discovered a single error of scientific fact*—their criticism is limited to the interpretation of historical events. Their claim that Hobbes's exclusion was related to his long-standing conflict with Wallis is not implausible; however, they present no evidence from either primary documents or secondary research to justify their claim.[15] Because secondary sources are not cited in their refutation, readers outside the field may be left with the impression that Gross and Levitt's claim of Hobbes's incompetence is an original insight which required specialized mathematical training. It is not: this claim is simply the received view found in elementary textbooks on the history of mathematics;[16] Schaffer, a historian of seventeenth-century science, is hardly unfamiliar with this claim. (But because the question of Hobbes's mathematics is not central to their study of the rise of experimental science, Schaffer's analysis of Hobbes's mathematical disputes is published in a separate research article, as is Shapin's work on Boyle's mathematics.)[17]

Historians, on the other hand, will recognize Gross and Levitt's assertion of Hobbes's incompetence as little more than Whig historiography in both content and approach—conclusions based on the claims of the victors and modern standards instead of the analysis of historical documents. Even a glance at secondary research articles would have helped Gross and Levitt avoid merely rehearsing this received view.[18] And readers familiar with *Leviathan and the Air-Pump* will recognize that this book was formulated as a critique of precisely this Whig historiography and the assumption that Hobbes was simply "incompetent" in particular (again, these arguments are all detailed in their first chapter [4–14]).

What, then, is the significance of *Leviathan and the Air-Pump* for Gross and Levitt? It cannot just be academic disputes about the history of seventeenth-century science: they have evinced little interest in either primary historical materials or secondary research literature; they do not share Shapin and Schaffer's interest in philosophical issues, which are ignored throughout their synopsis and refutation; and they have overlooked the central arguments that make *Leviathan and the Air-Pump* "an exercise in the sociology of scientific knowledge."[19] Instead, for Gross and Levitt this arcane historical treatise on seventeenth-century experimental philosophy represents the political threat of historians and sociologists taking control of science:

> [Hobbes] can be made to stand for the voiceless and excluded masses—and the intellectuals without serious scientific training—to whom science is an inaccessible mystery, seemingly beyond human control. Thus we are forced in our reading of the book to see it as a parable, whose fulsome celebration of Hobbes conveys the implication that "philosophers" who are not professional scientists (for which we must read "historians" and "sociologists") should have the authority to pronounce, or even to prescribe, on scientific questions. (69)

In this, their most rigorous attempt at refutation, Gross and Levitt's criticisms turn out to be focused on the tangential point of Hobbes's exclusion, oblivious to Shapin and Schaffer's own explicit explanation for his exclusion, unsupported by evidence from primary historical documents or scholarly research, and ultimately little more than a return to precisely the Whig histories that Shapin and Schaffer sought to criticize in the first place. Gross and Levitt's refutation provides no evidence that they have understood any of the central theses of *Leviathan and the Air-Pump*—or even the arguments presented in the eight pages that parenthetically discuss Hobbes's exclusion from the Royal Society.

Sandra Harding's The Science Question in Feminism. In their refutation of Sandra Harding's *The Science Question in Feminism*, Gross and Levitt fail to mention that she was trained in the philosophy of science and wrote her dissertation on Quine.[20] Her book, written in the clear prose and style of argument of the Anglo-American philosophical tradition, examines the tensions between the developments in three current feminist epistemologies—empiricist, standpoint, and postmodernist—and the assumptions of the philosophical systems from which they originally derived; she argues that these tensions indicate a need to formulate new questions and approaches that are not limited by the original philosophies.

Gross and Levitt's refutation focuses on *two sentences* from her book. These two sentences, along with two others that Gross and Levitt quote, come from a section in Harding's second chapter entitled "Paradigmatic Physics." It is only six pages in length, but Gross and Levitt provide no summary of her argument. Here Harding argues that although physics has been generally accepted by physicists, chemists, and others as the paradigm of science because of the successes of Newtonian mechanics, physics is in many ways atypical of the sciences in general: it explains phenomena less complex than biology; and it minimizes factors such as human action, the role of the observer, and the irrational which are important in the social sciences and psychology.

The two sentences that Gross and Levitt quote and criticize make no specifically feminist claim, nor is there any mention of feminist philosophy or epistemology:

> Would not physics benefit from asking why a scientific world view with physics as its paradigm excludes the history of physics from its recommendation that we seek [critical] causal explanations of everything in the world around us? Only if we insist that science is analytically separate from social life can we maintain the fiction that explanations of irrational social belief and behavior could not ever, even in principle, increase our understanding of the world physics explains. (107; and Harding, 47)

Instead of presenting arguments to counter Harding, they propose to decipher Harding's two sentences: "What can the quoted statement possibly mean? What can be said in defense of it?" (127). They follow with two pages of misreadings—approximately one page for each of Harding's sentences—contrived by quoting Harding out of context. These two sentences that Gross and Levitt quote are examples that Harding offers to explain her previous sentences (a fact that is further obscured by Gross and Levitt's removal of Harding's "For instance" from the beginning of the first sentence that they quote). The previous sentences read as follows:

> Why, then, should we take as the model for all knowledge-seeking a science [physics] that has no conceptual space for considering irrational behavior and belief? Moreover, possibly explanations even in physics would be more reliable, more fruitful, if physicists were trained to examine critically the social origins and often irrational social implications of their conceptual systems. For instance, would not physics benefit from asking why . . . [see above][21]

Gross and Levitt's deciphering proceeds by presenting feigned misinterpretations of the words *world-view, paradigm, everything, causal,* and *history.* First is Harding's use of *world-view*—a word very difficult to misunderstand since this commonplace philosophical term has only one accepted meaning.[22] But the appearance of *world* in *world-view* provides Gross and Levitt with a pretext to present more than an entire paragraph of refutation and asides purporting to demonstrate that the "scientific world-view" is not in fact a view of science shared by the entire world; the confusion finally ends with their admission that physics may indeed be considered paradigmatic by natural scientists:

> First of all, the [first] sentence is freighted with assumptions. There is supposed to be a regnant worldview, with physics as its paradigm. Given that only a vanishingly small fraction of the world's five billion souls know anything about physics, it cannot be that physics is any sort of demotic weltanschauung. If we limit our attention to what likes to think of itself as the Developed World, the situation is hardly different. Let us then narrow the focus to the microscopic subworld of professional intellectuals. At least half of those—shall we say most humanists, most historians, a good fraction of sociologists, a surprising number of philosophers—know virtually nothing about physics either. Such ignorance is not a particular virtue, but neither is it a vice: it has not precluded remarkable intellectual achievements. Humanists have not, traditionally, felt the need to apologize for this gap in background, *nor should they* (although a grain or two of embarrassment might well be felt for the propensity of a few colleagues to pontificate on the deep meanings of physics without having bothered to learn it).

We are left, presumably, with natural scientists and some of the social scientists who have been trying, with variable success, to introduce into their disciplines a measure of what is usually thought of as scientific rigor. It is to some degree justified to assert that this community takes physics for its paradigm, out of acknowledgment that physics is the most successful of empirical sciences. (127, emphasis in original)

We have returned to our starting point, and next is *paradigm.* Harding—by implication unaware of fundamental debates in the philosophy of science—is accused of vulgarizing because her single sentence fails to adequately address the question of whether physics *should* be the paradigm:

To say that [the last sentence in the previous extract], however, is to gloss over the important debates that engage scientists in all fields. How and to what extent *should* physics serve as a model? There is the issue of reductionism in the strict sense, the distinct question of whether a given discipline *ought* to rely on a logico-mathematical model for its theoretical structure, the question of the degree to which constructs appearing naturally within such a structure should be reified. These questions are as subtle as they are important. Leading thinkers in all fields of science confront them ceaselessly. Thus to speak of physics as a "paradigm" is to vulgarize the situation. (127, emphasis in original)

Has Harding vulgarized by glossing over debates over whether physics *should* be the paradigm? As noted above, Harding's two sentences come from a section entitled "Paradigmatic Physics" which Harding introduces with the sentence: "But why should we continue to regard physics as the paradigm of scientific knowledge-seeking?" (43). Her argument is that physics should *not* be the paradigm. It takes but three pages before Gross and Levitt quote Harding considering precisely this issue in reference to its consequences for feminism (Harding's statements that they quote begin: "If it is reasonable to believe that physics should always be the paradigm of science . . ." and continue "If physics ought not to have this [paradigmatic] status . . ." [130, interpolation in Gross and Levitt]).

Next is the term *everything* (Gross and Levitt point out physicists do not seek causal explanations of literally *everything*), followed by *causal* (Harding is again charged with being unaware of the subtleties of concepts from the philosophy of science), and then on to *history.*[23]

This is followed by a similar series of interpretive mishaps, sprinkled with ad hominem remarks, contrived for the second sentence: Gross and Levitt suggest that Harding's view is similar to that of Nazi physicist Philip Lenard and aver that physicists will find her "equally unpersuasive" (129).

In the end Gross and Levitt finally offer what appears to be their synopsis of Harding's book—they offer no other—but here there is no mention of any of the central philosophical issues that Harding has addressed:

> To the best of our ability to make out her program, as set forth in the book from which the quotation is taken, it is as follows:
> 1. Western society is in some measure rooted in assumptions about the differences in worth and ability between men and women.
> 2. These are irrational and harmful beliefs.
> 3. Such beliefs permeate all our social institutions and all aspects of our belief systems.
> 4. Therefore even physics is biased and distorted by the ineluctable influence of these irrational beliefs, and therefore
> 5. analysis of this root-and-branch unreason will lead eventually to clarification and rectification, even of the recondite world of physics. (129–30)

Even Gross and Levitt find their own synopsis too uninteresting to make the pretense of offering arguments against it, and all five assertions are swiftly refuted by dint of their following sentence: "Points (1) and (2) are unexceptionable; (3) is partially, but only partially true, and it is altogether too categorical; (4) is substantially *untrue* as regards the actual content and methodology of physics—there is no plausible body of evidence for it—and, consequently, (5) represents no more than the triumph of hope over logic" (130, emphasis in original). Gross and Levitt then offer one last misconstrual based on two quotations again stripped of context, and their refutation is complete.

Had Gross and Levitt outlined Harding's argument from the six pages in the section within which they found their quotations, they would have discovered that her views here are surprisingly close to their own view that "we accede in principle to what might be called the 'weak' version of cultural constructivism" (44).[24] Certainly they have much more substantive disagreements with Harding; they simply have failed to address them.

Before moving to their next refutation, Gross and Levitt repeat an

attack on Harding that commonly appears in the popular press, ironically under the pretense of reassuring the reader of their own fairness: "But these [sentences from Harding] are not atypical, nor are
they her most redolent remarks. Her stirring assertion to the effect that
Newton's *Principia Mathematica Philosophae Naturalis* is a 'rape
manual' may well have won her a lasting admiration in doctrinaire
feminist circles and even a place among physicists" (131). The remedy for this commonplace caricature of Harding's views is as simple
as looking up the passage. Harding is critiquing the opposing effects
attributed to metaphors in science by historians and philosophers that
are apparently based on little more than a consonance with modern
values: metaphors offensive to modern sensibilities (e.g., Nature as a
woman, probing her secrets) are dismissed as insignificant accidents
of historical circumstance; other metaphors (e.g., the universe as a
machine) are elevated to be historically causal and even necessary
conditions for the rise of modern science. Arguing that a consistent
approach to the role of metaphors in the development of science
would require treating the metaphor of rape as similarly important,
Harding highlights this inconsistency by asking, "In that case, why is
it not as illuminating and honest to refer to Newton's laws as 'Newton's rape manual' as it is to call them 'Newton's mechanics'?" (113).
Gross and Levitt, unconcerned with correcting the widespread misrepresentation of her argument, instead use this as a further opportunity for vulgar caricature: "We pity coming generations of freshmen
physics students who, titillated by this famous remark, will spend
long hours thumbing through that magisterial work, looking for the
dirty bits" (131).

Donna Haraway's Interview. Their criticism of Donna Haraway
offers the most stark example of the logic and evidence that for Gross
and Levitt suffices as refutation. They derisively note that Haraway—
a biologist, historian of science, and central figure in the feminist
studies of science—is "one of the greats of the business" (132). The
refutation of a representative work would seem crucial for their claim
that feminist studies of science are flawed; but instead of choosing
one of her books or one of her numerous research articles, they settle
for a single interview. Here Haraway discusses in simpler, more con-
versational language some of the ideas from her two books. Gross and
Levitt choose a single section of Haraway's views not on biology, history, postmodernism, or feminism, but on social constructivism; more
surprisingly, the section they choose for their refutation outlines Haraway's arguments *against* social constructivism as leading ultimately

to relativism. Once again, Gross and Levitt offer no summary or analysis of her arguments. For passage after passage that they quote—four paragraphs, three in their entirety—their criticism is quite simple: they do not understand. Their central concern is Haraway's position on relativism, and at one point they find that Haraway seems to be a "full-fledged relativist" (133). But they make no attempt to locate her remarks in the extensive literature surrounding the debates on relativism,[25] nor do they even look at her books or articles. (Her books are listed on the second page of the interview [66]; her article "Situated Knowledges: The Science Question in Feminism and the Privilege of Partial Perspective"—frequently cited, reprinted in her most recent book, and arguably one of her most important articles—includes her criticisms of relativism and outlines her alternative.) In their final paragraph on Haraway, instead of evidence they invoke the opinions of anonymous authorities, instead of analysis they offer their own confusion, and ultimately, instead of a refutation they offer only a gratuitous insult:

> Let the confused reader not worry: we are confused, too; and so were the five thoughtful people, all strongly sympathetic to feminism, to whom we applied for clarification. They don't know, and we don't know, whether Haraway is for or against relativism; but a vote indicates that whatever "limitations" she sees in relativism, they are not fatal "at every level of the onion." For this "discourse"—and the quoted passages are entirely typical—on what is, after all, a quite fundamental question of scientific epistemology, we have been able to find only one *signifier:* it is Peter Mayle's term, invented originally to describe certain goings-on in Provence, especially in the season of wine tasting: "delusions of adequacy." (134, emphasis in original)

Higher Superstition as an Exposé of Errors of Scientific Fact

In these refutations, not only have Gross and Levitt failed to identify the central theses and supporting arguments of the works they criticize, they have proved unable to understand even the short passages they quote. Contrary to their stated claim, then, they are not presenting synopses of crucial arguments or refutations that undermine entire scholarly works. More plausibly, they are instead searching to find within these works *something* they can refute. If this is the case, to what extent does *Higher Superstition* provide evidence of widespread errors on science in humanist scholarly publications, errors possibly damaging to the theses presented?

Scientific Fact and Historical, Sociological, and Philosophical Interpretation. Higher Superstition does contain several examples of misunderstanding and misstatement of scientific fact; given the length of their book and the number and range of sources criticized, these errors are remarkably few.[26] In rare cases the errors they discover are important—for example, N. Katherine Hayles misdates Logical Positivism, which would require revision of part of her argument. In other cases the errors of scientific fact they discover are irrelevant to the argument presented, as in their quotation from Derrida on the Einsteinian constant.[27] In other instances, the errors they claim to discover are in fact a result of their own misreading: for example, in the single mathematical mistake they claim to discover in Latour, they attribute to Latour a view that he is in fact attributing to Cantor.[28] However, most of the criticisms Gross and Levitt present are not even questions of scientific fact but rather criticisms of interpretations of historical or philosophical issues.[29]

Evidence in Historical, Sociological, and Philosophical Interpretation. If Gross and Levitt are criticizing interpretations in history and philosophy—fields in which they have neither special competence nor training—what kind of evidence do they present against the views they critique? In their discussions on history they do not cite primary historical documents and only rarely secondary research;[30] in their discussions on sociology they do not cite other research and offer only occasional commonplace criticisms of methodology.[31] Most of their criticisms are philosophical and directed against constructivism, feminism, and in particular relativism. However, they appear to be unaware of the extensive debates on these subjects: they neither cite nor make arguments in reference to these debates. And only rarely do they offer even a sketch of their own philosophical views: they appeal to common sense and occasionally make reference to Ayer's (amended by Popper) "no-nonsense logical positivism" (86); in a footnote, they cite with approval an article suggesting that the universe may be deterministic (261–62, n. 9); and they offer an outline of a naive weak constructivism (43–44). To avoid the charge that they simply rehearse naive philosophical views of science—and these views are precisely what the studies they sought to refute were written against—would require that they systematically present their own views with arguments instead of merely stating their dislikes.

Because Gross and Levitt are unfamiliar with the fields they criticize—history, philosophy, anthropology, and sociology—they depend in their refutations on citations of the works of other scholars, but their citations are often indiscriminate. In place of critical evaluations of arguments, they frequently present as decisive refutation views similar to their own by another scholar who is involved in an ongoing debate; sometimes these citations are from sources of lesser scholarly quality than their own. One ironic example is their quotation of the philosopher Argyros whose work, by their criteria, might be expected to be precisely the kind that they would indict with scathing sarcasm;[32] but apparently because they are unfamiliar with postmodernism, they quote Argyros against Lyotard (80), deconstruction (83), and Hayles (99; this quotation is discussed below). Elsewhere, instead of citations they invoke anonymous authorities: against Haraway, "five thoughtful people, all strongly sympathetic to feminism" (134); and against Latour "earnest and responsible philosophers of science" driven by Latour's style into "paroxysms of disgust" (58). And their failure to provide proper citations for relevant work of other scholars often leaves the impression that their criticisms are insights that only a trained scientist could offer, when in fact they are commonplace arguments from precisely the fields that they seek to criticize.[33]

Higher Superstition as the Ad Hominem Indictment of Individual Scholars

As several examples from the above analyses have suggested, an even more serious problem with their evidence is their quotation of material out of context. Not all of Gross and Levitt's misquotations are simply errors of omission, and not all result in just the obfuscation of academic arguments: their substantive misrepresentations play a crucial role in their attacks on the character of individual scholars. Three examples from chapter 4—Ross, Hayles, and Derrida—will suffice to show how omission, misrepresentation, caricature, ad hominem remarks, and altered quotations combine to form these attacks.

Andrew Ross's Strange Weather. In their refutation of Ross's *Strange Weather*, Gross and Levitt note that "this work studies contemporary popular subcultures [including New Age, cyberpunk, and hackers] that vulgarize standard science while, to some degree, challenging its authority" (90);[34] but the remainder of their synopsis simply inverts his thesis. Although Ross claims that his purpose is neither to speak for nor against these subcultures,[35] his portrayals are often extremely ironic (for example, he describes how "the epithet 'Western' [which] is like the mark of Satan for the holistic health community" is used in a New Age "technical" paper "The Hidden Crime of the [Western] Porcelain Throne" to sell alternative toilets).[36] Gross and Levitt's claim that he "celebrates them [these subcultures] as possible nuclei of resistance to a monolithic, global and capitalist techniculture" (90) is precisely the opposite of Ross's statement that "this book is not, however, a celebration of the 'resistant' qualities of these subordinate cultures" (9). Indeed, Ross's thesis is that despite their apparent opposition, these cultures appropriate many of their norms from mainstream science and technology (one example offered by Ross is his analysis of how metaphors borrowed from high-tech motorcycles, bodybuilding, and the Yuppie work ethic are used to sell "wacky brain-machine technologies" [30–32]).

Given that Ross's subject is not science, it is not surprising that Gross and Levitt's criticisms are not of errors of scientific fact. Instead, Gross and Levitt introduce their refutation with a lengthy misconstrual of a paragraph taken out of context.[37] Their only critique of Ross's science is their claim that "Ross parrots all the New Age mystifications of quantum mechanics" (91); in fact, Ross is describing the incongruities in New Age appropriation of the philosophical views of quantum physicists.[38] And their refutation contains the usual ad hominem remarks — for example, likening Ross's views to those of Stalin.

Gross and Levitt apparently are eager to be misled by Ross's extravagant subtitle — "Culture, Science and Technology in the Age of Limits." It is apparently on this basis — for they offer no other justification — that Gross and Levitt assert that *Strange Weather* "enlarges his claim to be expert on matters of science and technology" (91). It is unlikely that anyone who has read Ross's book will think that he claims to be an expert on science and technology: his introduction begins with the travails of using photocopiers, and there he describes himself as "a cultural critic and educator who does not consider himself especially technoliterate"; he acknowledges that he lacks training in science and is not able to critique or understand "official" science.[39] Gross and Levitt then develop their charge — again without offering any justification — into an indictment of Ross's character as a scholar: "Strange it is that a well-known scholar at one of the world's most distinguished universities should write a lengthy book upon a subject about which he knows, evidently, virtually nothing. It is stranger still that he can boast of his ignorance in the very first words" (91). In their kettle-logic criticism, Gross and Levitt show no interest in reconciling their first sentence stating that he has written "a lengthy book upon a subject about which he knows, evidently, virtually nothing" with their earlier description of the subject of his book as a study of "contemporary popular subcultures," or reconciling their second sentence stating that he "boast[s] of his ignorance in the very first words" with their earlier charge that Ross claimed he was an expert in science and technology.

Ironically, had Gross and Levitt read this book, they could have found many original criticisms of sublegitimate scientific subcultures missing in their own polemic. Instead, their only interest in *Strange Weather* was to impugn Ross's character through misrepresentations and ad hominem attacks.

N. Katherine Hayles's Chaos Bound. In their refutation of Hayles's *Chaos Bound*, Gross and Levitt again offer very little synopsis of her argument, noting only that she argues for parallels between developments in scientific and literary theories.[40] Instead of outlining her reasoning, Gross and Levitt repeatedly lampoon the very assertion of the possibility of parallels: "Why should the theory of dynamical systems be more closely related to the gyrations of literary exegetes than it is to major league baseball or Jane Fonda's workout tapes?" and "Why not compare Derrida to Charles Manson, or the Feigenbaum number to Roger Clemens's earned-run average?" (99). This is not a question of scientific fact but rather of intellectual history, and their caricature is in fact predicated for its effect precisely on omitting Hayles's arguments—including the basis for her assertions in philosophy and history, her criticisms of Prigogine's extravagant claims for chaos theory, and her assertions of important differences in the development of literary and scientific theories.[41] Their indictment of her integrity is based on these caricatures and misrepresentations and, in particular, the omission of her analysis of the differences between scientific and literary theories: "She is one of those who are eager to tell you, earnestly and at length, precisely why a raven *is* like a writing desk—especially if a publication can be got out of it" (99, emphasis in original).

Despite their ad hominem attacks against Hayles for suggesting the possibility of parallels between scientific and literary theories, one paragraph later, we find Gross and Levitt admitting that it may not be so unreasonable to see some parallels, after all:

> Even philosophers who see some parallels between deconstructive literary theory and the mathematics and physics of chaos theory are loath to push the comparison as far as does Hayles.

Alexander Argyros, commenting on Hayles's work, notes: "I suspect that this apparent compatibility may be implying to literary
theorists that chaos is a validation of deconstruction. My own
view is that such a claim is, for the most part, wrong. . . . While
it is certainly true that deconstruction and chaos are both interested in highlighting [the importance of] non-linearity, to claim
that they are fellow travelers is, I believe, to make an unwarranted assumption." (99; ellipsis in Gross and Levitt)

Gross and Levitt's choice of Argyros as an authority against Hayles
might seem at first to be just another case of indiscriminate citation:
Hayles received some training in science and Argyros—as Gross and
Levitt note—does not understand basic mathematics.[42] However, as
a philosopher, Argyros analyzes arguments for their cogency and evidence; and here the evidence he presents in support of his assertion of
differences between scientific and literary theories is the work of *N.
Katherine Hayles*. What Gross and Levitt remove from the middle of
the passage (marked by their ellipsis) is the following twenty lines of
text, including *sixteen lines* of text that Argyros *quotes* from Hayles:

Basically, I agree with Hayles when she describes many of the differences between deconstruction and chaos.
There are, of course, also significant differences between them
. . . One measure of these differences is disagreement among
deconstructionists and scientists on how extensive chaos is.
For Derrida, textual chaos is always already in Rousseau and
in every other text. By contrast, scientists acknowledge that
ordered, predictable systems do exist, although they are not
nearly as widespread as classical science had supposed . . .
Where deconstructionists see an apocalyptic break with logo-
centrism, scientists are likely to think of their work as a continuation of what went before. To a deconstructionist, to say someone is a recuperator is a damning comment; for most scientists
recuperation is not an issue, because they see their work as
enhancing rather than discrediting scientific paradigms. These
differences are symptomatic of the different values the two disciplines place on chaos. For deconstructionists, chaos repudiates
order; for scientists, chaos makes order possible. (1989, 316–17)
The gist of Hayles's article, and of her *Chaos Bound*, is that such
differences are not as important as the significant parallels between
chaos and deconstruction.[43]

Once again, Gross and Levitt simply do not wish to isolate specific academic issues—such as the relative extent of parallels and differences in the development of literary and scientific theories—where they disagree with Hayles. By misrepresenting her argument, by omitting mention of her critique of the exaggerated claims of chaos theorists, and by manipulating quotations to obscure her assertions of these differences, Gross and Levitt are able to devote entire paragraphs to lampooning Hayles and attacking her integrity as a scholar.

Post-Structuralism: Foucault and Derrida. Despite their repeated deprecation of what they term postmodernism,[44] there is little evidence that they have understood any of these works either; their overviews are apparently based on misreadings of uncredited popularizations. For example, their overview of Foucault's work fails to even mention his central concepts—genealogy, archaeology (a term that appears in the title or subtitle of three of Foucault's major works), or the constitution of the subject; instead, Foucault is described as presenting "ideas of consciousness and domination" (77).

Their summary of Derrida is similarly from uncredited popularizations; their only original criticism consists of accusing him of an "eagerness to claim familiarity with deep scientific matters" (79). Here they offer but two examples from more than a dozen books Derrida has authored. In the first example, the passage they quote from Derrida on the Einsteinian constant makes little sense, as they claim. But nowhere do Gross and Levitt inform the reader that this passage is not from the article but from a discussion following the paper.[45] Nor are we informed this was simply a response to a similarly worded question posed by Hippolyte; and—in a pattern now too familiar—this is further obscured by their removal of the beginning of the paragraph they quote, "Concerning the first part of your question. . . ." Nor do Gross and Levitt inform the reader that Derrida's paper—probably the most influential critique in the turn from structuralism associated with Levi-Strauss to post-structuralism—depends in no way on the physical sciences. The source that Gross and Levitt credit for this quotation—an article in the journal *Skeptical Inquirer*—quotes the discussion properly.[46]

Immediately following, Gross and Levitt offer their second and final example:

> This [the preceding example] is not, we assure the reader, an isolated case. In various other Derridean writings there are to be found, for example, portentous references to mathematical terms

such as "differential topology," used without definition and without any contextual justification. Clearly, the intention is to assure readers who recognize vaguely that the language derives from contemporary science that Derrida is very much at home with its mysteries. (79)

Reading the article would have been enough to assure them that Derrida is not discussing mathematics: the article they cite, "Before the Law," is an analysis of a Kafka fable of the same name in reference to the question, "What is literature?"; an editor's introduction to the article makes this point very clearly. There are no references to mathematics or science in the entire article, only to thinkers such as Freud, Heidegger, and Hegel. But here—instead of quoting a passage out of context—Gross and Levitt offer no quotation. Derrida's passage indicates why: "Guardian after guardian. This differantial topology [*topique différantielle*] adjourns, guardian after guardian, within the polarity of high and low, far and near (*fort/da*), now and later."[47]

Those even vaguely familiar with Derrida's work will recognize that the English term *differantial* (with an *a*) derives from *différance*, one of his central concepts, which indeed itself appears repeatedly throughout this article. The French gloss immediately following is *topique différantielle*, not the French for differential topology, *topologie différentialle*—a term that Levitt, who has written a book on topology and cites French work in his bibliography, should be familiar with.[48] So much for evidence: here the content of the article, the editor's introduction, the English term, and a French gloss together prove inadequate to dissuade Gross and Levitt from their ad hominem attack.[49]

Refutations and Rhetoric. In summary, *Higher Superstition* refutes very little. This is not of course to suggest that every argument in *Higher Superstition* is incorrect; nor is this intended to be a blanket affirmation of the works Gross and Levitt selected for their refutations. These works are hardly so incontrovertible that important academic criticisms cannot be presented, as reviews in scholarly journals amply demonstrate—some of the most incisive criticisms of science studies come from within the field itself.[50] However, Gross and Levitt were not interested in academic debates: their refutations were so carelessly written that in their haste they failed to identify the assertions they probably disagree with most strongly. The ad hominem approach they chose allowed them to complete their book without either the study required to understand the fields that they criticized or the research necessary to contest these views. Instead, they filled page after page with caricature, misreadings, and insults. More importantly, it was their attack on the honesty and scholarship of those they criticized that served as their fundamental rationale to justify their political intervention in the humanities.

This then accounts for the rhetoric in their ad hominem attacks. Gross and Levitt call others' views "unalloyed twaddle" (43), "vaporous pontifications" (98), "a howler" (98), "philosophical styrofoam" (98), and "asinine" (228). They characterize postmodernism as a "painful bolus" in the "costive bowels of academic life" (106). They liken deconstructionists to "con men" (85), one scholar to a "crackpot" (103), and their opponents to braying asses (6). They even stoop to grade-school taunts, as when they parenthetically state of a graduate student whose work they criticize: "It [the development of topological science] is also, we note, coincident with the era in which people played football without a helmet. Would it be too cruel of us to wonder whether Dean is overly fond of some similar activity?" (266, n. 12).

The "Academic Left" and the Academic Left

We must now briefly return to their definition of the *academic left.*[51] It would not in fact seem difficult to offer an imprecise but workable definition of the academic left—as journalistic accounts have done—and then proceed to analyze their views on science; in such a reformulated study some of the works included in *Higher Superstition* would undoubtedly reappear. This is not, however, the approach Gross and Levitt have chosen.

Their usage of the term *academic left* changes substantially throughout their book. In a section of the first chapter entitled "Distinctions," Gross and Levitt offer their most rigorous (and narrow) definition:

> When we use the phrase *academic left* we do *not* refer merely to academics with left wing political views. There are plenty of such people with whom we have no quarrel. There are countless academics who do excellent and penetrating work, in appropriate fields, from a left-wing viewpoint. There are countless left-wing scientists—although we are stodgy enough to insist that there is no such thing as left-wing science. We are using *academic left* to designate those people whose doctrinal idiosyncrasies sustain the misreadings of science, its methods, and its conceptual foundations that have generated what nowadays passes for a politically progressive critique of it. (9, emphasis in original)

Here they present two criteria that, in addition to political views, further narrow the scope of their "academic left": disciplinary affiliation and research program. Excluded at the outset from their definition of the "academic left" are scientists (and apparently at least engineers and physicians) with leftist political views. What disciplines count as "appropriate fields" is never delineated, but later in the book they also exclude many humanists with leftist views. In fact, Gross and Levitt go so far as to explicitly state that they are not against perspectivist, multiculturalist, or leftist scholarship if it is not taken to an anti-science extreme: "perspectivism points to some genuine issues and may lead to some valuable observations" (40); "there are honorable thinkers on the left—and elsewhere—who advocate strongly a multicultural approach" (259, n. 5); "historians with a left-wing perspective . . . have certainly made their respectable mark" (69).

When we include their criteria for research programs—without disputing here their sensibilities regarding how to define those "whose doctrinal idiosyncrasies sustain the misreadings of science"—all that remains of the "academic left" is what they term "left-wing criticism of science" (13) or, more precisely, the constructivist, postmodernist, and feminist studies of science (along with radical environmentalism), represented by the works they criticize in chapters 3 to 6. These constitute three separate research agendas within the history, sociology, and philosophy of science—a program itself so small that it does not even constitute a department in most large research universities (and radical environmentalists number considerably fewer still). In this narrow definition, what is missing from Gross and Levitt's "academic left" is the overwhelming majority of the academic left—engineers, scientists, physicians, and most humanists.

Narrowly defined, their "academic left" does not just exclude most of the academic left; not all of what they do include is either academic or left. Radical environmentalism is so tenuously associated with academia that Gross and Levitt offer but one academic representative. And to the extent that they identify the "academic left" not with individual political beliefs but instead with entire research programs, not all of what they criticize is leftist. In fact, Gross and Levitt evince so little interest in limiting their criticisms to the political left that they wind up having to offer caveats, admitting, for example, that Hayles's *Chaos Bound* "is not primarily concerned with leftist political agenda" (100). Thus defined, their term "academic left" refers not to political views but rather to fields of research.

Having narrowly defined "academic left," Gross and Levitt are unable to provide a consistent answer to what they believe these science critiques have in common. It is not, they repeatedly emphasize, a set of shared views, ideology, or doctrines: "although we have been speaking of an academic left critique, it must be stressed—and we are compelled to stress it throughout the discussion to follow—that this is not a self-consistent body of doctrine" (5); they note that there is "little to connect any two of them in terms of language or philosophy" (10); they further note the "rank incompatibility" (10) of these works. At various places throughout their book, they offer incongruous characterizations of what their "academic left" is to have in common: a uniform hostility toward science (2); a commitment to "fundamental political change" through the revolutionary transformation of cultural categories (3); a "shared sense of injury, resentment, and indignation against modern science" (5); "an overriding common purpose, which is to *demystify* science" (11, emphasis in original); advocacy of a form of perspectivism which questions the universality of the ideas upheld by Western culture (38); and a political moralism (220)—resorting "constantly and shamelessly to *moral one-upmanship*" (8, emphasis in original). But contrary to their assertion that the hostility toward science is uniform, Gross and Levitt wind up repeatedly explaining how to interpret the explicit admiration of science as hostility: while admitting that "we do not claim that all of the named writers [authors they criticize in *Zone 6/Incorporations*] are hostile to science: many of them, indeed, profess to admire it greatly," Gross and Levitt allege that these authors "help to set the stage for a kind of hostile science-criticism" (80); although Hayles's *Chaos Bound* "regard[s] science as, on the whole, liberatory and politically progressive," it becomes, by dint of its alleged misreading of science, "hardly distinguishable from hostility" (100); cultural critics who seem to be "celebrating science" are termed hostile because of their alleged "moralizing undertone" (105). What, then, of their claim that "it is actually this [leftist political] moralism, rather than any solid philosophical commonality, that unites the various critiques we have examined"? (220; see also pp. 6, 8, 41, 246). Discovering moralism in *Leviathan and the Air-Pump* was possible only by their extravagant misreading of it as a "parable"; and in chapter 4 they must repeatedly explain how to interpret postmodernism, which they have denounced throughout their book as relativistic and perspectivist, as being moralistic—"hidden beneath the relativistic pose, there lurks a stiff-necked moralism" (96).

Their definition of the "academic left" has so many problems that—

in contrast to the confidence they exude for their refutations—it is introduced with repeated disclaimers: they admit to using the term with "great misgiving" (2), find it a "troubling term" (3), and concede that "the term is not felicitous" (37). They offer five alternatives, along with their rationale for rejecting them: "postmodern left," "Post-Marxist left," "hypertheoretical left" (37), "perspectivist left" (40), and "New Rage Academics" (260, n. 13). And in response to Steve Fuller's criticism that "the range of opponents addressed by Gross and Levitt is truly staggering" and that "these groups do not have much in common,"[52] Gross and Levitt reply, "Yes, this is a 'staggering' range of 'opponents'; but the argument is that they have something important in common, and that the commonality is cause for worry"[53]—but exactly what comprises this "commonality" is never consistently explained.

"Science" and the "Academic Left"

The "academic left" is the central concept in Gross and Levitt's formulation of their thesis, and they have good reasons for insisting on a definition this bad.

"Science" against the "Academic Left." Narrowly defined as research programs—the constructivist, postmodernist, and feminist studies of science, with nonacademic radical environmentalism thrown in—the "academic left" stands refuted: chapters 3 to 6 were to show the work of their "academic left" to be flawed and fatuous. At this point, excluding leftist humanists and even multiculturalists, feminists, and Marxists is essential for Gross and Levitt: it obviates even the pretense of offering refutations on questions of politics, culture, art, and literature that Gross and Levitt can claim no special competence—arguments that Gross and Levitt are at pains to concede a validity to in general.

Excluding leftist scientists pits "science" against their "academic left." On the one hand, by excluding scientists by fiat, Gross and Levitt avoid the necessity of presenting arguments against the kinds of claims they criticize when they are made by scientists themselves: critiques of realism and positivism by physicists and philosophers—including the philosophical questions arising from quantum physics and the Bell theorem; the claims for chaos theory made by Prigogine and Stengers; the claims by scientists for a "postmodern science"; and claims by eminent figures in quantum physics of links to Eastern religions—claims that have been highly influential for New Age. But most importantly, it is through this dissimulation that Gross and Levitt ignore left-wing political attacks by scientists on science.[54] And on the other hand, it is precisely this same fiat that permits Gross and Levitt to invoke their personal views as a unanimous "science" against their "academic left" whose assertions were purported to result from nothing more than ignorance, hostility, and crude political motivation.

The "Academic Left" against "Science." Unfortunately, their "academic left" (narrowly defined) is minuscule, in general trained in the sciences, no more leftist than other academics and seldom politically active (with the possible exception of feminists), often admiring of science, and "earnestly, even pathologically, concerned with 'getting the science right.'"[55] Changing the definition of their "academic left" from research programs within the discipline of science studies to an essentially political definition (limited to academia), including multiculturalists, Marxists, feminists, and postmodernists outside of science studies acquires for their "academic left" a more significant representation in academia, and one often untrained in science and possibly politically active; popular polemics can be appropriated as evidence of partisan political meddling in academia. Broadened further to include the essentially nonacademic ecoradicalism, New Age, AIDS activism, and animal rights, the "academic left" becomes a threatening political movement that scholarly works from science studies hardly present. Further broadened by their historical contextualization to include the leftist political movements of the 1960s, their term "academic left" serves to link science studies to a series of unflattering traits.[56] In their final chapter, their term "academic left" loses any semblance of definition, and their contrived conflict becomes transfigured into all of the sciences and engineering battling against not just the "academic left" as science studies or even as leftist humanists but all of the humanities and social sciences—left, right, and center.[57] Only pages before they had noted the relative success of teaching in the humanities, and the comparative difficulties that scientists face—the time pressures of research and the difficulties of teaching a technical subject to students that are "wretchedly prepared" (245). Sacrificed here to their need to fabricate political threats posed by science studies is their repeatedly professed concern for education.

Conclusions

As Gross and Levitt prepare in chapter 2 to return to the refutations that comprise chapters 3 to 6, they again remind the reader:

> We thus come round to our own announced intention—to ana-
> lyze and refute the critiques of science—its methods, assump-
> tions, conclusions, and social aspects—that have arisen among
> left-wing scholars, or, more precisely, that specialized subset we
> have styled "the academic left." (37)

These refutations were the logical core of *Higher Superstition*; based
on these refutations entire research programs were to be shown to be
flawed and even fatuous, Gross and Levitt's polemics were to be vin-
dicated from charges of being ad hominem, and political intervention
in the humanities was to be shown to be imperative.

Gross and Levitt's refutations never even began: they were unable
to identify or summarize the central arguments of the works they
criticized, even when these were explicitly outlined in the works'
introductions and conclusions; several of their refutations failed to
raise even a single valid criticism. Nor was *Higher Superstition* a
compilation of errors in scientific fact: most of their criticisms were
questions not of fact but of interpretation in history, sociology, and
especially philosophy in which Gross and Levitt had no special
competence and against which Gross and Levitt proved unable to
offer either reliable evidence or cogent arguments. Instead, their
refutations often turned out to be little more than pages filled with
haphazard citations, misconstruals of simple English, caricature,
misrepresentation, quotations out of context, altered quotations, ad
hominem attacks, and malicious rhetoric hastily assembled to impugn
the character of individual scholars whose works they never understood and, in some cases, never even read.

The political threat they claimed the "academic left" posed to "science" was concocted by their manipulation of their term "academic
left." As the object of refutation by "science," their "academic left"
appeared small and isolated: leftist scientists were excluded to pit
"science" against their "academic left"; defined narrowly as research
programs in science studies—the constructivist, postmodernist, and
feminist studies of science along with nonacademic ecological radi-
calism—the "academic left" was refuted by "science." But as a political threat to "science," their "academic left" quickly grew: broadened
beyond the minuscule science studies to include multiculturalism,
Marxism, and feminism, their "academic left" acquired significant
academic representation and was linked to political meddling in
scholarship; further broadened to include New Age, animal rights and
AIDS activism, their "academic left" finally became a threatening

political movement. Their contextualization of its historical roots pieced together popular political polemics to further link their "academic left" to the legacy of an "informal *conspiracy of the heretical*" from the 1960s (222, emphasis in original). In describing real historical persons and events without any attempt to collect, analyze, synthesize, and evaluate historical materials, *Higher Superstition* can only be called historical fiction; in warning of future dangers based on undocumented claims of common purpose between groups without political, organizational, or ideological connections, *Higher Superstition* is simply conspiracy theory.

The controversies that *Higher Superstition* addresses are very real; their evidence, refutations, historical analysis, and ominous warnings of the threat of the "academic left" to "science" are not. The solutions to these controversies do not lie in further political sloganizing. If one's principal concern is to defend science against those who assert that Newton's *Principia* is a "rape manual," instead of caricature, misrepresentation, and ad hominem attacks, a more effective response might be simply to refer them to Sandra Harding's original statement. In this sense, perhaps at least some of the apparently intractable political controversies in which academia is currently embroiled may have rather simple academic solutions.

Notes

I wish to thank Mario Biagioli, Jed Buchwald, Alix Cooper, Benjamin Elman, Sam Gilbert, and Ted Porter for detailed criticisms.

1   I will follow their usage of the term *academic left* without quotation marks or further comment until the last two sections, where I will examine in detail the distinctions that Gross and Levitt make between the academic left and their own definition of the academic left (9).

2   Several academic reviews of *Higher Superstition* commend Gross and Levitt for the evidence they present: for example, Paisley Livingston asserts that "Gross and Levitt amply document the fact that the arguments and homework are simply missing in a lot of prominent work in contemporary humanistic theorizing about science" (review of *Higher Superstition* and *Evidence and Inquiry*, *MLN* 109 [1994]: 1025); Donald Kennedy calls it "a well-constructed and valuable polemic" ("Muddleheadedness exposed," *Nature* 368 [1994]: 410). Even academic reviews more critical of *Higher Superstition* seem to take for granted its basic scholarship: Michael Ruse states "I have serious reservations about *Higher Superstition*, but on balance I believe Gross and Levitt have done a fine job, covering an enormous amount of literature with a care that must have taken a huge investment of time," noting only that they "occasionally depict their opponents inaccurately" ("Struggle for the Soul of Science," *The Sciences* 34 [1994]: 40–41); although Steve Fuller finds the book "mean-

spirited," he continues, "nevertheless, Gross and Levitt have done their home-work. They simply don't like what they see, and they spare no punches" ("A Tale of Two Cultures and Other Higher Superstitions," *History of the Human Sciences* 8 [1995]: 115–116); while noting that they often fail to address more substantive arguments, Helen Longino states "Gross and Levitt's method of critique is to quote or summarize passages that exhibit arrant nonsense or ignorant misrepresentation of relevant scientific matters. And certainly they cite enough foolishness to make almost anyone blush for its authors and their associates" (review of *Higher Superstition, American Scientist* 83 [1995]: 202); Bennett Berger questions their selection of material but not their accuracy ("Taking Arms," *Science* 264 [1994]: 985–89); Richard Lewontin's scathing review does not question their evidence but instead criticizes them for attack-ing the "most vulnerable and easiest targets" ("À la recherche du temps perdu," *Configurations* 3 [1995]: 262). Only Michael Flower observes that *Higher Superstition* "dismisses its opponents with more caricature, misread-ing, and condescension than argument," noting that "their errors of impeach-ment bespeak a lack of detailed knowledge of constructivist, postmodern, and feminist literatures" (review of *Higher Superstition, Contemporary Sociology* 24 [1995]: 113–14).

3   Paul R. Gross, Letter to the Editor, *The Chronicle of Higher Education*, 20 October 1995, B3, emphasis in original.

4   This article will be strictly limited to their criticisms of academic research in the fields of the cultural, gender, and social studies of science, and thus will not address their criticisms of AIDS activism, Afrocentrism, animal rights, or ecoradicalism. This article makes no attempt to evaluate the works that are criticized—reviews of these works can easily be located within their fields. In particular, this article is not intended as a defense of these works; summaries of these works are limited to several sentences intended to describe only the scope of the work.

5   The three groups in science studies that they claim to refute could be more accurately categorized as the cultural, gender, and social studies of science; however, in this article it will sometimes remain necessary to use Gross and Levitt's terminology. Also, it should be noted that they use the term *postmod-ernism* to include poststructuralism (259, n. 2).

6   Steven Shapin and Simon Schaffer, *Leviathan and the Air-Pump: Hobbes, Boyle, and the Experimental Life*, 1989 ed. (Princeton, N.J.: Princeton Univer-sity Press, 1985). Among the works criticized in *Higher Superstition*, this is the one that Gross and Levitt take the most seriously: here, in contrast to their crit-icisms of other works, they occasionally offer a positive remark, noting that it raises questions that are "serious and genuine" (65) and that it is "exhaustively and meticulously researched" (68); and here we find an extended attempt to offer a synopsis of the central arguments of the book (63–65).

7   *Leviathan* is mentioned only once: " . . . the philosopher Thomas Hobbes, author of *Leviathan*" (63). Their only discussion of the air-pump makes no references to Shapin and Schaffer's analysis except its scarcity: "The air pump of the title was not a common device. Only a handful existed during the 1660s, and thus the possibility of investigating experimentally the emerg-ing theories of the weight and pressure of gases, now associated with Boyle's

name, was limited to the corresponding handful of people who had access to one" (64).

8   The chapter titles in *Leviathan and the Air-Pump* are: chap. 1, "Understanding Experiment"; chap. 2, "Seeing and Believing: The Experimental Production of Pneumatic Facts"; chap. 3, "Seeing Double: Hobbes's Politics of Plenism before 1660"; chap. 4, "The Trouble with Experiment: Hobbes versus Boyle"; chap. 5, "Boyle's Adversaries: Experiment Defended"; chap. 6, "Replication and Its Troubles: Air-Pumps in the 1660s"; chap. 7, "Natural Philosophy and the Restoration: Interests in Dispute"; chap. 8, "The Polity of Science: Conclusions."

9   Shapin and Schaffer, 332.

10   Ibid., 19, chaps. 3–4.

11   Ibid., 131, 139.

12   Quentin Skinner, "Thomas Hobbes and the Nature of the Early Royal Society," *Historical Journal* 12 (1969): 217–39.

13   Shapin and Schaffer, 139.

14   Gross and Levitt write "Was it true that 'the social order implicated in the rationalistic [i.e., a prioristic] production of knowledge threatened that involved in the Royal Society's experimentalism?' It's hard to believe! Recall the roster of thinkers . . ." (65, interpolation in Gross and Levitt). Gross and Levitt then present the varied religious beliefs, political views, and social rankings of Descartes, Spinoza, Halley, and Newton as examples; they make no mention of anti-experimentalism, Shapin and Schaffer's point.

15   The only relevant secondary source they cite—James R. Jacob, "The Political Economy of Science in Seventeenth-Century England," *Social Research* 59 (1992): 505–32—confutes their argument and is thus rejected as ideological.

16   For example, Boyer and Merzbach discuss Hobbes but once: "Thomas Hobbes (1588–1679) was foremost among those who criticized Wallis' arithmetization of geometry. . . . Hobbes, however, had more mathematical conceit than ability, insisting that he had squared the circle and had solved the other ancient geometric problems. Wallis could well afford to disregard Hobbes and go on to further discoveries" (Carl B. Boyer and Uta C. Merzbach, *A History of Mathematics*, 1989 ed. [New York: John Wiley and Sons, 1968], 427).

17   Simon Schaffer, "Wallifaction: Thomas Hobbes on School Divinity and Experimental Pneumatics," *Studies in History and Philosophy of Science* 19 (1988): 275–98. See also Steven Shapin, "Robert Boyle and Mathematics: Reality, Representation, and Experimental Practice," *Science in Context* 2 (1988): 23–58.

18   In addition to Schaffer's article cited in n. 17, research literature on Hobbes's mathematics includes the following: William Sacksteder, "Hobbes: The Art of the Geometricians," *Journal of the History of Philosophy* 18 (1980): 131–46; Helena M. Pycior, "Mathematics and Philosophy: Wallis, Hobbes, Barrow, and Berkeley," *Journal of the History of Ideas* 48 (1987): 265–86; Paolo Mancosu, "Aristotelian Logic and Euclidean Mathematics: Seventeenth-Century Developments of the *Quaestio de Certitudine Mathematicarum*," *Studies in History and Philosophy of Science* 23 (1992): 241–65; Douglas M. Jesseph, "Hobbes and Mathematical Method," *Perspectives on Science: Historical, Philosophical, Social* 1 (1993): 306–41; Siegmund Probst, "Infinity and Creation: The Origin of the Controversy Between Thomas Hobbes and the Savilian Profes-

sors Seth Ward and John Wallis," *British Journal for the History of Science* 26 (1993): 271–79.

19   Shapin and Schaffer, 15.

20   Sandra Harding, *The Science Question in Feminism* (Ithaca, N.Y.: Cornell University Press, 1986). See also her introduction in Sandra Harding, ed., *Can Theories Be Refuted? Essays on the Duhem-Quine Thesis* (Boston: D. Reidel, 1976).

21   Harding, *The Science Question*, 47.

22   *The Oxford English Dictionary*, 2d ed., states "world-view [G. *weltanschauung*], contemplation of the world, view of life" (vol. XX, 560); *Webster's Third New International Dictionary of the English Language, Unabridged* states "world view *n* [trans. of G *weltanschauung*]: WELTANSCHAUUNG" and provides two examples (2636).

23   Their argument against Harding—that physicists do accept the history of physics as an explanation—demonstrates only the extent to which Gross and Levitt do not. Here Gross and Levitt redefine history to exclude history as an explanation through the admitted tautology that "the history of physics as a collection of ideas is largely explained by the objective nature of the phenomena it describes and schematizes" (128).

24   Harding develops Quine's critique of the dogmas of empiricism to argue that physics is not completely isolated from the social but rather it is the science least affected—in this sense physics is not paradigmatic of the sciences but atypical. She agrees that it is unlikely that a feminist critique of physics could result in changes to the fundamental equations (though it is not in principle impossible); but Harding also argues that physics is more than just the set of paradigmatic equations of Newtonian mechanics or of relativity contained in textbooks. The "weak form" of cultural constructivism that Gross and Levitt accept includes in principle the possibility that "in scientific debate and in the process by which a preference for one paradigm over another emerges, attitudes of mind come into play that are in some measure dictated by social, political, ideological, and religious preconceptions"; they argue however that this idea is "oversold" and that "the areas of science in which such direct intrusion of ideology becomes possible are few" (44). The major difference is that for Harding this "in principle" is not meant to discourage cultural explanations; Gross and Levitt seem to admit the possibility in principle only.

25   Collections of essays include Martin Hollis and Steven Lukes, eds., *Rationality and Relativism* (Cambridge, Mass.: M.I.T. Press, 1982); Michael Krausz, ed., *Relativism: Interpretation and Confrontation* (Notre Dame, Ind.: University of Notre Dame Press, 1989). Gross and Levitt mention the term *situated knowledge* in their discussion of Aronowitz without attribution (50), apparently unaware of Haraway's use of the term.

26   Examples that Gross and Levitt offer of the misstatement of scientific fact include the following: Best's misunderstanding of the term "linearity" (98; Steven Best, "Chaos and Entropy: Metaphors in Postmodern Science and Social Theory" *Science as Culture* 2 [1991]: 188–226); and several careless analogies in *ZONE 6* (267–68, n. 17; Jonathan Crary and Sandford Kwinter, eds., *ZONE 6—Incorporations* [Cambridge, Mass.: MIT Press, Zone Books, 1992]).

27   This example is discussed in the next section, *"Higher Superstition* as the Ad Hominem Indictment of Individual Scholars."

28   Gross and Levitt state of *Science in Action*: "The one reference to an actual piece of mathematical research manages to misunderstand an anecdote utterly" (62). They assert that "Latour regards this result [the one-to-one correspondence of the unit interval to the unit square] as 'scarcely conceivable'"; this leads to an ad hominem attack on Latour (263, n. 23). Latour's claim in this section is that scientific objects are defined through "performances" which assign to these objects "competence"; he asserts that "mathematics also defines its subjects by what they *do*" and cites Cantor's claimed incredulity at the one-to-one correspondence as an example that Cantor "creates his transfinites from their performance in these extreme, scarcely conceivable conditions" (Bruno Latour, *Science in Action: How to Follow Scientists and Engineers through Society* [Cambridge, Mass.: Harvard University Press, 1987], 89–90).

29   For example, in their refutations of Shapin and Schaffer, Harding, and Haraway discussed above, Gross and Levitt make no claim to have discovered even a single error of scientific fact.

30   They cite no relevant secondary historical works in support of their refutations of the historians they criticize—Haraway, and Shapin and Schaffer.

31   For example, they raise the problems of reflexivity (48–49) and of language fluency in the anthropological studies of science (62).

32   Gross and Levitt themselves note that "Argyros himself is concerned with the philosophical implications of contemporary mathematics and science, in relation to the postmodern ideological positions promulgated by Derrida and others" (266, n. 15); his work is indeed very speculative, much more so than Hayles's more conventional work on intellectual history. Argyros makes serious errors, but none of his work is the target for criticisms in the main text: his work is discussed in two footnotes (266–67, n. 15 and 270, n. 52). Only in these footnotes do Gross and Levitt note that "Argyros seems to think that continuously differentiable functions are, in general, linear, which is grossly untrue" (this is from first-year calculus). Yet in contrast to their characteristic sarcasm, Gross and Levitt are sympathetic throughout, explaining that "he is also *slightly guilty* of bluffing his way through mathematical points" (266–67, n. 15, emphasis mine).

33   This appears to be more an unwillingness to properly cite sources from the fields they criticize than a lack of familiarity with these arguments. Throughout their book Gross and Levitt are at pains to defend the originality of their arguments, but their refusal to credit the fields they criticize obscures which of these arguments are from the science-studies seminars Levitt attended. For example, they note that their principal argument against social constructivism (including here Latour) is a common criticism—that scientific fact is claimed to be constructed but sociological fact is not—without mentioning that this argument is most commonly raised within science studies itself (as "reflexivity"), and by Latour in particular.

34   Andrew Ross, *Strange Weather: Culture, Science, and Technology in the Age of Limits* (New York: Verso, 1991). As Ross phrases it, his "chief interest lies in describing how various scientific cultures—sublegitimate, alternative, marginal, or oppositional—both embody and contest these claims [of official sci-

ence] in their cultural activities and beliefs." His critique describes (1) "the ways in which technocratic elites . . . have molded and regulated public opinion about the role of science and technology in shaping our future," and (2) "challenges to these elite languages in popular and alternative cultures" (9).

35 Ibid., 28.

36 Ibid., 56.

37 Although Gross and Levitt note that the paragraph "despite its length, . . . is worth quoting in full" (89), what is for them not worth even mentioning is that the sentence immediately following states "*I do not want to insist on a literal interpretation of this analogy*, but *it is an analogy* that informs my own thinking as a cultural critic about some of the points I want to make in this chapter" (Ross, "New Age Technoculture" in *Cultural Studies*, ed. Lawrence Grossberg, Cary Nelson, and Paula A. Treichler [New York: Routledge, 1992], 535, emphasis mine). Nor do Gross and Levitt note that in the context of his article, this analogy is primarily used to conceptualize these sublegitimate cultures.

38 Ross argues that linking New Age Orientalism to science only further mystifies science as a "secularized Western priesthood," thus making science even less accountable to the nonexpert (Ross, *Strange Weather*, 42–44).

39 Ross further insists in his introduction that it is important "to acknowledge the contexts and limitations of my own speculations," stating that because of his lack of training "my point of identification, even as a demystifier, could not be with official or 'high' scientific culture" (Ibid., 3, 8).

40 N. Katherine Hayles, *Chaos Bound: Orderly Disorder in Contemporary Literature and Science* (Ithaca, N.Y.: Cornell University Press, 1990).

41 One of the most influential assertions of parallels in the development of apparently disparate fields is Michel Foucault's controversial study of natural history, general grammar, and the analysis of wealth in *The Order of Things: An Archaeology of the Human Sciences* (New York: Vintage, 1973). Hayles states of Ilya Prigogine and Isabelle Stengers's *Order Out of Chaos: Man's New Dialogue with Nature* (Boulder, Colo.: Shambhala, 1984) that it is conjectural, "an ambitious synthesis that goes well beyond what many scientists working in chaos theory would be willing to grant are legitimate inferences from their work" (Hayles, 91). For Hayles's comparison of parallels and differences between the literary and scientific theories, see Hayles, 183–84.

42 See n. 32.

43 Alexander J. Argyros, *A Blessed Rage for Order: Deconstruction, Evolution, and Chaos* (Ann Arbor: University of Michigan Press, 1991), 237–38; Argyros's quote is from Hayles's "Chaos as Orderly Disorder: Shifting Ground in Contemporary Literature and Science," *New Literary History* 20 (1989): 316–17. Hayles's article is incorporated with some editorial revisions in Hayles, *Chaos Bound*, 183–84.

44 As noted above, Gross and Levitt use *postmodernism* to include poststructuralism. Their criticisms of poststructuralism are not explicitly claimed to be refutations.

45 "Discussion" following Derrida's "Structure, Sign, and Play in the Discourse of the Human Sciences," in *The Languages of Criticism and the Sciences of Man: The Structuralist Controversy*, ed. Richard Macksey and Eugenio Donato (Baltimore, Md.: Johns Hopkins University Press, 1970), 267. Thus their quo-

tation from Derrida does not appear in the version of the article "Structure, Sign, and Play" included in *Writing and Difference* (Chicago: University of Chicago Press, 1978).

46   Gross and Levitt credit their quotation to Ernest Gallo, "Nature Faking in the Humanities," *Skeptical Inquirer* 15 (1991): 371–75.

47   Jacques Derrida, *Acts of Literature*, ed. Derek Attridge (New York: Routledge, 1992), 208.

48   Norman Levitt, *Grassmannians and Gauss Maps in Piecewise-linear Topology* (New York: Springer-Verlag, 1989), 202.

49   The accompanying footnote states: "Defenders of deconstruction and other poststructuralist critical modalities will no doubt wish to point out that *topos* (pl.: *topoi*) is a recognized term within literary theory for a rhetorical or narrative theme, figure, gesture, or archetype, and that therefore it is permissible, without asking leave of the mathematical community, to deploy *topology* to designate the analysis of textual topoi. One's suspicions are reignited, however, when the term *differential topology* suddenly appears" (266, n. 11).

50   For two examples, see Latour's review of Haraway's *Simians, Cyborgs, and Women*, in *American Anthropologist* 94 (1992): 501–2; and H. M. Collins's review of Latour's *We Have Never Been Modern*, in *Isis* 85 (1994): 672–74.

51   From this point on I will use the phrase *academic left* to refer in an admittedly imprecise manner to academics with leftist political views and use the term "academic left" in quotation marks to refer to Gross and Levitt's definition, in order to clarify the differences between the two.

52   Fuller, 116.

53   Paul R. Gross and Norman Levitt, "A Higher Superstition? A Reply to Steve Fuller's Review," *History of the Human Sciences* 8 [1995]: 128.

54   One example is presented in Ruse's generally favorable review of *Higher Superstition*. He terms Gross and Levitt's reasoning here "dishonest," stating that Gross and Levitt know well that "some of the strongest critiques of science have come from the ranks of science itself." In Ruse's words, Harvard's Richard C. Lewontin and Stephen J. Gould have "accused their colleague the sociobiologist Edward O. Wilson of sexism, racism and the general support of unrestrained capitalism." As Ruse notes, Gross and Levitt praise Gould throughout *Higher Superstition* (56, 263, n. 15, and 269, n. 44). Sociobiology is one of four examples they offer of the assault by "left-wing antiscientism" against science (242–43), and Gross and Levitt are well aware of Gould's critiques of sociobiology—Ruse notes that "Gross was director of the Marine Biological Laboratory in Woods Hole, where most of the parties embroiled in such controversies take their summer holidays." Yet Gould's criticisms are never even mentioned in *Higher Superstition*. Ruse concludes that "Gross and Levitt have deliberately slanted the picture both to make humanists look bad and to avoid awkward questions about why serious scientists are part of the attack," Ruse, 41–42. One need not endorse Ruse's characterization of Gould's views to see that Gould's arguments can be subjected to precisely the same reductive political caricature that Gross and Levitt offer for Shapin and Schaffer, Harding, Haraway, and the rest of science studies.

55   This phrase is from Mark L. Hineline [mhinelin@bruin.bowdoin.edu], "Re: STS and Superstition," in SCI-TECH-STUDIES [sci-tech-studies@ucsd.edu], 7 May 94.

56   This is primarily accomplished by their description of the social and ideological roots of the "academic left" in chapter 8. In Gross and Levitt's hands, this
history involves no collection, evaluation, or analysis of historical documents,
but simply the indiscriminate citation of similar views from popular political
polemics; social contextualization involves none of the sociological methodology, hypotheses, or surveys they scorn, but instead simply attributes crass
motives to those they attack; and psychology similarly involves no case studies, no interviews, and none of the psychological theories they find unreliable,
but instead simply ascribes to their opponents unflattering personality traits—
such as resentment, hostility, revenge, and "physics envy."

57   In their "fanciful hypothesis," the humanities faculty at M.I.T. "walk out in a
huff," and Gross and Levitt suggest that scientists—"autodidacts" who have
learned humanities "not through formal or systematic study"—could put
together humanities courses that would be "no worse than operative." This
justifies their warning that "the notion that scientists and engineers will
always accept as axiomatic the competence and indispensability for higher
education of humanists and social scientists is altogether too smug" (243).

A la recherche du temps perdu: A Review Essay

Richard C. Lewontin

THE political movements in Europe and America in the 1960s that Americans identify primarily with opposition to the Vietnam War were not, at base, pacifist or anticapitalist or "countercultural" or simply a revolt of youth against age—although they were all those things. Rather, they were held together by a general challenge to conventional structures of authority. They were an attempt to create a general crisis of legitimacy. They were a "Call to Resist Illegitimate Authority" and were made in the image of 1792 and the revolt of the Paris Commune. The state, the military, the corporate holders of economic power, those over thirty, males, white—all were the sources of authority and legitimacy that maintained a social structure riddled with injustice. Those who were in the forefront of the struggles of the sixties knew what their revolutionary forebears knew, that a real crisis of legitimacy is the precondition of revolutionary change. But their attempt failed, and the main sources of authority and legitimacy for civil and political life remain what they have been for two hundred years, apparently unaltered in their stability or sense of permanence.

There is, however, one bit of the body politic whose sores from the abrasions of the sixties have never quite healed over, rather like a bloody heel that is perpetually rubbed raw by a new shoe that doesn't fit the old foot. It is the academy and its intellectual hangers-on who, while not themselves professors, depend on academics to buy, assign, review, and cite their works. No one was more troubled, hurt, and indignant than the professional intellectuals when their legitimacy was

Paul Gross and Norman Levitt, *Higher Superstition: The Academic Left and Its Quarrels with Science*. Baltimore, Md.: Johns Hopkins University Press, 1994. 328 pages $25.95.

Gertrude Himmelfarb, *On Looking into the Abyss: Untimely Thoughts on Culture and Society*. New York: Alfred A. Knopf, 1994. 192 pages $23.00.

challenged. The state and the corporations, after all, have long been the objects of attack. They are used to the fight, they know their enemies, and they have the weapons to hand. Their authority can always be reinforced when necessary by the police, the courts, and the layoff. Intellectuals, on the other hand, are particularly vulnerable, because professional intellectual life is the nexus of all strands of legitimacy, yet it has had no serious experience of opposition. Despite the centrality of authority in intellectual life, the academy has not, since the seventeenth century, been immersed in a constant struggle for the maintenance of the legitimacy of its methods and products; on the contrary, it seemed for a long time to be rooted in universal and unchallenged sources of authority. Then, suddenly, students began to question the authority of the older and the learned. No longer were genteel and civilized scholars allowed to propagate their political and social prejudices without rude challenges from pimply adolescents. The attack on the legitimacy and authority of the academy during the sixties was met by incredulity, outrage, and anger. It produced an unhealing wound that continues to be a source of pain to some intellectuals, who see nothing but an irrational nihilism in the rejection of traditional structures of academic authority.

Were it only the institutional authority of professors that was challenged, the hurt would be nearly forgotten. For the most part the control of the scholarly environment has returned to its former masters—although not without alteration: professors are no longer free to make racist and sexist remarks in class without challenge, and even quite innocent events may lead to serious struggles, making many academics long for the days when they could say anything they damn well pleased. But even more sinister developments have continued the crisis in the academy, long after the rest of civil and political society has restabilized. For the last three decades there has been a growing attack on the very intellectual foundations on which academic legitimacy is ultimately grounded. What was revealed even by the rather unsophisticated attacks of thirty years ago has encouraged a thoroughgoing foundational reexamination in every field. It is no longer obvious to all that the methods and problematic of natural science produce an "objective" picture of the world untainted by ideology and by the social and political predispositions of scientists, or that the Divina Commedia contains all that much of universal or lasting value to someone uninterested in the history of medieval and early Renaissance Italy (or without the ability to read fourteenth-century Italian). What makes this attack even more unsettling is that it comes from within. God grant us another Urban VIII!

The reaction to the foundational attack on the intellectual presuppositions of the sciences and the humanities, following so soon on the blows to the personal status of academics, has been the creation of a literature of indignation, characterized for the most part by the analytic coherence of a cry of pain. Among the most recent expressions of hurt and anger are *Higher Superstition: The Academic Left and Its Quarrels with Science* by Paul Gross and Norman Levitt, and *On Looking into the Abyss: Untimely Thoughts on Culture and Society* by the longtime protector of traditional values of the intellectual family, Gertrude Himmelfarb.

What suicidal impulse must have possessed Paul Gross and Norman Levitt when they produced, as the first line for their book, "Muddleheadedness has always been the sovereign force in human affairs"? While reading the book I thought it might be amusing to review it entirely through artfully arranged quotations from it, producing a kind of autophagous destruction, but then I decided it was not worth the considerable effort required to copy out all the passages. Yet it is impossible to resist totally: "This is a book that is content, in the main, to posture, rather than to argue. It is driven by resentment, rather than the logic of its ideas" (91). "Very few positions are analyzed at great enough length to make them coherent; names and phrases are simply run in and out of the text as props for [their] views" (51).

The argument of *Higher Superstition* is simple, although its rhetoric is rococo:

(1) There is a set of antiscientific critics who comprise the "academic left" and are the direct descendants of the Marxist or Marx-inspired new-lefties of the sixties. Their program to devalue science is the deliberate extension of the attempt to destabilize bourgeois society, an attempt that failed politically but continues to plague intellectual life.

(2) A great deal of nonsense has been written about science by the "academic left," who, in fact, hate science. The claims of these people are that the content and method of science are culturally biased— against feminine values, against non-Europeans—and are tools for the oppression of groups without power. Moreover, according to these critics, science is just another language, and like all texts, the texts of science can mean many different things at different times and in different contexts. Such people deny the objective reality of the material world that is described by science.

(3) Science is a set of practices that has been developed in order to produce an objective picture of the natural world. Scientists, of

course, make mistakes like anyone else, but the results of science that really last are those that are "written in nature." Moreover, science is good for you. It is the one methodology that is guaranteed to produce objective knowledge about the world, and it is the only way to solve the world's problems. "The wretched of the earth want science and the benefits of science."

The first problem with Gross and Levitt's thesis is that it is impossible to tell what is meant by the "academic left," although they spend a lot of energy trying to justify the term. It definitely does not mean academics who are politically left: they exclude all practicing scientists with leftist politics. Indeed, some of their best friends are lefties. They love Steve Gould. Nor does it include all leftist humanists and social scientists. They use, for example, an article in the *New Left Review* by an admirer of Marx, Elizabeth Wilson, to castigate the "academic left." On the other hand, the academic left includes such well-known lefties as Paul de Man! Nor does one have to be an academic to be included (Jeremy Rifkin is on the list). Their archetype of the "academic left" is Stanley Aronowitz, whose leftist credentials are for them that he is actually a member of the Democratic Socialists of America, the left wing of what used to be the Democratic Party. The hopeless muddle they make of the category renders the term *academic left* useless for any analytic purpose, yet it appears over and over, beginning with the subtitle of the book itself. What is revealed is the unbroken historical line that connects the present literature of indignation with the struggles for authority and legitimacy of the sixties and the still-present memories of clenched fists and cries of "Ho Ho Ho Chi Minh!"

It is certainly true, and Gross and Levitt provide some lovely examples, that some people have written nonsense about the method and content of natural science. What is not clear from their treatment is whether these examples of nonsense represent any significant or threatening attack on rationality, any more than their own vulgar six-page history of the Left in the United States threatens the profession of political history, or their one-liners out of Cliff Notes characterizing Blake, Wordsworth, Goethe, and Coleridge need worry those who study European literature. By deliberately choosing a few extreme examples—so extreme that they require only quotation and not analysis—the authors have created a bogeyman meant to frighten us so much that we will be distracted from considering the real critique of naive reductionism and positivism. The vulgarity of their approach prevents any serious analysis of the presuppositions, methods, and results of what goes on under the name of Science.

The "science" of Gross and Levitt is something out of a high school textbook. It is the Law of Combining Proportions, the motion of a falling body in a vacuum, the ratio of round to wrinkled peas in the second generation of a hybrid cross. They know that there are serious problems in epistemology, but they announce their intention to ignore these problems because they have already been disposed of by others: "This is a book about politics and its curious offspring, not about epistemology or the philosophy of science; we cannot therefore refute, *in abstracto*, the constructionist view. . . . Nor are we obliged to do so: serious philosophers have been at it for decades" (48). Decades, indeed! Since Plato's cave.

What Gross and Levitt have done is to turn their back on, or deny the existence of, some of the most important questions in the formation of scientific knowledge. They are scornful of "metaphor mongers," yet Gross's own field of developmental biology is in the iron grip of a metaphor, the metaphor of "development." To describe the life history of an organism as "development" is to prejudice the entire problematic of the investigation and to guarantee that certain explanations will dominate. "Development" means literally an unrolling or an unfolding, seen also in the Spanish *desarollo*, or the German *Entwicklung* (unwinding). It means the making manifest of an already predetermined pattern immanent in the fertilized egg, just as the picture is immanent in an exposed film, which is then "developed." All that is required is the appropriate triggering of the process and the provision of a milieu that allows it to unfold. This is not mere "metaphor mongering"; it reveals the shape of investigation in the field. Genes are everything. The environment is irrelevant except insofar as it allows development. The field then takes as its problematic precisely those life-history events that are indeed specified in the genome: the differentiation of the front end from the back end, and why pigs do not have wings. But it ignores completely the vast field of characters for which there is a constant interplay between genes and environment, and which cannot be understood under the rubric of "development." Nor are these characters trivial: they certainly include the central nervous system, for which the life history of the nerve connections of the roundworm is a very bad metaphor.

The study of evolution is filled with ideological prejudices whose influence is increasing. Notions of "optimality," "strategy," and "utility" have been taken over from economics and are the organizing metaphors of fields of biology, like sociobiology that Gross and Levitt so admire. Yet there is no "hard science" here. In its place is a collection of imaginative stories with no empirical test that can put them into the frame of analytic genetics on which evolutionary theory is claimed to be built. One of the most extraordinary developments in evolutionary studies has been the coming into dominance of metaphors of selective adaptation for explanations at the level of whole organisms, while, simultaneously, explanations in population genetics have become characterized by reference to historical contingency, "random walks," and "gamblers' ruin."

Even molecular biology, with its talk of "self-reproducing" genes that "determine" the organism, is ideological in its implications. DNA is certainly not "self-reproducing," any more than a text copied by a Xerox machine is self-reproducing; in fact, it is the machine that is interesting and needs to be understood. So it is the total cell machinery that needs to be understood if we are to understand both the production of new DNA and how the information in the DNA is, in fact, turned into flesh. *Higher Superstition* is not a serious book about the problems of understanding and constructing science. It is, instead, one long fit of bad temper, taking as its object the most vulnerable and easiest targets. Its authors remind one of the father who, having been told off by his wife and children, goes out and kicks the dog.

While *Higher Superstition* misses the real action, the author of *On Looking into the Abyss* knows the enemy and engages it directly. As Himmelfarb correctly perceives, the traditional bases for authority and legitimacy in questions of aesthetic, historical, and moral judgment are under direct attack. The claim for "contingencies of value," in Barbara Herrnstein Smith's resonant phrase, is the demand for a thoroughgoing revision in our arguments about what is good and bad in both the moral and aesthetic spheres. If the struggles of moral philosophers can only be judged in time and place, if Shakespeare was only a marvelous *English* poet and dramatist, then we are indeed adrift. Can it really be that Tupac Shakur and Ludwig van Beethoven are in some way on the same plane? After all, they both qualify as antisocial personalities. The problem of the source of authority and legitimacy of values is more than an academic issue, and its implications are far greater than just finding a good reason to make all undergraduates take a survey course in English literature.

Like *Higher Superstition*, Himmelfarb's book belongs to the genre of the literature of indignation. Her argument is simple and direct. If there is no universal and absolute source of value, then there is no basis on which we can hold in check the most destructive and inhumane behavior of individuals and nations. If Opus 131 is not "great" in some absolute sense, then we are doomed to an eternity of punk rock. If we cannot hold the Western ideas of freedom to be absolute, then we are doomed to be slaves or slave drivers. The claim of contingency *must* be rejected, because the alternative is the abyss.

Unfortunately, the seriousness of a project does not guarantee the coherence of its consideration. First, she is wrong about history. She makes many references to the Holocaust, all meant to warn us of the hideous consequences of a loss of commitment to absolute values. But, if there is one thing that characterized Nazism, it was not a nihilism of values but rather a psychopathic adherence to absolute principles of the right, the good, and the beautiful. Race purity, the morality of the *Volk*, the rescue of culture and civilization from the evils and corruption of Jews and other orientals were the cornerstones of justification for the Holocaust. No word was more important to Nazi cultural criticism than "degenerate." Does Himmelfarb think that the tortures of the Inquisition were in the name of cultural relativism? In fact, we do not have a single example of mass inhumanity that was the consequence of a rejection of value; on the contrary, institutions of human slavery and oppression have always been justified by an appeal to the highest principles. The question, alas, is not one of "freedom," but of freedom for whom and to do what. Himmelfarb is too well versed in political philosophy not to know the deep contradictions in concepts of liberty, but none of that surfaces in her discussion. For her, the philosophical questions of liberty were all definitively dealt with by Mill.

Second, there is no argument in *On Looking into the Abyss*, only alarm and indignation. Suppose it were true (and I will argue that it is not) that the abandonment of absolute cultural values would lead ineluctably into the depths. It would not follow that values do indeed have an absolute basis but only that a willing suspension of clear thought is a prudential necessity for the maintenance of a humane society. Himmelfarb does not present a single argument for the existence of an absolute standard either of morality or of "greatness." Indeed, there are only two positions she might take. One is religious: the good and the beautiful are given by God or by some equivalent source of value that is prior to human existence. The other is a Darwinian gloss on Kant: all human beings, as a result of the evolution of their central nervous systems, have in the structure of their brains a set of a prioris that dictate what appear to be universal values. Himmelfarb is too perspicacious to commit herself to either of these, at least in public. Curiously, she does not try to finesse the problem by the standard negative argument from the *Theaetetus*, namely, that to argue that man is the measure of all things is self-contradictory, and so there must be absolute values. This is a Socratic ploy, which cannot carry real weight. It is a feature of language that the statement "There is no absolute truth" is self-denying; but it does not follow that there is absolute truth, any more than it follows from Russell's dictionary paradox that we should burn all our dictionaries.

In the absence of a religious or natural historical claim for absolute standards, there remains only the evident fact that human beings have created values in the course of their varied histories. To the extent that some values have made possible stable social orders and others have not, there has been a convergence on values, and this prudential consideration is quite sufficient as an argument for adherence to them. But this instrumental argument leads to the possibility that prudence may lead in other directions in other times. It is this possibility that Himmelfarb wishes to deny, despite the fact that the intellectual circle to which she belongs was quite willing to support the murder of Vietnamese by Americans for the sake of freedom. It simply does not follow that if there are no absolute and transcendent values, then there are no values. If that were true we would still be on the gold standard. I do not need to believe in God or a universal human nature to know that I would find it intolerable to live in a society where I could not dissent publicly from received truth, but I also know that the members of the College of Cardinals do not share my view and that they have a better claim to represent the long sweep of European history than I do. The difference is that I am on the victorious side of the most recent serious crisis of legitimation.

One of the ironies of Himmelfarb's position is that she, like the school of cultural theorists to which she belongs, makes the same indissoluble linkage between absolute moral values and absolute cultural standards that was made by the National Socialists. The most common attribute in *The Abyss* is "greatness," but it is never quite clear what one needs in order to qualify as "great" in a universal sense, beyond the historical approbation of people who are in a position to know. But how are we to tell who is in a position to know, except that they have read and approved of the "great"? The claim that "great" authors have provided deep and novel insight into the general human condition that necessarily speaks to all, irrespective of class and culture, is patently false. When I reply to a friend who has twitted me about taking myself too seriously, "He jests at scars that never felt a wound," I am not providing a deep philosophical insight, unknown to the rudest groundling, but I am quoting a superb bit of English poetry. Anyone who is in any doubt that Shakespeare was an English poet should try André Gide's translation of *Hamlet*: "Thou wretched, rash, intruding fool, farewell!" comes out as "Pauvre sot, brouillon, indiscret, bon voyage!" Nor can Pushkin's dancing tetrameter,

> Onegin, dobryi moi pryatel',
> Rodilcya na bregakh Nevy,
> Gde, mozhet bit', rodilic' vy

be carried into any English translation of *Evgenii Onegin*, not to speak of creating any cultural resonance with the upbringing and love life of a late eighteenth-century dandy who "was born on the banks of the Neva, where, perhaps, you too were born, my dear reader." Sorry, wrong river, wrong century, wrong social class, wrong language.

The body of writing to which *Higher Superstition* and *On Looking Into the Abyss* belong, while appealing to transcendent standards, is, ironically, the product of a particular historical moment in the development of European culture. In a movement that began with the growth of the *noblesse de robe* in prerevolutionary France, technical and intellectual competence has increasingly become a pathway to upward social mobility. More secure and, from all attitudinal surveys, more prestigious than entrepreneurship or state service, intellectual activities increasingly have provided status, material well-being, and some forms of social power. Professional intellectuals, chiefly academics, have only relatively recently found themselves to be a major source of authority and legitimacy in European bourgeois society. An important part of that power is the image that intellectuals speak for no special interest, time, or group but are the conduits into society of the eternal verities. Thus, they have not appreciated the degree to which they, like any other source of legitimacy, necessarily become identified with the general structures of authority, and so they are unprepared for the attack on their authority that periodic crises of political legitimation must bring. In reading these books I saw before me Masaccio's bathetic image of Adam and Eve, faces screwed up in anguish, shedding bitter tears and covering their genitals as they are expelled from Paradise.

Note

This essay previously appeared in *Configurations* 3, no. 2 (spring 1995): 257–65.

Science Skirmishes and Science-

Policy Research

Les Levidow

THE Science Wars of the United States have rather marginal counterparts on the other side of the Atlantic; at most, we have some "science skirmishes." However, the implied comparison to the United States may be misleading. Unlike the maneuvers around the National Science Foundation (NSF) and the American Association for the Advancement of Science (AAAS) in the United States, few scientists in Britain attempt to discredit the social studies of science, much less to restrict our research funding. Indeed, at the European and British levels, more funds are being earmarked for social science research on technology-related issues.

As government recognizes the policy relevance of science studies, this discipline encounters new opportunities and dilemmas. On what terms do we design research for policy? What are the implicit politics of science studies? What dilemmas do we face? In discussing these questions, I will mention technological controversies in Britain—in particular, biotechnology examples that I have studied. I will also draw on European debates about the proper role of science studies.

Science Skirmishes in Britain

In spring 1994 *New Scientist* asked me to review *Higher Superstition*, which had received little attention in Britain. My review focused on the authors' main argument—namely, that a detailed knowledge of science is a prerequisite for criticizing it, especially for analyzing its value-laden content (Gross and Levitt 1994; Levidow 1994). Only much later did I realize that the book had become a focus of prolonged, heated debate between science studies and its opponents in the United States.

Later I was reminded of that argument when a particularly obtuse version came from Lewis Wolpert, a media-friendly missionary for scientific progress. The Third World Network had coordinated a petition by scientists advocating a moratorium on large-scale releases of genetically engineered organisms. The petition challenged the exclusive expertise of biotechnologists to judge the environmental safety of their own activities. In response, Wolpert argued that concerned scientists should sign the petition only if they are molecular biologists: "Scientists should spell out dangers and difficulties on the basis of their expertise—but only if they are experts." With this quotable quote, he helped a journalist turn a low-key event into a good "story" (Irwin 1995).

Meanwhile, it seemed that no one cared to ask the following questions: Who in fact is responsible for this policy issue? What kinds of scientific experts are already involved? And why? No one—not the petitioners, nor Wolpert, nor the journalists—seemed to notice how Britain's Department of the Environment was already regulating genetically engineered organisms. In fact, the government had recruited ecological expertise to help justify its safety judgments and, implicitly, to help legitimize biotechnology (e.g., Levidow and Tait 1993). My main point here is that ideologues such as Wolpert remain marginal to policymaking, while regulatory experts have been in dialogue with science studies—as represented by me, for example.

Another example comes from Wolpert's appearances in Channel 4 programs about the social role and content of science. In a panel discussion, a medical worker argued that genetic screening raises new ethical issues and that our society faces a potential "tyranny of genetic knowledge"; for example, new knowledge broadens the definition of a "genetic defect." Wolpert retorted, "Knowledge is liberating" because it gives people greater freedom of choice; he also denied that genetic advances raise new ethical issues—at least not for the laboratory scientists who produce the knowledge. Wolpert's denial may have confirmed viewers' worst suspicions about scientists' capacity to regulate themselves (as one panelist noted).

Meanwhile, in response to a wider ethics debate in Britain, a medical foundation prepared a report which acknowledged new ethical issues arising from human genome research. The report recommended setting limits on how such knowledge may be applied (Nuffield 1993), though it implied that the knowledge itself was ethically neutral. In effect, such bioethics sets terms of reference for treating social problems as genetic deficiencies (as analyzed by King 1995; Levidow 1995a; Stemerding and Jelsma 1996). At the same time, its legitimacy depends on "drawing the line," even if the line often shifts. Wolpert marginalized himself from this subtle debate when he celebrated new genetic knowledge as inherently beneficial.

In my two examples, state-funded procedures serve to accommodate public unease into an apparently expert-based consensus. Such a procedure characterizes Britain's political culture, whose regulatory style has been theorized as "consultative" (Jasanoff 1986). For its social legitimacy, the outcome needn't depend on a particular account of science, much less a positivist account of scientific "truth." Indeed, the relevant expertise is politically negotiable. In Britain's political culture, then, science skirmishes seems a more apt description than Science Wars.

It would be tempting for us to focus our critical efforts on ultra-positivist scientists such as Wolpert, who invoke their expertise to limit public debate on science. Certainly such a critique is necessary, given the mass-media attention such individuals attract. Yet this focus could divert science studies from analyzing the links between science and policy.

What kind of science informs policy—and vice versa? What policy-relevant research can obtain funding?

Research for Competitiveness?

In both Britain and the European Union, research funding is being more closely linked to market criteria. In particular, the "competitiveness" criterion serves to promote a self-fulfilling account of objective imperatives for throwing us into yet more intense competition with each other and for eliminating all values that defy productivity calculations. As the marketization process continues unabated in Britain, we continue to be haunted by the ghost of Margaret Thatcher, the rhetorical founder of "the enterprise culture"—which is now praised by Tony Blair, the progenitor of "New Labour."

Within higher education, marketization operates at several levels. Student grants and academic research become increasingly dependent on industrial sponsorship. Scientific research proposals are expected to demonstrate relevance to enhancing technological innovation, even industrial competitiveness. Until recently, critical studies of technology were funded through links between research councils (ESRC-SERC), but their joint board has been abolished.

Market language pervades the European Union as well as Britain. In a document on research policy, the European Commission redefined both the state and the environment in such terms:

It is these [public authorities] which must bring about the creation and maintenance of an overall economic "environment" and a respect for free competition, which is necessary so that firms can effectively develop supply policies. And this is very much the Community's task. The Treaty [of Rome] clearly confirms this, especially in the version adopted at Maastricht. (CEC 1992, 16)

This perspective was soon elaborated by diagnosing our problems as follows: Europe must overcome the obstacles to an optimum exploitation of new technologies which could enhance its industrial competitiveness (CEC 1993a).

In that vein, the Commission drafted the Fourth Framework Programme for Research and Technological Development, or RTD. It is funded to $15 billion (12 billion ecu). Its language was partly a compromise with the European Parliament: "In particular, efforts will be made to identify the science and technology options with the most favourable impact on growth, competitiveness and job creation in Europe." The document largely equated "favourable" with "competitive," though it also mentioned environmental, health, and ethics problems (CEC 1994).

The RTD's priorities illustrate the tendency to define society's problems in genetic terms. According to an adviser for the Life Sciences component, Europe must develop bioinformatic technology for processing gene sequences and achieve better "understanding of how genes work." According to this adviser, European policy could offer "added value" to the life sciences, via inter-European teams, research infrastructure, and training (DGXII 1994, 8–9).

By attributing our problems to genetic deficiencies, these research priorities promote biotechnology's molecular reductionism (as theorized by Kay 1992; see also Levidow 1995b). This perspective in turn lends multiple meanings to "value-added research" (e.g., market criteria for selecting biological qualities). While supporting genetic research, the European Parliament declared "that human life is not a marketable commodity and that there can be no commercial competition in this sphere" (CEC 1993b, 304). This proviso indicated a broader wish to set limits on the commodification of life.

At the behest of the European Parliament, the RTD also included a component on "targeted socio-economic research," to the tune of $130 million (100 million ecu). The parliament intended this research to explore "relations between the development of scientific and techno-

logical policies and the prevailing social order, the structures underlying the perception by experts and citizens of the risks and dangers
linked to technology" (CEC 1993b, 311). This research was to emphasize social marginalization and exclusion as a potential impact of economic change. In describing the research, however, an adviser to the
European Commission hardly linked social exclusion to the "competitiveness" that drives technological change toward reducing and dis-
ciplining the labor force. Moreover, he implicitly equated exclusion
with unemployment, as if waged labor were an unproblematic means
of social inclusion (DGXII 1994, 7–8).

In all those ways, the RTD has an implicit politics of science and
technology. Although the research guidelines may sound broad,
their practical content depends on whether researchers resist marketization pressures, for example, by encompassing the perspectives
of oppositional forces. Yet academics have largely abandoned their
links to NGOs, especially the labor movement. According to
Britain's Socialist Action Research Network, university-trade union
cooperation has had setbacks in recent years. Moreover:

> Great problems had occurred for those seeking to re-invigorate
> this cooperation under the EU's Fourth Framework Programme
> for Targeted Socio-Economic Research. There were very few
> openly committed, socialist researchers now inside the academic
> institutions; indeed, they were perceived by some as "dinosaurs"
> in a period when all the trends were to work away from unions
> and class issues, and towards business interests and concerns.
> (SAR 1995)

British and E.U. research programs carry a double-edged potential:
new funding can help science studies to challenge and broaden the
present terms of policy debate. Alternatively, funding can help
accommodate science studies to the prevalent policy language. How
have such researchers responded?

Following the Money

The aforementioned dilemmas have been debated somewhat
within the European Association for the Study of Science and Technology (EASST). Some EAAST members refer to their field as *science and technology studies* (STS) or the *sociology of scientific knowledge* (SSK), terms I will use below.

As British government policy pushed academia toward greater
dependence on industry funding, university-industry relations be-

came both a practical problem and a new object of study for STS. Having coordinated a NATO-funded study of such relations, one academic reiterated long-standing STS criticisms of science policy, in particular its assumption that science is a "black box" which can be managed by society. At the same time, he concluded that the recent "process of commercialization of academic work has not been to its scientific detriment" (Webster 1991, 89).

The latter's approach was criticized by a fellow EASST member, the late Paul Hoch. In his view, Webster aimed "not to discredit the managerialist stance of elite science policymaking, but rather to encourage sociologists (especially of SSK persuasion) to participate." He criticized Webster for evading the central issue of political power, that is, the design of innovation for exploiting and dominating people (Hoch 1992). Subsequent research diagnosed universities' difficulties in earning royalties from their biological patents (as reported by Patel 1995 and Coghlan 1995). Hoch's questions could be extended, for example, by asking whether STS ends up trying to outcapitalize the would-be academic capitalists.

However, such criticism somewhat begs the question of how STS could promote a different kind of policymaking. That question is bound up with alternative ways to criticize the stereotypical "linear model" of innovation. With such a model, government policy tends to assume that successful technologies simply "apply" scientific knowledge, which comes ready-made from the laboratory. Many STS researchers have criticized the model, though their political basis and implications may be ambiguous. For example, if we emphasize how innovation requires constantly recasting and accommodating technical knowledge for market relations, then STS may beg the question of power, or even help to link lab work with commodification processes. If we emphasize how technical knowledge embeds and promotes a reordering of society, then we may help to open up technological choices for debate as social choices, despite the appearance of technology as a force beyond social control (e.g., Winner 1977). Although those two emphases may be complementary, they tend to be polarized in STS debate.

Such tensions arose in proposals to extend Constructive Technology Assessment (CTA) toward policy issues. CTA exercises were developed to anticipate potential impacts of a new technology, so that its design could be reconsidered at an early stage. A leading STS writer has interpreted that experience as follows: "The key point for CTA is that a collusion of actors and factors is needed to

have a reliable and otherwise 'good' technology." He proposed that
STS develop a professional service role, for example, by its "identification with the problem of the client," so as "to provide added
value to what the actors themselves can do already" (Rip et al. 1994,
11–16).

As my rejoinder argued, his historical account of CTA downplays
fundamental social conflicts about the implicit purposes and power
relations embedded in a technological design (Rip et al. 1994, 18).
Indeed, often the actors have incompatible ways of defining the
problem which technology might solve. A consultancy language of
"clients," however, may narrow the range of problem-definitions that
researchers consider. From the start, we face the issue of whether or
how incompatible views of "good technology" could be accommo-
dated. Would such an attempt "subtract" consultancy value from the
exercise?

Influencing Policy?

In his account of debates in Britain, Steve Fuller (1995, 21) notes
that science studies colleagues encounter a conflict between offering
advice to scientists and remaining autonomous. That is, some
STSers seek to persuade scientists that our analysis will help them,
for example, by relieving them of the political expectations that are
often put upon science. As he argues, such an approach may not
only lack credibility but may also dull the critical edge of our analysis. There lies a pitfall of expecting that scientists will seek, or at
least accept, more modest claims for their specialist knowledge. On
the other hand, "autonomy" may entail some illusion: autonomy
from what? For STS, as for science itself, there can be no neutral reference point—no independence from someone's definition of the
problem for science to address, even solve.

There may be a somewhat different tension in a "professional-
client relation" between STS and policymakers. On the one hand,
government policy need not necessarily claim a basis in some
unique scientific truth. On the other hand, policy often conceals and
legitimizes implicit political purposes, which STS may help make
more accessible to public debate—not an aim of most policymakers.
Moreover, research bids undergo pressure to promise that STS will
help government solve its own problems, for example, public resentment of technological foreclosures.

Here lies a dilemma for STS, especially now in Europe: despite
the high-profile "science skirmishes," our main problem is not how

to counter accusations that we are promoting some "higher superstition." Rather, our problem is how to develop policy-relevant research, even to obtain government funding, yet somehow use the resources to challenge marketization pressures.

References

EASST membership, which includes the *EASST Review*, is available @ £14 from: SPSG, 11 Hobart Place, London SW1W 0HL. I would like to thank the following people for helpful editorial comments on an earlier version of this article: Steve Fuller, Andrew Ross, Andy Samuel, Peter Taylor, and Dave Wield.

CEC. 1992. Research after Maastricht: An assessment, a strategy. *Bulletin of the European Communities*, supplement 2/92. Brussels: Commission of the European Communities.

CEC. 1993a. Growth, competitiveness, employment: The challenges and ways forward into the 21st century. *Bulletin of the European Communities*, supplement 6/93. Brussels: Commission of the European Communities.

CEC. 1993b. Scientific research. *Official Journal of the European Communities* C329 (6 December): 282–329.

CEC. 1994. EC Fourth Framework Programme, *Official Journal of the European Communities* C228 (17 August): 177–87; excerpted in *EASST Review* 13, no. 4: 17–19.

Coghlan, Andy. 1995. Waiting for the gold rush. *New Scientist*, 8 July, 14–15.

DGXII. 1994. *RTD Magazine* 1: 8–9.

Fuller, Steve. 1995. Two cultures II: Science studies goes public. *EASST Review* 14, no. 1: 21–24.

Gross, Paul, and Norman Levitt. 1994. *Higher superstition: The academic left and its quarrel with science*. Baltimore, Md.: Johns Hopkins University Press.

Hoch, Paul. 1992. A social construction of science policy? *EASST Newsletter* 11, no. 2: 3–4.

Irwin, Aisling. 1995. Scientists call for genetic freeze. *Times Higher Education Supplement*, 20 January.

Jasanoff, Sheila. 1986. *Risk management and political culture*. New York: Russell Sage Foundation.

Kay, Lily. 1992. *The molecular vision of life: Caltech, the Rockefeller Foundation, and the rise of the new biology*. New York: Oxford University Press.

King, David. 1995. The limits of bioethics. *Science as Culture* 5, no. 2: 303–13.

Levidow, Les. 1994. The left and right of it. *New Scientist*, 11 June, 45.

———. 1995a. Whose ethics for agricultural biotechnology? In *Biopolitics: Biotechnology, feminism, and ecology*, edited by V. Shiva and I. Moser. London: Zed.

———. 1995b. Scientizing security: Agricultural biotechnology as clean surgical strike. *Social Text* 13, no. 3: 161–80.

Levidow, Les, and Joyce Tait. 1993. Advice on biotechnology regulation: The remit and composition of ACRE. *Science and Public Policy* 20: 193–209.

Nuffield. 1993. *Genetic screening: Ethical issues.* London: Nuffield Council on Bioethics; reissued 1995.

Patel, Kam. 1995. Scientists patently poor at capitalising on new ideas. *Times Higher Education Supplement,* 23 June, 335–51.

Rip, Arie. 1994. STS and CTA (plus commentaries). *EASST Review* 13, no. 3: 11–19.

SAR. 1995. Socialist Action Research Network, notes of meeting at the Conference of Socialist Economists, Newcastle-upon-Tyne, 7–9 July; see also *The trade union action-research guide 1995,* from CAITS, 104 Southgrove Road, Sheffield S10 2NQ.

Stemerding, Dirk, and Jaap Jelsma. 1996. Compensatory ethics for the Human Genome Project. *Science as Culture* 5, no. 3.

Webster, Andrew. 1991. *Sciences, technology, and society: New directions.* London: Macmillan.

Winner, Langdon. 1977. *Autonomous technology.* Cambridge, Mass.: MIT Press.

A Few Good Species

Andrew Ross

I F there has been one constant in the history of science, it is the relationship of applied research and technology to military force. Nothing belies the myth of pure science more than the evidence that it has served as the handmaiden of warfare or, in the period of the national security state, as a central component of the permanent war economy that continues to sustain elite interests among the major powers and their clients. We all know something about science's utility to the military trade of destruction, but what happens when the military is charged with utilizing science to repair the destructive consequences of that trade? The euphemism of the "peace industry" took on new life after the cessation of the Cold War at a time when the security establishment, deprived of its staple of Manichaean ideological conflict, turned in the direction of environmental considerations and elevated the concept of environmental security to the forefront of its global overviews. The result, by no means conclusive, is the outcome of a messy encounter between the functional ethics of ecological science and the institutional mentality of war making.

A Salvation Army

The recently publicized plan to reconstruct the Pentagon building according to ecological, energy-efficient principles is an outward sign of a much heralded makeover of the military's image. New acquisition regulations for weapon systems are being geared to "integrating environmental considerations" into assessments of their "life cycles." Native Americans may soon be allowed access to religious and sacred sites on Department of Defense (DOD) lands. The military's mighty surveillance networks are being revamped as ecological early warning systems. What's next? The reforestation of Vietnam? In an apparently humor-free gesture, the Marine Corps introduced a recruiting poster in the fall of 1994 which shows an amphibious landing at Camp

Pendleton, California. Strutting in the foreground is a Western snowy plover, a bird that made it on to the endangered species list in 1993, and whose (along with that of twelve other endangered species) coastal scrub sage habitat is now apparently protected through its use by the Marines as a combat training area. The poster's smugly captioned graphic—"We're Saving a Few Good Species"—is, among other things, a droll reminder of the Marines' overseas role as military executors of social Darwinism for an entire century now (if we take the Hawaiian coup in 1893 as an official point of origin). But it also captures one of the most incongruous spectacles of our times—the so-called greening of the armed forces or, from another perspective, the emergence of a military-industrial-environmental complex.

Since the cessation of the Cold War and the onset of the half-hearted conversion of the permanent war economy, the U.S. Armed Forces and their allies have taken on many of the public relations functions of a "health-care provider" in their attention to Operations Other Than War (OOTW). Things got so soft for the leathernecks in the new peacekeeping industry that Secretary of Defense William Perry complained to Congress: "We are an army, not a Salvation Army." As for the military's new ventures into environmental preservation and restoration, an intermittent PR campaign has focused on ecosystem management and nature conservation at high profile bases like Pendleton, Maryland's Fort Meade, Illinois's Joliet Arsenal, Florida's Eglin Air Force Base, and Massachusetts's Fort Deevens.[1] The less laudable task focuses on the military's "iatrogenic" problems, which, in medical terminology, are illnesses induced by exposure to medical institutions themselves. In common with many of the new developments in technoscience, the Pentagon is fighting a battle against threats generated by its own industrialization. The bill for cleaning up the military's own Cold War legacy is already enough to sustain a sizable defense industry. The DOD's latest cleanup estimate begins at $30 billion and rises as high as $200 billion. Thomas F. Grumbly, environmental supremo at the Department of Energy (DOE), who presides over the largest environmental budget in the world, estimates the cleanup cost of the nuclear weapons complex (at Hanford, Rocky Flats, Savannah River, and thousands of other DOE facilities) at $1 trillion. For the job of mopping up at the more than ten thousand military sites at home and abroad, his respective counterparts at each of the armed services hustle for their massive budgets with the same technorationality as they sue for new weapons systems. As Lieutenant Colonel Sherman Forbes, chief of the U.S. Air

Force's acquisition pollution prevention program, shrewdly puts it, his people "are environmental managers, not environmentalists," ensuring environmental compliance in the same way as they ensure combat readiness.[2]

Do these developments represent an ethical awakening of the military mentality as we know it? Or are they just another PR greenwashing campaign of the sort pioneered by the giant chemical and oil companies? No one needs to believe that the Pentagon's mission is really being reshaped along humanitarian lines to see that attention to environmentalism has become one of the doctrinal cornerstones for the future of the defense establishment. For the first half of the 1990s, the concept of "environmental security" (introduced at the 42d session of the U.N. General Assembly in 1987) was touted as one of the more likely successors to the "Communist threat" as an organizing principle of military-industrial policy. When the aftermath of the 1989 revolutions in Eastern Europe revealed the full effects of environmental degradation in the Black Triangle of Poland, Czechoslovakia, and the GDR, the politics of post-Communist Europe provided a crucible for putting the concept into action. "Threats without Enemies" was the watchword of the day.[3] When "enemy" warfare returned in 1991 with the Gulf War, it was perceived, from first to last, as a struggle over environmental resources, despite the attempt to politicize Allied involvement. As for the wars in Somalia, Rwanda, and Bosnia, the chief concerns for the Western powers seem to have revolved abstractly around the logistics of refugee movements. Under Clinton and Gore's watch, environmental security was accepted into realpolitik at the heart of the policy community. With the 1994 Republican takeover's new hawkish temper, it has been joined, but not displaced, by the more bellicose concept of the two-war strategy.

Consider the testimony of Sherri Wasserman Goodman, in her recently created position as Deputy Undersecretary of Defense for Environmental Security, speaking in the U.S. Senate before a fiscal appropriations meeting of the Armed Services Subcommittee on Military Readiness and Defense Infrastructure, on 9 June 1993:

> I was just down last week at Fort Bragg and I had a chance to see how the Army is doing in protecting the Red Headed Woodpecker at the base. And let me say, I think it is doing a terrific job there. It has protected the pine trees in which this bird lives. It is one of the few homes for this bird remaining. In fact, the Army now even builds homes for these birds in the trees because the bird

requires three to five years to actually peck its home, and the
Army goes out and builds some additional homes for it.[4]

This kind of talk, when it was first heard on Capitol Hill, must have
given Pentagon lifers the creeps. It has already provided ammo for
many of the Gingrich "tree-hugger" haters now in control of the
House committees (some substantial cuts have already been legislated
in the DOD's Environmental Restoration Account, although nothing
as sweeping as the assault on major federal environmental legislation). As it happens, Environmental Protection Agency (EPA) violations concerning the woodpeckers had caused stoppages at base in the
past, and Goodman was underlining the economic sense of environmental compliance. In addition, the publicity has come in very handy
since the Pentagon sits on some of the better preserved lands in the
Federal system. But listen to Goodman as she goes on to brief this, and
many subsequent committees, on the threats to environmental security faced by the national interest. Broken down into the Pentagon's
customary theaters of operation, these threats are either global (such
as "ozone depletion, global warming and loss of biodiversity") re-
gional (such as "environmental terrorism, conflicts caused by scarcity
or denial of resources, and cross-border or global commons contamination") or national (such as "risks to public health and the environment from DOD activities, increased restrictions on military operations, inefficient resource use, reduced weapons system performance,
and erosions of public trust"). Goodman's description meshes with
the plans for a Strategic Environmental Initiative (phrased as a Democratic counter to Star Wars) favored by the military's friends in the old
Democrat-controlled Congress as well as with the vision of a global
Marshall Plan to meet the new environmental threats that Al Gore
advanced in his book, *Earth in the Balance*. Not only were these initiatives intended to help preserve the massive research infrastructure
and budget of the military and intelligence agencies, they would also
provide a virtuous defense of just wars everywhere and an opportunity to speak the language of free market environmentalism, vying to
become the new lingua franca of the global economy.[5]

With the evaporation of the communist threat, the business of international security is no longer hedged around by ideological values:
"free" versus "totalitarian." Increasingly, security overviews and war-
game scenarios focus on the new tensions and conflicts caused by
environmental "threats": shortages of natural resources, water, and
oil; cross-border pollution, including radioactivity and acid rain; the

environmental underbelly of North-South trade; resource degradation; and population control in the new migrant economy generated by economic restructuring. To cite one symptom, the mapmakers at the CIA no longer highlight military installations and intercontinental ballistic missiles sites. They compile demographic maps, detailing where ethnic, national, and religious groups live, and environmental maps, pinpointing sites of nuclear power plants, especially in Russia and the Third World, where chances of meltdown are highest and where plutonium smuggling originates. "We at the agency are in the predicting business," says one cartographer, "and when we think about what will explode, demographic factors and environmental factors are what we look at first now. What policy makers want to know most from us when they ask for maps is the ethnic mix, because that is what determines the hot spots."[6] Many other government agencies that are officially nonmilitary have followed suit. NASA now sends the space shuttle on environmental missions: its space radar equipment, designed for espionage by national security agencies, is now used to detect changes in vegetation, land movement, and water flow that could cause earthquakes and other catastrophes.

Risk Calculus

The global character of environmental problems today has little regard for national boundaries. Food chains connect all over the earth, pollution crosses borders, and, while wealth, power, and class geography can differentiate the social impact of many of these threats, nuclear, chemical, and genetic hazards tend not to distinguish hierarchically between or within nation states. Even the North-South toxic trade is not a one-way flow; its effects are returned to the North through their largely invisible presence in consumer commodities and food produce. This is the "boomerang effect" cited by Ulrich Beck in his influential book *The Risk Society*. He argues that a society characterized by the overproduction of risks, from which no one is ultimately immune, no longer corresponds to the paradigm of a class society: hunger is hierarchical, nuclear contamination is not.[7] From a military perspective, what is most threatened by these transboundary activities and conditions is the old nation-state alliance system, which lives on informally in the fault lines of the new security transstate symbolized by global coalitions such as the Montreal Protocol on Substances that Deplete the Ozone Layer and the Basel Convention on the Control of Transboundary Movements of Hazardous Waste and Their Disposal.

Conceptually, "environmental security" is a strange bird, especially when invoked by corporate-military powers against threats generated by the very industrialization from which their interests have traditionally profited. Old hands at Cold War doublespeak will recognize a similar circular logic in the new rhetoric of risks. Most favored is the administrative rationality known as "risk assessment" and all its related techniques in McNamara whiz-kid jargon: risk management, risk communication, risk prioritization, and the like. Risk assessment, a form of cost-benefit analysis that conceals social interests in the form of mathematical possibilities, evolved as a managerial strategy in response to the massive Cold War public anxiety about hazardous technologies. Recently, it has become the new interface language between Congress, the Pentagon, and defense contractors. Both the DOD's Goodman and DOE's Grumbly, have become expert rhetoricians, rationalizing all environmental expenditure in the name of Goodman's military rubric of $C^3 P^2$—cleanup, compliance, conservation, and pollution prevention. Sometimes, the tone is abstract, almost philosophical. Grumbly, for example, is prone to quoting at length to congressional committees from the likes of Pericles of Athens:

> "We Athenians . . . take our decisions on policy or submit them to proper discussion. . . . We are capable at the same time of taking risks and of estimating them beforehand. Others are brave out of ignorance; and when they stop to think, they begin to fear. But the man who can most truly be accounted brave is he who best knows the meaning of what is sweet in life and what is terrible, and then goes out to meet what is to come."[8]

(When military hacks on Capitol Hill choose to be Greeks instead of Romans then we know some sea change has occurred.) At other times, the risk calculus is used to undermine existing environmental regulation. In the process of returning military bases to productive civilian use, the stricter and nonconditional cleanup required by Superfund is being passed over in favor of standards tied to land use considerations under CERFA (the Community Environmental Response Facilitation Act of 1992). Why observe environmental standards appropriate for a day-care center when the community plans to convert the contaminated site into a parking lot? (But who knows what will replace the parking lot at some future date?) As budgets get tighter, bases may be forced to stay open to evade the high cost of decontamination.

Risk assessment may seem like a far cry from the Cold War mathematical heyday of nuclear megadeath projections, but it is much more than a fiscal mechanism of military policy making. The massive Republican bills on regulatory reform would have made the vast bureaucracy of risk assessment a budgetary requirement for the enforcement of all existing legislation relating to public health and environmental safety. These Contract with America bills were intended to paralyze the process of effective regulation, but they were basically an upgrading of Democrat-initiated policies whereby acceptable levels of risk such as the "maximum tolerated dose" (MTD) are assessed, that is, how many people's lives have to be endangered before it is cost-effective to legislate? For example, in the case of the 1993 decision to dismantle Hanford seventy-five years hence, it was estimated that while the natural disintegration of the reactors would result in twenty additional cancer deaths over ten thousand years, dismantling them now would result in an indefinite number of cancer deaths to workers and residents living near the facility. In seventy-five years, some of the waste will have decayed to harmless levels at a site officially classified as one where exposure is occurring right now, and likely to increase as the earliest built holding tanks for radioactive waste give out.

Under the fiscal demand that environmentalism should pay its way, green accountability is now a *condition* for economic growth, not a *cause* of the limits to growth. As a primary technique of global economics, risk assessment is therefore central to the future of free market environmentalism. It also facilitates the classic task of·population management at state level, favoring the kind of governmentality that seeks to turn the welfare state against its clients while gutting the core of social citizenship. Through an assortment of technical appraisals, individuals and groups can be predetected as social risks or threats and assigned destinies in the technostructure according to their predicted level of competitiveness and social investment potential. In this manner, the "risk" to society is contained or is dispersed across society, just as the social guarantees of a welfare state are already broken down and scattered around by a highly rationalized process managed by the huge insurance companies that dominate corporate medicine.[9] In our medical institutions, and especially in the genetics fields, this recipe for social eugenics has already been partially implemented. But predetected individuals who can genetically pass on their risks to society are not the only ones affected by the new preventive social medicine. Prevention of risk by any means necessary is becoming the principle of administration in a liberal society that no longer wants to assist its poor citizens nor be perceived as visibly repressive but needs to guarantee safe passage for its predictive capability. What remains of the welfare state will increasingly be redefined as an institutional opportunity to protect "society" *against* the risks posed by its costly clients. Notwithstanding the devolutionary barking of the Gingrich pit bulls, welfare institutions have a healthy future—as the nemesis, and not the guarantor, of the social democratic vision of society. In their classic study of public welfare, *Regulating the Poor*, Frances Fox Piven and Richard Cloward show that the historical purpose of relief programs for the poor has been to regulate marginal labor and contain political unrest. Once civil order has been restored, the programs contract and are to degrade their clients so as to instill resentment (or fear of pauperization) on the part of working poor.[10] Welfare recipients are then in a position to serve conveniently as society's scapegoats.

The Leopard's Spots

As a prime example of the new fiscal wisdom of risk, the greening of the military is an early model for this tendency to reformat existing institutions. "The new partnership between the DOD and the EPA," in Goodman's words (the agencies have testified jointly), means that the watcher and the watched are now in bed together. The Pentagon is being touted as the reliable steward of its 25 million acres of public land, hosting one hundred thousand archaeological sites and three hundred listed and candidate endangered species; its Corps of Engineers are the new ecopolice, and its surveyors and cartographers are the new guardians of biological diversity.

What is wrong with this picture? Mainstream environmentalists have come to see eye to eye with big business. Why not with the military? Should the Pentagon be appointed to direct its own wing of the ecology movement? Can the military, which has been such a large part of the problem, be part of the solution? The purity tradition of environmental ethics makes it especially difficult to address questions that involve such a range of compromise. It is not enough to say that a leopard cannot change its spots. On the one hand, environmental compliance simply means that the Pentagon is abiding by the laws of the nation, or what's left of them. Taxpayers might also be happy to see the defense budget (or 2.5 percent of it) being put to progressive uses once in their lifetime. So, too, the principle that the polluter pays sets a good example for the corporate sector. Since the military, unlike most industries, is not beholden to corporate pressure (in many respects, it occupies the "commanding heights" of the economy), it is in position to act in an autonomous and progressive manner, as it did in its wartime integration of African Americans. And, who knows, we may see some reforms in the nature of military masculinity as a result of taking on the traditionally gendered role of cleaning up.

On the other hand, we are confronted with tough political questions about the moral status and behavior of institutions. The relationship between the U.S. military and the corporate state has seen the most profound distortion of the workings of democracy in modern times. This resulted in a truly psychotic consciousness that has proved structurally incapable of thinking and acting in a fashion consistent with the common or public good that is ecology's bailiwick. The dementia runs so deep that it cannot see the folly of funding hard-energy weapons production *and* environmental programs from the same defense budget that researches and executes environmental warfare. In its attempts to identify a new global enemy in the existence of environmental threats, the defense establishment displays yet again its structural need for all forms of planetary life to mirror its own bellicose mentality. Risk assessment, among other things, is emerging as the new managerial language for preserving this mentality. If the Pentagon proceeds any further in its kinder, gentler missions, the result may not be the greening of the military but the militarization of greening. The history of science's relations with institutions of warfare will have advanced to another level.

Notes

1   See William K. Stevens, "Wildlife Finds Odd Sanctuary on Military Bases," *New York Times*, 2 January 1966, B9.

2   Cited by Stacey Evers, "The Green Air Force," *Air Force Magazine*, September 1994, 23.

3   Gwyn Prins, ed., *Threats without Enemies: Facing Environmental Insecurity* (London: Earthscan, 1993).

4   Sherri Wasserman Goodman, U.S. Senate, Subcommittee on Military Readiness and Defense Infrastructure, Committee on Armed Services, 9 June 1993.

5   See Andrew Ross, "Earth to Gore, Earth to Gore" in *Social Text*, no. 35.

6   Thomas L. Friedman, "Cold War without End" *New York Times*, 22 August 1993, sec. 6, p. 28.

7   Ulrich Beck, *Risk Society: Towards a New Modernity*, trans. Mark Ritter (London: Sage, 1992).

8   U.S. Senate, Subcommittee on Nuclear Deterrence, Arms Control and Defense Intelligence of the Committee on Armed Services, 28 July 1993, 126.

9   Robert Castel, "From Dangerousness to Risk," in *The Foucault Effect*, ed. Colin Gordon (Chicago: University of Chicago Press, 1991).

10   France Fox Piven and Richard Cloward, *Regulating the Poor: The Functions of Public Welfare* (New York: Pantheon, 1971).

Contributors

Stanley Aronowitz teaches sociology and cultural studies at CUNY Graduate Center. His latest book is *The Jobless Future: Sci-Tech and the Dogma of Work* (Minnesota, 1994).

Sarah Franklin lectures in anthropology and cultural studies at Lancaster University in England. She is the coeditor (with Celia Lury and Jackie Stacey) of *Off Centre: Feminism and Cultural Studies* (HarperCollins, 1991) and the coauthor of *Technologies of Procreation: Kinship in the Age of Assisted Conception* (Manchester, 1993). Her ethnographic monograph addressing new conceptive technologies will be published by Routledge U.K. in 1996.

Steve Fuller is professor of sociology and social policy at the University of Durham, U.K. He is the executive editor of the journal *Social Epistemology* and the author of three books: *Social Epistemology* (Indiana, 1988), *Philosophy of Science and Its Discontents* (Westview, 1989), and *Philosophy, Rhetoric, and the End of Knowledge* (Wisconsin, 1993).

Sandra Harding is professor of philosophy at the University of Delaware and adjunct professor of philosophy and women's studies at UCLA. She is the author or editor of many books and essays on science and epistemology issues, including *The Science Question in Feminism* (Cornell, 1986), *Whose Science? Whose Knowledge?* (Cornell, 1991), and *The "Racial" Economy of Science: Toward a Democratic Future* (Indiana, 1993). She is writing a book on decolonizing science and technology studies.

Roger Hart is a Ph.D. candidate in history at the University of California, Los Angeles. Currently he is a Visiting Fellow in the Department of the History of Science at Harvard University, where he is completing a dissertation on the history of Chinese mathematics.

N. Katherine Hayles, Professor of English at the University of California, Los Angeles, holds advanced degrees in both chemistry and literature. Before beginning work in the interdisciplinary area of literature and science, she worked as a research chemist for Beckman Instrument Company and Xerox Corporation. Her books include *Chaos Bound: Orderly Disorder in Contemporary Literature and Science* and *Chaos and Order: Complex Dynamics in Literature and Science*.

Ruth Hubbard is professor of biology at Harvard University. Her most recent books are *The Politics of Women's Biology* (Rutgers, 1990), *Exploding the Gene Myth* (Beacon, 1993), *Profitable Promises: Essays on Women, Science, and Health* (Common Courage, 1995), and, coedited with Lynda Birke, *Reinventing Biology: Respect for Life and the Creation of Knowledge* (Indiana, 1995).

Joel Kovel is Alger Hiss Professor of social studies at Bard College. His most recent books include *History and Spirit* (Beacon, 1991) and *Red Hunting in the Promised Land* (Basic Books, 1994).

Les Levidow is a research fellow at the Open University, Milton Keynes, where he has been studying the safety regulation of agricultural biotechnology, colloquially known as "the real Jurassic Park." He is the editor of several books, including *Science, Technology, and the Labor Process* (Humanities, 1981), *Anti-Racist Science Teaching* (with Dawn Gill) (Free Association, 1987), and *Cyborg Worlds: The Military Information Society* (Free Association Books, 1987, 1989), and has been managing editor of *Science as Culture* since its inception in 1987.

George Levine is Kenneth Burke Professor of English at Rutgers University and director of the Rutgers Center for the Critical Analysis of Contemporary Culture. Author of *Darwin and the Novelists* (Harvard, 1988), recent editor of *Realism and Representation* (Wisconsin, 1993) and *Aesthetics and Ideology* (Rutgers, 1994), he has published widely on the relations of science to culture. His *Lifebirds,* a book of birding memoirs, was published in September 1995 by Rutgers University Press.

Richard Levins is a veteran of Science for the People, an evolutionary ecologist who has worked in the areas of agro-ecology, mathematical biology, international development, and public health. He coauthored *The Dialectical Biologist* with Richard C. Lewontin (Harvard, 1985) and *Humanity and Nature* with Yrjö Haila (Pluto, 1992). Levins teaches at the New York Marxist School and the Harvard School of Public Health and is a long-term research collaborator with the Cuban Institute of Ecology and Systematics.

Richard C. Lewontin holds the Alexander Agassiz Chair in Zoology at Harvard University. He is the author of *Biology as Ideology* and *The Genetic Basis of Evolutionary Change* and is the coauthor of *Not In Our Genes* and *The Dialectical Biologist.*

Michael Lynch is Professor in the Department of Human Sciences at Brunel University in West London. He has published widely in the fields of social studies of science, ethnomethodology, and social theory. His most recent books are *Scientific Practice and Ordinary Action* and (with David Bogen) *The Spectacle of History: Speech, Text and Memory at the Iran-contra Hearings.*

Emily Martin is professor of anthropology at Princeton University. Her work on ideology and power in Chinese society was published in *The Cult of the Dead in a Chinese Village* (Stanford, 1973) and *Chinese Ritual and Politics* (Cambridge, 1981). Beginning with *The Woman in the Body: A Cultural Analysis of Reproduction* (Beacon, 1987), she has been working on the anthropology of science and reproduction in the United States. Her latest research is described in *Flexible Bodies: Tracking Immunity in America from the Days of Polio to the Age of AIDS* (Beacon, 1994).

Dorothy Nelkin is a university professor at New York University teaching in the department of sociology and the school of law. She is the author of *Selling Science* (Freeman, 1987) and coauthor (with Susan Lindee) of *The DNA Mystique: The Gene as a Cultural Icon* (Freeman, 1995).

Hilary Rose has been active, initially as a socialist and later as a socialist feminist, in the politics of science since the radical science movement came into existence as part of the opposition to the Vietnam War. She has published a number of books and articles in this area, beginning with the coauthored (with Steven Rose) *Science and Society* (London, 1969). Her most recent book is *Love, Power, and Knowledge: Towards a Feminist Transformation of the Sciences* (Indiana, 1994).

Andrew Ross is Professor and Director of the American Studies Program at New York University. His books include *The Chicago Gangster Theory of Life: Nature's Debt to Society* (1994), *Strange Weather: Culture, Science, and Technology in the Age of Limits* (1991), and *No Respect: Intellectuals and Popular Culture* (1989). A columnist for *Artforum* and coeditor of the journal *Social Text*, he is also the editor of *Universal Abandon?* (1988) and the coeditor of *Microphone Fiends* (1994) and *Technoculture* (1990).

Sharon Traweek is associate professor in the history department and director of the Center for Cultural Studies of Science, Technology, and Medicine at UCLA. She studies variations in the international high energy physics community's craft knowledge, research styles, learning and pedagogic practices, disputing processes, social structures, and political economy. Her work is situated at the intersection of cultural, feminist, and rhetorical studies.

Langdon Winner is professor of political science in the department of science and technology studies at Rensselaer Polytechnic Institute. He is the author of *Autonomous Technology* (MIT, 1977) and *The Whale and the Reactor: A Search for Limits in an Age of High Technology* (Chicago, 1986), and editor of *Democracy in a Technological Society* (Kluwer, 1992). Winner is past president of the Society for Philosophy and Technology. His column, "The Culture of Technology," appears regularly in *Technology Review*.

Index

Absolute standards in science. *See* Universals in science

Academic environment: asymmetry in, 249–52; hostility against science studies in, 106–7, 110, 165, 252–54, 285; questioning legitimacy of, 293–95. *See also* "Academic left"; Scientific communities

"Academic left," 10, 239, 248, 253, 259–60, 279–85, 295–97

Academic Questions (journal), 63, 65

Accountability. *See* Scientists: self regulation of

ACT–UP, 223

Adorno, Theodor, 221

Agassi, Joseph, 206

AIDS, 68, 117, 222

Alcoff, Linda, 73–74

Alternative medicine, 11, 62, 180–82

Alternative movements, 115, 117, 185–86, 214, 273–74. *See also* New Age movements; Radical movements of the 1960s and 70s

Althusser, Louis, 213

American Association for the Advancement of Science (AAAS), 204

American Chemical Society, 112, 120

Analogy. *See* Metaphors: use of in science

Anderson, Perry, 213

Anthropology, 142–43, 251

Antiabortionists. *See* Right-to-Life movement

"Antidemocratic right," 16, 17, 18, 26

Antiscience phenomenon, 127, 129, 134, 142, 295. *See also* Science Wars

Arendt, Hannah, 53–54

Argyros, Alexander, 273, 276

Aronowitz, Stanley, 248, 260, 296

Atkins, Peter, 97

Bachelard, Gaston, 213

Bacon, Francis, 140

Baltimore, field work in, 67–69

Banks, Robert, 95

Battersby, Christine, 158

Beck, Ulrich, 2–3, 315

Benveniste, Jacques, 92–93

Bernal, John Desmond, 208–9, 216

Bernal, Martin, 96

Berreby, David, 130

Best, Steven, 260

Bhaskar, Roy, 211–13

Binary male/female model, 173–78

Biological psychiatry, 82–84

Biology, 97; developmental, 297; and ecology, 206–7; molecular, 10, 206–7, 221–22, 298, 303; and social theory, 210

Biotechnology. *See* Genetics

Blair, Tony, 304

Bloom, Amy, 177

Bockris, John, 50

Bodmer, Walter, 85, 88

Body: health of the, 19, 193–95; knowledge about the, 64–65, 169. *See also* Gender identity

Bohm, David, 211–12

Bohr, Niels, 211, 221

Boyle, Robert, 213, 220, 261–63
British Association for the Advancement of Science, 84
Brush, Stephen, 29
Butler, Judith, 73

Cancer, rise in, 192
Capitalism, 5, 9–10, 35–36, 50–51, 75–76, 110, 184, 186; and the environment, 5, 198–99. *See also* Funding for science and technology
Carlsen, Elizabeth, 195
Carpenter, Clarence Ray, 228
Carson, Rachel, 198, 209
Cetina, Karen Knorr, 12
Chaos Bound (Hayles), 128, 133–34; analyzed in *Higher Superstition* (Gross and Levitt), 275–77, 280–81
Chemical industry, 194–95
Chemical Manufacturers' Association, 197
Chinese and Islamic sciences, 22, 184
Choreographing History (Foster), 139
CIA map makers, 315
Cloward, Richard, 318
Club of Rome report, 2
Cold War, 10, 46, 52, 87, 117, 119, 316; end of, 7, 31, 234, 311, 312
Collins, Harry, 12, 84–85, 89–94, 97
Colonialism. *See* European expansion
Committee on the Public Understanding of Science (COPUS), 84–86, 88, 96–97, 254
Computer simulation programs, 230
Comte, Auguste, 36, 39, 41
Confucianism, 41–42
Conservative organizations, 63
Contract with America, 1, 109, 317
Copernicanism, 41
Creationism, 9, 11, 48–49, 115, 117, 184–85
Crew, Frederick, 81–82
Crick, Francis, 206, 222
Critics of science, 16, 34; antiracist, 19; conservative, 184–85, 187, 190–91, 207; diversity of, 8, 10–11, 20, 53, 80–81, 105, 109, 115, 151, 184–87, 248, 281–85; feminist, 6, 53, 81, 208,

222; liberal, 19, 185–87, 190–91; radical, 53, 186–90, 209, 213–14, 280, 282–83 (*See also* Marxism). *See also* Science studies; Science Wars
Crystals, Fabrics, and Fields (Haraway), 154
Culliton, Barbara, 120
Cultural constructivism. *See* Social constructionism
Culture. *See* Society
Culture Wars, 7, 114, 151, 202; in Britain, 80

Danish Environmental Protection Agency, 192
Danish Hydraulic Institute, 230
Darwin, Charles, 31, 125, 128, 146, 206, 299
Darwin and the Novelists (Levine), 125, 128–30
Darwinians: social, 41, 210
Dawkins, Richard, 90, 92, 97, 156
Decade of the Brain, 82
Decline of the West (Spengler), 219
Deconstructionism, 20, 45, 90, 104, 276
Defoe, Daniel, 159–60
De Lauretis, Teresa, 74
De Man, Paul, 296
Democratic Socialists of America, 296
Department of Defense (DOD), 7, 205, 214, 311–19
Department of Energy (DOE), 312
Department of the Environment (Britain), 303
Derrida, Jacques, 259, 272, 277–78
De Santillana, Georgio, 206, 220–21
Dickson, David, 108, 205, 215
Dilthey, Wilhelm, 210–11
DNA, 298; discovery of, 206–7, 222
Dominican Republic: third sex in, 172
Douglas, Mary, 115, 120
Duden, Barbara, 162
Durham conference of 1994, 95, 96, 98

Earth in the Balance (Gore), 314
Eastern Europe: environmental degradation in, 313

Economy: market–driven. *See* Capitalism

Education: in Britain, 52, 151–52; decrease in funding for, 151–52; in France, 41; in Germany, 39–41; in Japan, 39–41; and land–grant universities, 40; legislation, 132; liberal arts, 40; and religion, 47–48; science, 2, 12, 16, 19–20, 30–31, 39–44, 52–53, 84–89, 120, 132, 136–38. *See also* Public understanding of science

Einstein, Albert, 37, 38, 42, 221, 223

Einsteinian constant, 277

The End of History and the Last Man (Fukuyama), 35–37

Engelhart, Tom, 120

Enlightenment, 39, 41, 209

Enola Gay exhibit (Smithsonian Institution), 111, 120–21

Environment: degradation of in Eastern Europe, 313; and the economy, 2, 6, 197–98, 314, 317–18; education about, 136–38; effects of technology on, 2, 17–18, 192–200, 207, 214; global, 315; legislation, 1–2, 198–99, 317–18; and the military, 311–19

Environmental Protection Agency (EPA), 1, 314

Environmental sciences: conflict with technosciences, 17–18, 206–7

Environmentalists, 260, 280

Eurocentricism. *See* Multiculturalism

European Association for the Study of Science and Technology (EASST), 306–7

European expansion, 17, 22–24

Evolution, 48–49, 159, 188–89, 297–98

Fact and Feeling (Smith), 126, 132

Fausto–Sterling, Ann, 171

FBI investigation of the Unabomber, 202–3

Federation of American Scientists (FAS), 204

Fee, Elizabeth, 222

A Feeling for the Organism (Keller), 232

Feminism, 73–76, 169; alliances, 80–81, 90; backlash against, 81; in Britain, 81, 88, 93; diverse types of, 71; as instrument of change, 66–67; recognition of, 93–94; and social science, 69–71. *See also* Science studies: feminist

Feuerbach, Ludwig, *The Essence of Christianity*, 47

Feyerabend, Paul, 46, 51, 218

Fieser, Louis, 10

"Flight from Science and Reason." *See* New York Academy of Sciences: June 1995 conference

Forbes, Sherman, 312–13

Foreman, David, 260

Forman, Paul, 218, 221

Forrester, John, 82, 84

Foster, Susan, 139

Foucault, Michel, 277

Fraasen, Bas van, 125

Fukuyama, Francis, 35–37, 39

Fuller, Steve, 9, 90, 282, 308

Funding for science and technology, 7, 50–51, 92, 107, 110, 121; in Britain, 31, 85, 304–6; from conservative organizations, 63; military and corporate, 14, 49–50, 118, 205, 215, 303–8, 311; state, 9, 31, 46–47, 85, 116–19, 304–6. *See also* Capitalism; Science: funding cut–backs

Galileo, 41–42, 43, 140, 190, 220–21

Galton, Francis, 210

Garfinkel, Harold, 177

Gellner, Ernest, 82

Gender identity, 25–26, 73–76, 170–78, 232–33

Genetics, 10, 82, 97, 117, 214, 303; in Britain, 88, 303; in Europe, 305–6. *See also* DNA

The German Ideology (Marx), 47

Germany: education in, 39–40, 41; science in, 43. *See also* Weimar Republic

Gibson, William, 209

Ginsburg, Faye, 161
Glassy mirror ideology (GMI). *See*
  Scientific realism
Goals 2000 (education legislation), 132
Goldstein, Sheldon, 62
*The Golem: What Everyone Should
  Know about Science* (Collins and
  Pinch), 89–94, 97
Goodman, Sherri Wasserman, 313–14,
  318
Good society, vision of, 35–36, 70
Gore, Al, 314
Gould, Stephen J., 65, 97, 296
Greek and Roman ideas about concep-
  tion, 158–59
Gross, Paul, 7–8, 11, 14, 80, 102–3,
  107, 110, 123–26, 131, 133, 152,
  154–57, 203, 226–29, 233–36,
  239–40, 247–53, 259–85, 295–98
Grumbly, Thomas F., 312
Grünbaum, Adolf, 82, 84

Haberer, Joseph, 115
Habermas, Jürgen, 82, 146
Hacking, Ian, 92
Haldane, J. B. S., 86, 208–9
Haraway, Donna, 10, 90, 93, 152, 154,
  222, 228, 259, 260, 270–71, 273,
  285
Harding, Sandra, 6, 10, 44, 45, 93, 222,
  229, 232, 259, 260, 265–70
Hart, Roger, 227
Harvey, William, 160
Havel, Vaclav, 115
Hayles, N. Katherine, 128, 133–34,
  259, 260, 272, 275–77, 280–81
Hegel, G. W. F., 39, 41
Heisenberg uncertainty principle, 211,
  218, 221, 247
Herbal medicine. *See* Alternative
  medicine
Herdt, Gilbert, 172–73
Hermaphroditism. *See* Sexual
  ambiguity
Hermeneutics, 82, 84
Herschbach, Dudley, 65
Hess, David, 330
Hessen, Boris, 5–6, 209

Higher Superstition (Gross and Levitt),
  7–8, 11, 14, 80, 102–3, 107, 110,
  123–26, 131, 133, 152, 154–57, 162,
  203, 226–29, 233–36, 239–40,
  247–53; analysis of, 259–85,
  295–99; reproductive imagery in,
  157, 160
Hilbert, David, 221
Himmelfarb, Gertrude, 62, 295–301
Hirobumi, Ito, Prime Minister, 43
Historiography, 33–36; Whig,
  264–65
History: predetermined, 35–38. *See
  also* Evolution
History of science, 21–24, 29, 32–38,
  51, 96, 120–22, 204–5, 244
History of Science Association:
  membership records subpoenaed,
  202
Hobbes, Thomas, 219–20, 261–65
Hoch, Paul, 307
Hogben, Lancelot, 208
Holton, Gerald, 9, 11, 44, 105, 127–28,
  129
Human Genome Project, 10, 88, 303

Imperial University of Tokyo, founding
  of, 40
Industrialization. *See* Technology

Japan: education in, 39–41; medicine,
  42; and Western science, 29, 35,
  38–44, 48
Japanese language, 42, 147

Kant, Immanuel, 41, 299
Kantrowitz, Arthur, 115
Keller, Evelyn Fox, 10, 93, 222, 232,
  260
Kenney, Martin, 205, 215
Kessler, Suzanne, 174–75
Klausner, Richard, 132
Knowledge, scientific. *See* Scientific
  knowledge
Komesaroff, Paul A., 125
Kreiger, Nancy, 188
Kuhn, Thomas, 34–37, 39, 51, 52, 116,
  125, 140, 217, 218, 240, 245

Laboratory environment: studies of, 217
Laboratory Life (Woolgar), 94–95
Lang, Marvin, 112
Language: and gender identity, 170, 208; influence of on scientific thought, 157–62, 204; Japanese, 42, 147; translations, 301; use of singular generics in, 146–49
Latour, Bruno, 10, 12, 44, 45, 54, 93, 124, 127–28, 217, 221, 248–49, 260, 272, 273
Lawrence, Ernest, 222
Lenard, Philip, 269
Lessing, Gotthold, 46
Lévi–Strauss, Claude, 18
Leviathan and the Air-Pump (Shapin and Schaffer), 219–20, 229; analyzed in *Higher Superstition* (Gross and Levitt), 261–65, 281
Levins, Richard, 206–7
Levitt, Norman, 7–8, 11, 14, 80, 102–3, 107, 110, 123–26, 128–29, 131, 133, 152, 154–203, 195–98, 226–29, 233–36, 239–40, 247–53, 259–85
Lewontin, Richard, 62, 97, 206–7
Liebig, Justus von, 40
Life of Jesus (Strauss), 47
Lincoln, Abraham, 40
Linnaeus, 146, 189
"Literature of indignation," 289, 295, 298
Logical positivism, 31, 51, 206, 213, 272
Logic of Scientific Discovery (Popper), 216–17
Lynch, Michael, 98

Mach, Ernst, 42
Maddox, John, 82, 92–93
Madison Center for Educational Affairs, 63
Marcellino, John (community leader in Baltimore), 67–69
Marx, Karl, 39, 47
Marxism, 25, 87, 180, 186–90, 199–200; in Britain, 208–9, 213
Maturana, Humberto, 230–31

Mazur, Eric, 132
McClintock, Barbara: Keller biography of, 93, 232
McKenna, Wendy, 174–75
McLaren, Anne, 88
Medawar, Peter, 86
Memory transfer experiments, 89, 91
Merchant, Carolyn, 260
Merton, Robert, 207, 209, 216
Metaphors: use of in science, 270, 297–98
Mettler's Woods (New Jersey): preservation of, 136, 138
Military forces: and environmental security, 311–19
Military research and development, 6, 13, 31, 108, 205, 210, 215, 311
Mills, C. Wright, 210
Modern Language Association Conventions, 233–34
Montague, Peter, 195, 199
Moore, John, 163
Muchie, Mammo, 96
Mulkay, Michael, 93
Multiculturalism, 18–22, 19–25, 148–49, 152
Mumford, Lewis, 105
Muslim culture, 39

National Association of Scholars, 7; November 1994 conference, 63, 123, 125, 127, 129, 130–31
National Science Foundation, 107
Native Americans: access to sacred sites, 311; third sex among, 173
Nature: concepts of, 168–69, 187–90; diversity of, 190
Nature (journal), 92–93, 204
Nazism, 299–301
Needham, Joseph, 22, 208, 209, 216
Nelkin, Dorothy, 91–92, 205
Neo–Kantians, 210–12
New Age movements, 8, 49, 50, 129, 134
New Guinea, third sex in, 172–73
New Left. *See* Critics of science: radical
New Left Review (journal), 296
New Scientist (journal), 302

Newton, Isaac, 6, 31, 37, 38, 41, 140, 270, 285; and science of navigation, 216; translations into Japanese, 42

New York Academy of Sciences: June 1995 conference, 8, 61–63, 65, 72, 123, 203

Noble, David F., 107

Nuclear weapons, 95, 208, 209, 222

Objectivity in science. *See* Science: objectivity in

Office of Technology Assessment, 108

Ohio State Medical School, 64

Olin (foundation), 7

Onians, Richard, 158–59

On Looking into the Abyss (Himmelfarb), 295, 298–301

Open Science and Its Enemies (Popper), 127

Operations Other Than War (OOTW), 312

Origin of Species (Darwin), 48

Overdetermination, 34–35

Peirce, Charles Sanders, 218

Pentagon building, 311

Perry, William, 312

Perspective. *See* Scientific knowledge: influence of viewpoint on

Petchesky, Ros, 162

Phelan, Peggy, 155, 156, 165

Philosophy: and science, 51, 207–9, 212–13, 240, 244–47

Physics: all sciences reducible to, 51, 266–69; and Japan, 42; and measuring instruments, 211–12; solid state, 222

Pickering, Andrew, 217

Pinch, Trevor, 12, 84, 89–94, 97

Polemics. *See* Science Wars

Popper, Karl, 82, 86, 127–28, 216–17, 221, 223, 240, 245

Popular subcultures. *See* Alternative movements; New Age movements

Population: density of, 187–88; management of, 193, 317

Postmodernism, 74–75, 80–81, 90, 94, 104, 124–25, 199–200, 277

Power of science. *See* Science: authority of

Price, Don, 117

Principia (Newton), 6, 270, 285

Probability, 211–12, 221

Proctor, Robert, 6

Protestant Reformation, 29, 46

Psychoanalysis: attack on, 80–84

Psychopharmacology. *See* Biological psychiatry

Public understanding of science, 30–31, 62, 72, 89, 96–97, 119, 127, 131, 134, 136–38, 242; in Britain, 84–89, 239, 254; in Denmark, 86. *See also* Education; Science, public participation in

Puerto Rico: oral contraceptives trials in, 64

Quantum theory, 211–12, 218, 274

Quine, W. V., 240, 244–45

Racism, 81, 184; in Britain, 87

Radical movements of the 1960's and 70's, 87, 293–94

Rapp, Rayna, 71

Raymond, Janice, 176

Realism: scientific. *See* Scientific realism

Reality: nature of, 133, 155–56, 206; science as means of coping with, 65–66

Reductionism, 38, 51, 186–87, 212, 268

Relativism, 3, 4–5, 17–18, 23–24, 34–36, 272

Reliable Knowledge (Zinman), 231

Religion: and education, 47–48; and science, 22, 46–48, 141

Religious wars in Europe in sixteenth and seventeenth centuries, 46

Reproductive imagery, 157–62, 165

Research and Technology Development (RTD), 305–6

Richter, Burton, 111

Rifkin, Jeremy, 260, 296

Right–to–Life movement, 117; fetal imagery, 161–62, 165

The Risk Society (Beck), 315
Rorty, Richard, 65
Rose, H. J., 159
Rose, Mark, 159–60
Ross, Andrew, 134–36, 259, 273–75
Ross, Dorothy, 69
Royal Society, 40, 229, 261–65
Russell, Denise, 83
Ryle, Gilbert, 238–40, 250, 253–54

Safe, Stephen, 195, 197
Schaffer, Stephen, 206, 220, 229, 259,
    260–65, 281
Science: assumptions about, 140–41,
    143–44, 229, 246; authority of, 3–5,
    9, 53, 62, 64–65, 86–87, 92, 95, 109,
    144, 156, 191, 298–301 *(See also*
    Scientists: self–regulation of);
    autonomy of, 12, 84, 116, 197, 210,
    213–16, 222, 308; benefits of, 153,
    182–83, 193, 204–5, 296; in Britain,
    30; and Christianity, 22; consensus
    in, 231–32; as discourse, 126, 137,
    219; disputes about *(See* Science
    Wars); education, 2, 12, 16, 19–20,
    30–31, 39–44, 52–53, 84–89, 120,
    132, 136–38; elitism of, 4, 10, 13, 16,
    19, 62, 86, 89, 109, 229; expertise in,
    4, 12–13, 72, 89, 135, 142, 163, 203,
    227, 249, 271–78, 274; funding cut-
    backs in, 7, 9, 31, 85, 116–18, 125,
    234 *(See also* Funding for science
    and technology; Scientific research:
    influenced by funding); in Germany,
    43, 44–45, 219, 221; history of *(See*
    History of science); and literature,
    125, 127, 128–29, 133–35, 275–77;
    and the media, 15, 84–87, 91–92,
    98–99, 119; as myth, 89–90, 92;
    nature of, 31–33, 84, 228, 254; nine-
    teenth century, 40; objectivity in,
    19–20, 99, 109, 152–54, 169, 183,
    212–13, 232, 294, 295–96; and other
    disciplines, 32, 76, 103–4, 114,
    126–27, 164, 247, 249–51; and
    philosophy, 51, 207–9, 212–13, 240,
    244–47; and politics, 89, 93, 107,
    214–15; and progress, 35–36, 37–38;
    public participation in, 1–2, 4, 13,
    18, 30–31, 41, 72, 75, 108, 191, 223
    (See also Public understanding of
    science); and public policy, 51,
    104–5, 107–8, 116, 205, 302–9; and
    religion, 22, 46–48, 141; seculariza-
    tion of, 29, 46–48, 51; seventeenth
    century, 261–65; and society, 16–18,
    20–22, 23–24, 32, 37, 42, 45, 72, 75,
    86–88, 92, 95, 98–99, 107–8, 110,
    114, 120, 124, 125–26, 136, 144,
    168–69, 183–84, 186–87, 197, 205,
    212, 219–20, 229, 232–33; special-
    ization in, 32–33; universality *(See*
    Universals in science); and values,
    4–6, 10, 12–13, 37, 41, 53, 64–65,
    69, 70, 88, 104–5, 107, 119, 122,
    151, 210, 229, 231, 298–301, 319;
    Western monopoly on, 4–5, 11, 21,
    39, 43–44. *See also* Scientific
    knowledge; Scientific research;
    Scientists; Technology
Science (journal), 93, 132, 204
Science and Anti–Science (Holton),
    127
Science and technology studies (STS).
    See Science studies
Science as a Way of Knowing (Moore),
    163
Science in American Life exhibit
    (Smithsonian Institution), 111–12,
    120
The Science Question in Feminism
    (Harding), 265–70
Science studies, 16, 17–22, 53–54,
    162–63, 203, 278; academic envi-
    ronments hostile to, 106–7, 110, 165,
    285; compared to secularization of
    religion, 46–48; expectations of,
    45–46; feminist science studies, 6,
    11, 18–20, 24–25, 34, 152, 208, 222,
    228, 265–66, 270; history and devel-
    opment of, 21–24, 52–54, 102–7,
    124–25, 207–10; impact of, 29, 47,
    104, 105–7, 226; purpose of, 72,
    125–27, 163–64, 197, 302–9; rela-
    tionship to natural sciences, 61–63,
    75–76, 98–99, 102–4, 128, 164,

Science studies (*continued*)
233–36, 281; scapegoating of, 7, 30, 106, 116, 131, 234–35; as science appreciation, 106, 110. *See also* Critics of science; Science Wars; Sociology of Scientific Knowledge (SSK)

Science Wars, 7–12, 16, 26, 29, 61, 63, 102, 114, 123–25, 151–52, 156, 199–200, 202–4, 222, 239–40, 285; in Britain, 80, 84–85, 98–99, 151–52, 239–40, 302–9; common ground in, 65–66, 76, 90–91, 124, 126, 189–90, 233–36; in Europe, 302–9; polarizing effect of, 121–22, 239–40, 254–55; and reproductive rights, 161–62, 165; as "turf" battle, 125–27, 130–31, 265. *See also* Critics of science

Scientific communities, 107, 203–4; British seventeenth century, 261–64; and the military/industrial complex, 210, 215–16; role of power and influence in, 217–18

Scientific knowledge: accepted, 64–67, 217–18; access to, 143–44; ambiguity in, 195–96; changes in, 34, 36–38, 181, 218; and common sense, 96–97, 138–39, 220, 240–47; consistency in, 189–90, 231–32; influence of outside factors on, 218–22; influence of viewpoint on, 154–56, 162–63, 182–84, 188–89, 206–7, 209–10, 228, 232; nature of, 152, 156, 162–63, 180–81, 210–11, 241–47; organization of, 140, 145–49, 189–90, 241–47; questions in, 102–3, 228, 297; relation to all knowledge, 210–11. *See also* Science; Scientific research

Scientific methods, 30, 143, 204, 211–12, 220, 229–30, 243

Scientific realism, 34–38, 94–98, 240, 243

Scientific research, 141; in Britain and the European Union, 304–9; distinction between discovery and justification in, 48, 215, 228–29; experiments in, 89, 91, 143–44, 230;

ideological motivation for, 216–17; influenced by funding, 50–51, 92, 108, 118, 119, 130, 144, 203, 205, 215, 304–9; influenced by political and cultural environment, 216, 302–9; instruments of, 153, 211, 231; and the military, 31, 311; process of, 187–90; rejected theories of, 217–18; variety of, 142; writings about, 143. *See also* Science; Scientific knowledge

Scientists: self–regulation of, 12, 43, 116–19, 203–4, 303; and unemployment, 30

Scott, Joan, 73

Selling Science (Nelkin), 91–92

Separation of church and state, 47–48

Sexual ambiguity. *See* Gender identity

Sexuality in Western nations. *See* Binary male/female model

Shapin, Steven, 12, 206, 220, 221, 229, 259, 260–65, 281

Silent Spring (Carson), 198, 209

Skeptical Inquirer, 277–78

Skinner, Quentin, 263

Smith, Barbara Herrnstein, 298

Smith, Jonathan, 126, 132

Smithsonian Institution, 111–12, 120–21

Snow, C. P., 52, 95, 157

Social constructionism, 61–63, 69, 72–76, 92, 115–16, 129, 132–33, 138, 183–84, 212, 226, 233–36, 248, 250–51; analyzed in *Higher Superstition* (Gross and Levitt), 260, 269, 270–71, 272

Social Relations of Science Group, 208–9, 213

Social sciences, 66–67, 69–72; public contributions to, 67–69, 75; and science, 103–4

Social Studies of Science (journal), 93

Social Text (journal), 14, 139–40

Society: relation of science and technology to, 16–18, 20–24, 32, 37, 42, 45, 72–76, 92, 95, 98, 107–8, 110, 114, 124, 125–26, 136, 144, 183–84, 205, 212, 229, 232–33

Society for Social Studies of Science, 108

Sociology of Scientific Knowledge (SSK), 10, 12, 13, 80, 90, 94, 97, 306–8

Sofia, Zoë, 157

South. *See* Third World science

Spencer, Herbert, 39, 41, 210

Spengler, Oswald, 219

Sperm counts: decline in, 192–98

Sputnik, 117, 131

Ston, Sandy, 176

Strange Weather (Ross), 134–36; analyzed in *Higher Superstition* (Gross and Levitt), 273–75

Strategic Environmental Initiative, 314

Strauss, David Friedrich, 47

The Structure of Scientific Revolutions (Kuhn), 34–37, 52, 116, 125, 140

Superconducting supercollider project, 7, 50, 118, 125, 130

Tadao, Shizuki, 42

Technology: economic aspects of, 6, 9–10, 14, 31, 49–51, 110–11, 197, 203; and environment, 2, 6, 17–18, 115, 181–82, 194–200, 207; as means of social improvement, 75–76, 104–5, 108, 117; and the military, 108, 210, 215, 221–22; public relations for, 197; risks of, 2–3, 6, 316–17, 319; skepticism about, 1–3, 12, 85–86, 105–6, 115, 118, 194, 207, 209. *See also* Science

Thatcher, Margaret, 80, 88, 304

Third World Network, 303

Third World science, 4–5, 13–14, 18, 21, 31, 96

Times Higher Education Supplement (THES), 81–82, 85, 95, 153

Totalitarian regimes, 53–54

Traditionalists. *See* "Antidemocratic right"

Transsexuals, 175–78

Travis, David, 91

Traweek, Sharon, 12, 217

Trefil, James, 62, 132

Treichler, Paula, 75

Truth: belief in absolute, 62–63, 157, 205, 231

The Two Cultures (Snow), 95

Underdetermination, 35, 244

Universals in science, 23–24, 70–74, 95–97, 146–49, 183–84, 189–90, 198, 230–31, 298–301. *See also* Truth: belief in absolute

University of Chicago, School of Social Sciences, 66, 72

The Unnatural Nature of Science (Wolpert), 84, 94–98, 239–47

Value–free science. *See* Science: and values

Varela, Francisco, 231

Vietnam War, 87, 209

Viewpoint. *See* Scientific knowledge: influence of viewpoint on

Wallis, John, 262, 264

Walter, Paul, 112

Warnock Committee, 88

Watson, James, 206, 222

Webster, Andrew, 307

Weimar Republic: antiscience sentiment in, 44–45, 219, 221

Weinberg, Steven, 32–33, 125

Weissman, Gerald, 62

Welfare state, 317–18

Williams, Raymond, 213–14

Wilson, Elizabeth, 296

Wilson, E. O., 152

Winner, Langdon, 202–3

Wittgenstein, 246

Wolpert, Lewis, 32–33, 84–85, 88, 94–98, 239–47, 252, 254, 302–4

Woolgar, Steve, 12, 93, 94, 217

Working-class issues, 67–69, 111

World-view, 267–68

World War II, 221–22

Worm runners experiments. *See* Memory transfer experiments

Young Hegelians, 47

Ziman, John, 231

Library of Congress Cataloging-in-Publication Data
Science wars / Andrew Ross, editor.
Includes index.
ISBN 0-8223-1881-4 (alk. paper). — ISBN 0-8223-1871-7
(pbk. : alk. paper)
1. Science—Social aspects. 2. Science and state.
I. Ross, Andrew, 1956– .
Q175.55.S294 1996
303.48'3—dc20 96-22506 CIP